Savannas cover approximately half of the African land surface and one fifth of that of the world. They are one of the most important, but least studied terrestrial ecosystems. They are the basis of the African livestock industry and the wildlife they support is of key importance in bringing in tourists.

The Nylsvley site in South Africa is one of the most intensively studied savannas in the world, and as such, it is a key source of data and theory relating to this important tropical biome.

The South African Savanna Biome project was set up to develop the understanding necessary to predict changes in the ecosystem's stability, induced by both natural and man-made stresses. This book provides a synthesis of the programme's sixteen years of research at Nylsvley and aims to develop a unified vision of the ecology of the dry savanna.

An African savanna: synthesis of the Nylsvley study

Cambridge Studies in Applied Ecology and Resource Management

The rationale underlying much recent ecological research has been the necessity to understand the dynamics of species and ecosystems in order to predict and minimise the possible consequences of human activities. As the social and economic pressures for development rise, such studies become increasingly relevant, and ecological considerations have come to play a more important role in the management of natural resources. The objective of this series is to demonstrate how ecological research should be applied in the formation of rational management programmes for natural resources, particularly where social, economic or conservation issues are involved. The subject matter will range from single species where conservation or commercial considerations are important to whole ecosystems where massive perturbations like hydro-electric schemes or changes in land use are proposed. The prime criterion for inclusion will be the relevance of the ecological research to elucidate specific, clearly defined management problems, particularly where development programmes generate problems of incompatibility between conservation and commercial interests.

Also in the series

Graeme Caughley, Neil Shephard & Jeff Short (eds.) *Kangaroos: their ecology and management in the sheep rangelands of Australia*
P. Howell, M. Lock & S. Cobb (eds.) *The Jonglei canal: impact and opportunity*
Robert J. Hudson, K. R. Drew & L. M. Baskin (eds.) *Wildlife production systems: economic utilization of wild ungulates*
M. S. Boyce *The Jackson elk herd: intensive wildlife management in North America*
Mark R. Stanley Price *Animal re-introductions: the Arabian oryx in Oman*
K. Homewood & W. A. Rodgers *Maasailand Ecology: pastoralist development and wildlife conservation at Ngorongoro, Tanzania*
D. Pimentel (ed.) *World soil erosion and conservation*

AN AFRICAN SAVANNA

SYNTHESIS OF THE NYLSVLEY STUDY

R. J. Scholes
Department of Botany, University of the Witwatersrand, Johannesburg, South Africa

B. H. Walker
Commonwealth Scientific and Industrial Research Organisation, Canberra, Australia

PUBLISHED BY THE PRESS SYNDICATE OF THE UNIVERSITY OF CAMBRIDGE
The Pitt Building, Trumpington Street, Cambridge, United Kingdom

CAMBRIDGE UNIVERSITY PRESS
The Edinburgh Building, Cambridge CB2 2RU, UK
40 West 20th Street, New York NY 10011–4211, USA
477 Williamstown Road, Port Melbourne, VIC 3207, Australia
Ruiz de Alarcón 13, 28014 Madrid, Spain
Dock House, The Waterfront, Cape Town 8001, South Africa

http://www.cambridge.org

First published 1993
First paperback edition 2004

A catalogue record for this book is available from the British Library

ISBN 0 521 41971 9 hardback
ISBN 0 521 61210 1 paperback

CONTENTS

PREFACE

This book presents the major findings of a sixteen-year study of savanna ecology undertaken at Nylsvley, South Africa. Savannas are a tropical vegetation type in which both trees and grass are an important component. They are the basis of the African livestock industry and their wildlife is a key tourist draw-card. The development of the genus *Homo* in Africa has largely taken place in savannas, and they are currently home to tens of millions of people. One half of the land surface of Africa is covered by savanna, making it the most extensive African biome. It is also one of the biomes which has received the least ecological study.

Scientists worked in savannas for many years before these became regarded as a distinct biome, rather than a special case of grassland or forest. It became apparent that certain classes of management problem associated with savannas, such as bush encroachment and multispecies herbivory, could not be addressed with conceptual models based on temperate grasslands or tropical forests. One of the results of this realisation was the establishment of the South African Savanna Biome Programme in 1974. The Nylsvley Nature Reserve, 200 km north of Johannesburg, was its principal study site (Figure 0.1).

The aim of the Savanna Biome Programme was to 'develop the understanding necessary to predict changes in the ecosystem's stability induced by various natural and man-made stresses' (Anon. 1978). The organisational approach was to encourage scientists in universities, research institutes and government departments to collaborate in a multi-disciplinary study of the ecology of a broad-leafed savanna. When the programme officially ended in 1990, a recommendation of the review committee was that the extensive published and unpublished research which had resulted from the Savanna Biome Programme be brought together in an easily accessible form. This book is the result.

The book aims to be a synthesis of research rather than a summary. A large amount of work was done in the Savanna Biome Programme, and if each topic was to be justly treated, this book would be unacceptably long, and probably rather boring. Our main purpose has been to draw together the findings of diverse projects into an integrated view of the structure and function of one particular savanna ecosystem. Where possible we relate the findings to savannas in general, and other African savannas in particular. Although our review of the Nylsvley research is not exhaustive, we believe that the major conceptual thrust and findings of the South African Savanna Biome Programme are represented. Most of the studies undertaken at Nylsvley get at least a mention. Readers seeking more methodological detail than is presented here are directed to the original sources, cited in the bibliography.

In writing the book, we have tried to keep the needs and backgrounds of several types of reader in mind. Each chapter begins with a brief background

Figure 0.1. The location of the Savanna Biome research site at Nylsvley, South Africa.

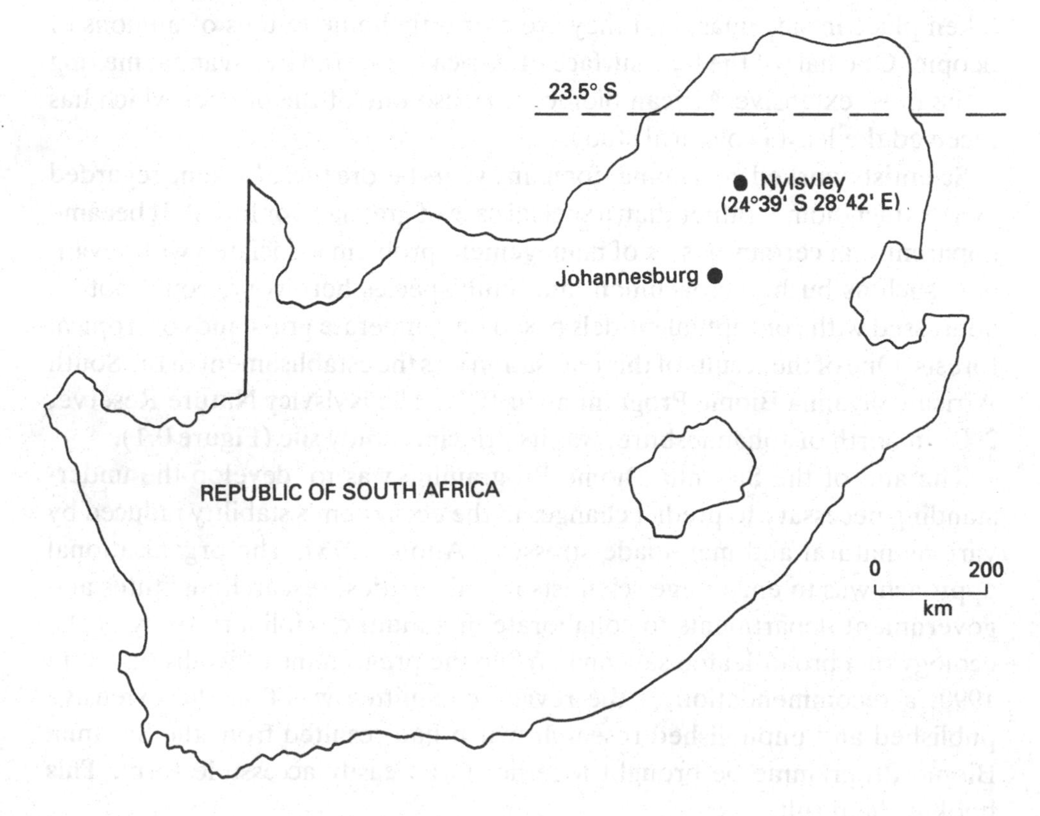

to the topic and its relevance to savanna ecology. These paragraphs are aimed mainly at undergraduate students, land managers and newcomers to the field. The Nylsvley research results on that subject follow, and will be of most interest to advanced students and specialist research scientists. A short summary of the research highlights concludes each chapter. The overall structure of the book follows a similar strategy: basics in the beginning, detail in the middle, and an overview at the end.

Savanna managers, be they traditional pastoralists, commercial cattle ranchers or nature conservators, are the group which the Savanna Biome Programme was ultimately designed to serve. Although much of the research conducted at Nylsvley was of a fundamental nature, and unlikely to be immediately transferable to the day-to-day operations of a farmer in the field, most of the pressing practical questions which savanna managers have were addressed to some degree. The task of translating the understanding of ecological processes gained at Nylsvley into farming practices in savannas falls mainly to the professional staff of the agricultural and nature conservation authorities. It is hoped that this synthesis of the Nylsvley findings will stimulate them to do so. Some pointers to issues relevant to savanna management are contained in Chapter 17.

The findings reported in this book result from the work of a great many people, including scientists, administrators, students and technical staff. As far as possible the discussion is based on published data (although some of it is not readily accessible), and citation is the main form of acknowledgement of the data source. In most of the topic areas, however, a few individuals played a leading role, and are acknowledged in the introduction to the section.

This book is dedicated to the late Prof. A. J. 'Pine' Pienaar, Director of Agriculture, who was the driving force behind the initiation of the Savanna Biome Programme. He was supported in this effort by Dr S. S. du Plessis of the Transvaal Nature Conservation Division, and Dr van der Merwe Brink of the Council for Scientific and Industrial Research.

Every research programme needs a champion: few are fortunate enough to get one as able as Brian Huntley, the manager of the Ecosystems Programmes at the CSIR. His energetic commitment to the Savanna Programme was crucial to its success. Other key personnel were the CSIR liaison officers Gudran Schirge, Tisha Greyling and Marie Breitenbach; Peter Frost, scientific coordinator of the programme for many years; Ernie Grei, the on-site senior technician; and field assistants Kazamula Baloyi, Fanie Baloyi, Caine Mashishi, Katrina Nkgumane, Ephraim Masunga, John Hlongwane, Samuel

Hlongwane, Johannes Ngobeni and Simon Molekoa. The writing of this book would not have been possible without the help of Mary Scholes.

Savannas cover one third of South Africa, but this bald statistic cannot begin to express their importance in the national sense-of-place. For South Africans returning from abroad, the sight of a dry, tawny-coloured grass layer, with a broken canopy of trees, is the essence of homecoming.

Present address:
R. J. Scholes, Division of Forest Science and Technology, CSIR, Pretoria, South Africa.

Part I

Nylsvley in an African savanna context

1

African savannas: an overview

Africa presents two dominant images to the world. One is of natural disaster, human suffering and environmental degradation. The other, paradoxically, is of boundless vistas, teeming with wildlife. The backdrop to both these images is a vegetation characteristically African, and offering a paradox of its own. Although it may have abundant grass, it does not seem to be a grassland, because it also has many trees; yet the trees are not sufficiently dense and tall to be called a forest. All efforts to bound it by precisely defining its limits end in pointless argument, since it grades imperceptibly into dry forest on the one hand, and desert shrublands and grasslands on the other. It is highly changeable, both over short distances and over the brief time for which it has been scientifically studied, and yet it is always there in some form; old as the landscape, older by far than us. These are the savannas of Africa, the cradle in which our species evolved.

This book is a synthesis of ecological knowledge concerning one African savanna site, Nylsvley, which has been studied in detail. However, the insights gained in that study have relevance to savannas in other areas of Africa and the world. The purpose of this chapter is to place the Nylsvley savannas within a global, continental and local context.

Savannas are second only to tropical forests in terms of their contribution to terrestrial primary production (Atjay, Ketner & Duvigneaud 1987: their categories 'Tropical Woodland' and 'Tropical Grassland' are largely savannas following the definition used in this book). They represent a substantial terrestrial organic carbon pool, which could act as either a net source or a sink of atmospheric carbon dioxide in future decades. They occupy about 20% of the land surface of the world, and about 40% of Africa. In Africa, they are home to most of the population and are the areas in which population growth is most rapid. In South Africa savannas make up 35% of the land area, and are the basis of two major industries: cattle ranching and wildlife-related tourism. In 1989 these industries had a turnover of around US$ 1000 million.

Savannas occupy the extensive areas between the equatorial forests and the mid-latitude deserts. Although the principal elements of the vegetation are trees and grasses, the ecology of savannas is neither that of a forest, nor that of a grassland. The strong and complex interactions between the woody and herbaceous plants give this vegetation a character of its own. In no other major terrestrial biome is the primary production shared so equitably and persistently between two such different life forms as trees and grasses.

There is no general consensus on the precise definition of savannas, despite much discussion of the issue. The central concept – a tropical mixed tree–grass community – is widely accepted, but the delimitation of the boundaries has always been a problem. Figure 1.1 places savannas within the continuum of vegetation types which have trees and grasses as their main constituents. We take a broad view of what constitutes a savanna, acknowledging that at the extremes, the distinction between savannas and woodlands and savannas and grasslands is unavoidably arbitrary. We contend that the vast extent of African rangelands is intermediate between woodlands and grasslands, and

Figure 1.1. Savannas occupy part of a continuum of vegetation types with varying proportions of woody plants and grasses. The boundaries between types on this continuum are arbitrary. Here they are represented by the lines of at least 5% contribution to the total primary production by trees and grasses, with most of the remainder contributed by the other type.

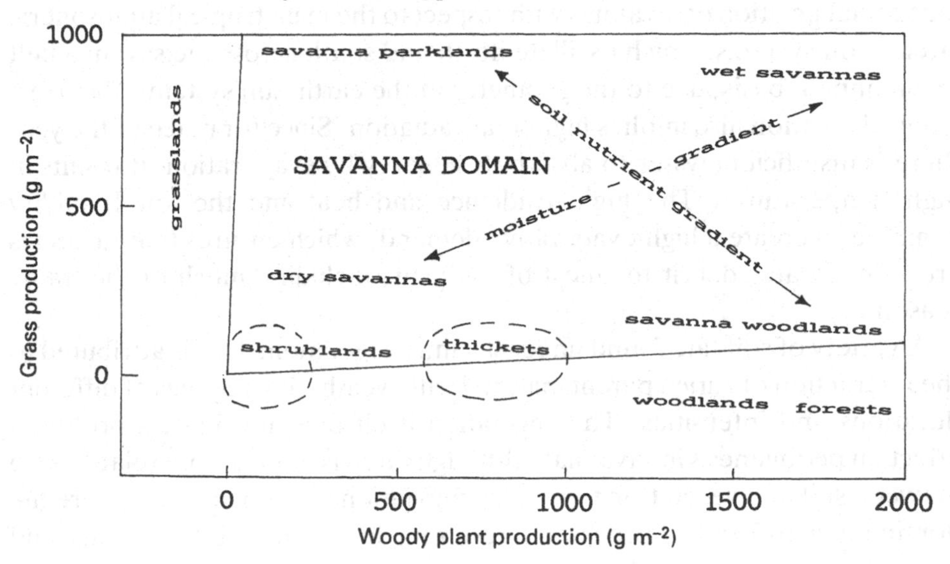

can be classified readily and usefully as savannas. To categorise them as grassy woodlands, or woody grasslands, is to dodge the issue and ignore their special attributes.

The core of the savanna definition used in this book is that it is a tropical vegetation type in which ecological processes, such as primary production, hydrology and nutrient cycling, are strongly influenced by both woody plants and grasses, and only weakly influenced by plants of other growth forms. This follows the spirit of the working definition used in the international Responses of Savannas to Stress and Disturbance (RSSD) programme (Frost *et al.* 1986). This definition is deliberately non-quantitative. A more rigorous definition is provided in Box 1.1, for use when unambiguous delimitation is necessary.

Many environmental and biotic factors correlate with the distribution of savannas. These include geomorphology, climate, soils, vegetation, fauna, and fire (Bourlière 1983; Cole 1986). It is not always clear which of these factors cause the main features of savanna structure and function, and which are consequences, or coincidences. Furthermore, many of these factors are not independent of one another. Figure 1.2 attempts to organise some of the many characteristics of savannas according to their relationships and interactions. These factors are briefly introduced below, and are dealt with in more detail for Nylsvley in the chapters which follow in this section.

The savannas of the world all occur in hot regions with a highly seasonal rainfall distribution. This results in a warm dry season (or two, in monsoonal climates) with a duration of three to eight months, and a hot, wet season for the remainder of the year. The rainfall seasonality occurs as a result of the latitudinal position of savannas with respect to the main tropical atmospheric circulation systems, which oscillate north and south across the savanna belt on an annual basis, due to the geometry of the earth–sun system. The near-tropical location also implies high solar radiation. Since for much of the year there is insufficient water to absorb this energy by evaporation, it results in high temperatures. The high irradiance and heat and the low humidity combine to create a high evaporative demand, which ensures that savannas are in net water deficit for most of the year, including much of the 'rainy season'.

A variety of soils are found under savanna vegetation. This is attributed to the interaction of varied parent material with weathering regimes of different durations and intensities. The vegetation itself does not have a profound effect on pedogenesis in savannas, although there is often a close relationship between soil and vegetation type. Clay illuviation and ion movement are the dominant soil-forming processes, resulting in distinct soil horizons and

BOX 1.1. A set of rules for deciding whether a given vegetation is a savanna or not. Note that the rules rely both on vegetation characteristics (such as tree canopy cover) and site characteristics (such as climate). This is because the vegetation parameters in savannas are subject to short-term variation. These rules were chosen to predict the presence of a savanna based on relatively easily collected data.

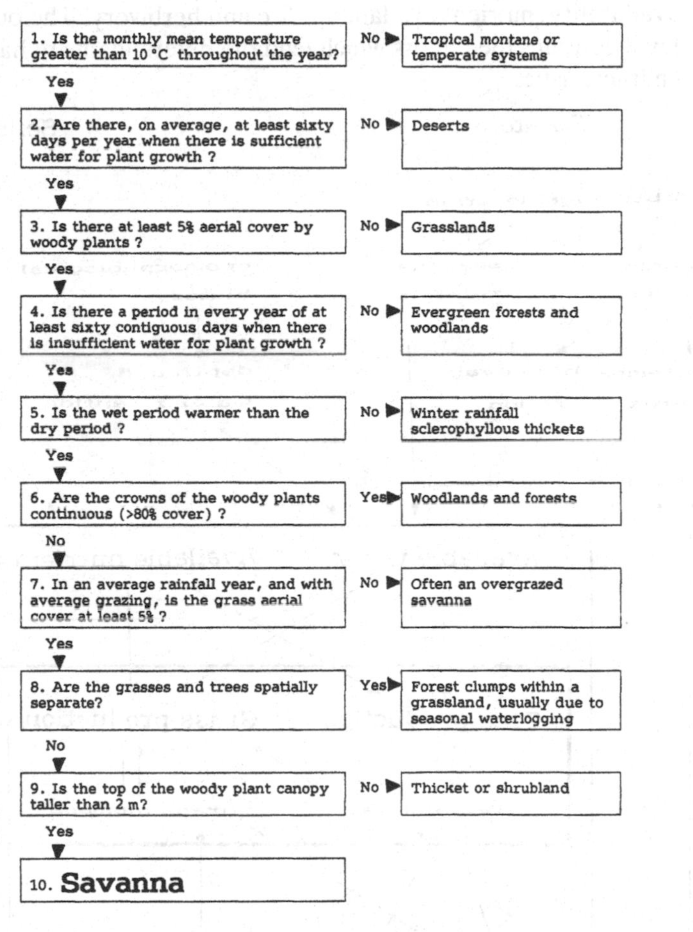

catenary sequences. The organic matter content of savanna soils is generally low. This has been attributed to the high temperatures, which lead to a high rate of organic matter decomposition. It is also due to the frequently sandy nature of savanna soils, and the predominance of low-activity clays, which do not encourage organic matter stabilisation.

1. Is the monthly mean temperature greater than 10 °C throughout the year?
No
Tropical montane or temperate systems
Yes
2. Are there, on average, at least sixty days per year when there is sufficient water for plant growth ?
No
Deserts
Yes
3. Is there at least 5% aerial cover by woody plants ?
No
Grasslands
Yes
4. Is there a period in every year of at least sixty contiguous days when there is insufficient water for plant growth ?
No
Evergreen forests and woodlands
Yes
5. Is the wet period warmer than the dry period ?
No
Winter rainfall sclerophyllous thickets
Yes
6. Are the crowns of the woody plants continuous (>80% cover) ?
Yes
Woodlands and forests
No
7. In an average rainfall year, and with average grazing, is the grass aerial cover at least 5% ?
No
Often an overgrazed savanna
Yes
8. Are the grasses and trees spatially separate?
Yes
Forest clumps within a grassland, usually due to seasonal waterlogging
No
9. Is the top of the woody plant canopy taller than 2 m?
No
Thicket or shrubland
Yes
10. Savanna

The soils of dry savannas are base-rich where the parent material is basic igneous rock such as basalt, or fine-grained sediments such as shale or

Figure 1.2. The factors which lead to the occurrence of savannas, and their interrelationships. Only the main linkages and feedbacks are shown. In the centre are the factors contributing to savanna structure and function, and in particular those required for the savanna definition we have adopted: a significant contribution to the ecosystem primary production by both trees and grasses. The next level represents the factors which determine savanna structure and function: water availability, nutrient availability, fire and herbivory. The outermost level contains the factors which lead the determinants to have their characteristics.

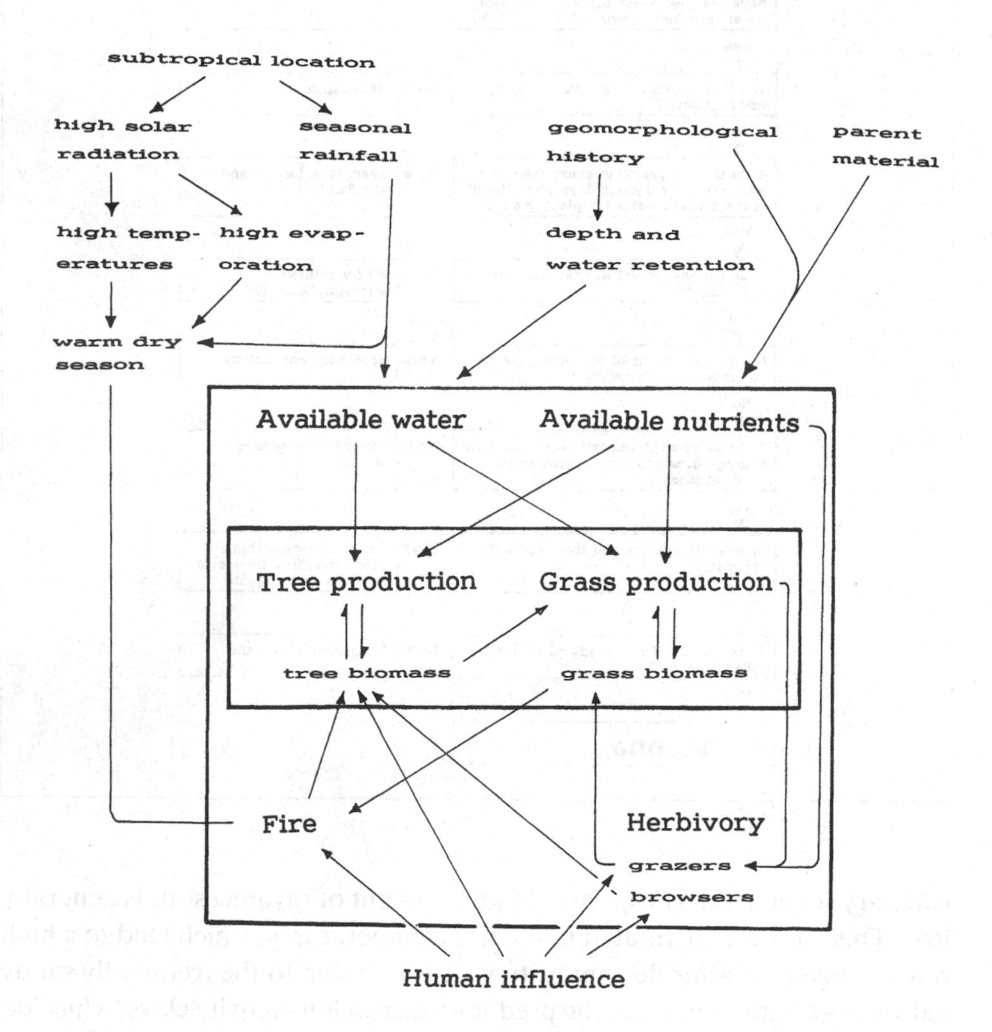

mudstone. The concentration of bases in dry savannas is sometimes so high that the soils are alkaline, and accumulations of free salts occur in the profile. Where the parent material is basalt or related basic lavas, clayey soils dominated by smectites (cracking clays) and rich in organic matter occur. Acid igneous parent materials (typically granites) in dry savannas result in a landscape with sandy, infertile uplands and clayey, fertile bottomlands.

The wet savannas mostly occur on the high-lying surface of the African shield, which is extremely old. The soils are acid and infertile as the result of prolonged leaching, almost regardless of the parent material. The soils can be very deep, and may not have a clear clay increase with depth. The clays are predominantly kaolinites and oxides of iron and aluminium.

In southern and Central Africa there are vast deposits of aeolian sand, which occur in both wet and dry areas. The soils they provide are deep, structureless and low in plant nutrients.

Savannas, by definition, could not have existed before the evolution of trees and grasses. The main savanna tree families had evolved by the end of the Cretaceous Period, 150 million years ago (Raven & Axelrod 1974). The first grasses probably appeared in the Palaeocene Epoch, about 60 My ago (Clayton 1981), perhaps as forest understorey plants. The breakup of Gondwanaland into the present continents was just commencing. By 30 My ago, Africa had moved 15° northwards, to its present position. As the gap between Africa, Antarctica and Australia grew, and the gap between Africa and Eurasia decreased, so the present pattern of oceanic circulation became established, and with it, the present climate distribution in Africa. The increasing areas of seasonal rainfall were ideal for grasses, which came into their own during the Oligocene Epoch (38–26 My). By the Miocene (25–15 My), grass pollen is conspicuous in the fossil record, and the first grazing mammals appear. The annual dry season and the presence of a continuous, easily dried fuel made regular fires inevitable, and savannas were born. Their distribution in Africa has been relatively constant since the late Miocene Epoch, 15 My ago.

The best-known feature of the African savanna fauna is the diversity and biomass of large mammals, particularly antelopes. It is seldom noted that other groups, such as the rodent fauna, are equally diverse (Bigalke 1978). During the period between the commencement of the breakup of Gondwanaland and the re-establishment of the connection with Eurasia in the Oligocene Epoch, the land animal fauna of Africa was effectively isolated from that of the rest of the world. Many of the rodents endemic to African savannas are thought to have evolved during this period. The Bovidae (antelopes) are thought to have arrived from Eurasia during the Oligocene, along with the Suidae (pigs), Giraffidae and Rhinocerotidae, while the Equidae and Hippopotamidae arrived during the Miocene. The Pliocene Epoch (7–2.5 My) saw a massive species radiation in the Bovidae and Suidae, which corresponded to speciation among the grasses, and the exploitation of the newly formed savanna environment. Apart from the coincident timing of bursts of speciation, there is morphological evidence for linking the evolution of grasses and ungulates (Stebbins, 1981). The high-crowned teeth needed to graze siliceous grass leaves, and the slender, hoof-tipped legs of the antelope, would be of no use outside fairly open, grassy environments. The hallmarks of the graminoid life form – basal tillering and intercalary meristems – provide grasses with a unique advantage in a heavily grazed environment. However, it could also be argued that they improved the fitness of grasses in a fire-prone environment, and any advantage they confer in a grazed environment is a happy accident. The parallel radiation of grasses and ungulates occurred in Asia and the Americas as well as in Africa, but the rich antelope communities survived only in the African savannas, possibly because they were less exposed to the Pleistocene climatic fluctuations there. On the other hand, only a few remnants of the diverse African Pliocene pig fauna remain.

The word 'savanna' first appears in Spanish writings in the sixteenth century and is thought to be a Carib Amerindian word meaning a treeless grassland. It is still used in this sense in South and Central America, but for at least a hundred years it has also been used to describe wooded grasslands, both in South America and elsewhere. The current scientific usage of the term generally implies a landscape which is conspicuously woody. To remain consistent with the functional savanna definition adopted above, it is necessary to exclude treeless tropical grasslands. In Africa at least, these are of relatively small extent, and are associated with characteristic soil conditions (Michelmore 1939; Whyte 1962). We believe that there are sufficient ecological differences between tropical grasslands and tropical savannas to justify treating them as separate biomes. For instance, tree–grass interactions are central to our concept of savannas. Montane tropical grasslands share many functional characteristics with temperate grasslands, and tropical seasonally flooded grasslands are a type of tropical wetland. Treeless tropical grasslands may more logically be grouped with these biomes, at least conceptually. In practice, just how many trees a grassland needs before it qualifies as a savanna is obviously an arbitrary decision. Sparsely wooded tropical grasslands that are neither cold nor occasionally flooded, but are kept open by frequent, intense fires represent a particularly troublesome case, since a small change in fire regime can result in their classification either as savannas or grasslands. We class them as savannas.

Seasonally waterlogged grasslands with included tree clumps are widespread in Africa. The tree clumps are often on slightly raised ground associated with termitaria. In South America these would be classified as hyperseasonal savannas (Sarmiento 1984), and probably conform to the original meaning of 'savanna'. We prefer to treat this type conceptually as a mosaic of islands of woodland or forest (the bush clumps) within a grassland matrix, rather than as a special type of savanna.

Another vegetation formation which does not fit comfortably into the savanna definition is the tropical thicket. These are dense, low-growing (typically 1.5–3 m tall) formations dominated by multi-stemmed woody plants, and are widespread in Africa. They occur in situations where the rainfall is low and temperatures are high. It appears that the water supply to woody plants in thickets is more evenly spread through the year than in typical savannas. This can be due to a less highly seasonal rainfall, or edaphic conditions such as deep, sandy soils. In extreme cases there is hardly any grass layer under the thicket, and they are therefore disqualified as savannas. However, many thickets do include a significant amount of grass. Furthermore, savannas can be induced to become thicket-like through poor management, and some thickets will, after a time, revert to savannas. The large, well-known thickets (for instance, the Valley Bushveld in South Africa, the Jesse Bush in Zimbabwe, and the Itigi thicket in Tanzania: White 1980), on the other hand, are apparently stable configurations, as is evidenced by the high species endemism associated with them. Thickets and savannas share many species, as do thickets and forests. In the absence of better understanding of the ecology of thickets, we treat them as a class of savannas rather than as miniature dry forests.

At the dry end of the savanna spectrum the woody plants become shorter, but not necessarily sparser. It is not clear whether low-growing, dry savannas are functionally distinct from summer-rainfall desert shrublands, or whether a single conceptual model will suffice for both.

Despite their tropical nature, savannas can be exposed to temperatures low enough to reduce the rate of processes such as growth and decomposition. A large portion of the African savannas occurs at elevations greater than 1000 m and at latitudes greater than 20°. Severe frosts occur infrequently in these savannas, but can have a marked effect on the structure and composition of the woody component. This raises the question of whether temperate mixed tree–grass associations can be considered to be savannas. The question is unanswered, but we intuitively feel that the annual occurrence of an extended period during which low temperatures are the major controlling factor over plant production removes temperate wooded grasslands from our savanna concept, in which lack of water during the 'winter' months is the unifying factor.

Fire is a crucial factor in the ecology of all savannas. It varies in frequency from annual in the wet savannas, to once every ten or more years at the dry extreme. If fire is excluded from savannas, the woody plant density and biomass increase. To argue from this observation that savannas are merely arrested forests is analogous to saying that if only they received more water they would be forests. The association between fire and savannas is as old as savannas themselves, since it follows inevitably from their climate and fuel characteristics. The perception that savannas are a fire-subclimax to some other vegetation is a prejudice introduced by ecologists trained in areas where fire is regarded as a natural catastrophe rather than a regular feature of the environment, like the occurrence of a dry season.

Another common misconception is that savannas are transitional between grasslands and forests. This is seldom true in a purely geographical sense in Africa, where the main vegetation gradient is one of increasing aridity. The vegetation sequence along this gradient leads from forest, through woodland and savanna to desert shrubland, not grassland. Pure grasslands are rather rare in tropical Africa. The temperature gradient associated with elevation in mountainous areas does sometimes place savannas between tropical lowland forest and montane grasslands, but can also proceed from montane grassland to montane forest, without a savanna intermediary, or to montane thickets. The other main class of African grasslands comprises those maintained by seasonal waterlogging, and therefore occur between riparian forests and savannas on the aridity gradient.

Furthermore, savannas do not represent an ecologically intermediate case between forests and grasslands. They lack the tendency towards organic matter accumulation in the topsoil characteristic of grasslands, and the microclimatic amelioration typical of forests. The conceptual problems posed by savannas have been sidestepped in some circles by treating the herbaceous layer as a grassland, separate and independent of the tree layer. However, it is precisely in the interaction between trees and grasses that savannas differ from grasslands or forests. Based on these observations, a working definition of savannas can be proposed. It is presented in the form of a set of rules in Box 1.1.

The plant families of the African savannas belong to the Palaeotropical Kingdom, which includes the tropical flora of all the continents. The next lower level of phytogeographical classification is the Sudano–Zambezian Region, which corresponds to an ecological (largely climatic) definition of African savannas (Werger & Coetzee 1978). White (1980) separates the

Sudanian and Zambezian floristic regions and defines a number of transition zones (Table 1.1). While he recognises that the Sudanian and Zambezian regions have a great deal in common, both in appearance and species composition, he justifies the division on the basis that the former falls in 'low Africa', where the geology is mostly sedimentary and the altitude is below 1000 m, and the latter in 'high Africa', above 1000 m and with a largely igneous geology. To some extent this mirrors the distinction, introduced later in this chapter, between the infertile and fertile savannas. Rough calculations from White's estimates indicate that there are about 13 000 plant species within the combined savanna regions of Africa (excluding Madagascar), of which about 8000 are savanna endemics.

Several classification schemes have been proposed for the division of African savannas into subtypes. Most are based on vegetation structure rather than species composition, owing to the unevenness with which species data are available on an Africa-wide basis. The Yangambi proposal, on which the first vegetation map of Africa was based (Keay 1959), recognised woodlands, savannas and steppes under our broad savanna definition. The difference between savannas and steppes was based on the growth form of the grasses. Savannas had tall, perennial grasses with leaves produced all the way up the stem, while steppes had shorter grasses, with leaves produced at ground level. Both growth forms occur in the same plant communities at Nylsvley. We feel that the term 'steppe' should be reserved for temperate grasslands with a strong temperature constraint.

Disagreements over the meaning of the word 'savanna' led to it being dropped as a vegetation category in systems such as that of Pratt, Greenway & Gwynne (1966) and Edwards (1983). They used structurally defined terms such as 'woodland', 'bushland', 'wooded grassland' and 'thicket' instead. The Nylsvley study site would be classified as a wooded grassland in these systems, since it has about 40% canopy cover by trees, which are about 6 m tall on average. Cole (1986) proposed a structural classification specifically for savannas, based mostly on the height and canopy cover of the woody plants. She defined five classes: savanna woodlands (trees more than 8 m high, grass more than 0.8 m high); savanna parklands (scattered trees less than 8 m high, grasses more than 0.8 m); low tree and shrub savannas (short grass, short trees); thickets (dense, low shrubs); and savanna grasslands (treeless, with tall grass). She classed the savanna study site at Nylsvley as a savanna woodland, on the basis that the *Burkea africana* trees were up to 15 m tall. In fact, few are taller than 7 m, and the woody cover is 30–40%, which places Nylsvley among the low tree and shrub savannas.

The most recent all-Africa vegetation description is White's (1980)

Table 1.1. *The proportion of savannas and species richness in various African phytochorological regions (after White 1980). All the values in this table are approximate*

Phytochorological region	Proportion of savannas (%)	Total area (km^2 × 1000)	Total number of species	Number of endemic species
Predominantly savannas				
II Zambezian Region	>95	3770	8500	4590
II Sudanian Region	>95	3731	2750	910
IV Somali–Masai Transition Zone	90	1873	2500	1250
X Guinea–Congolian/Zambezian Transition Zone	20	705	2000	few
XI Guinea–Congolian/Sudanian Transition Zone	30	1165	2000	few
XIII Zanzibar–Inhambane Coastal Mosaic	50	336	3000	200
XIV Kalahari–Highveld Transition Zone	75	1223	3000	few
XV Tongaland–Pondoland Mosaic	50	148	3000	200
XVI Sahel Transition Zone	50	2482	1200	40
Forests				
I Guinea–Congolian Region	20	2800	8000	6400
XII Lake Victoria Mosaic	30	224	3000	few
Sclerophyllous thicket				
V Cape Region	0	71	7000	3500
VII Mediterranean Region	0	330	4000	2900
Montane grassland with forest patches				
VII Afromontane & Afroalpine Archipelago	0	715	4000	3000
Deserts				
VI Karoo–Namib Transition Zone	0	661	3500	1750
XVII Sahara	0	7387	1620	188
Totals		27 619		

AETFAT/Unesco map. The geographical distribution of African savannas, as shown in Figure 1.3, is based on this map. White's classification uses Greenway's structural categories, but incorporates floristic (species composition) information as well. He divided Africa into seven broad regional centres of endemism, and a number of transition zones and special cases. Although he avoided the use of the word savanna in his classification, he recognised it as an embracing term for a range of ecologically similar

Figure 1.3. The savannas of Africa. Core savanna areas (1) are defined as having > 80% of the vegetation as savanna. The transition zones between savannas and forests (2), and savannas and arid shrublands (3), contain 20–80% savanna by area.

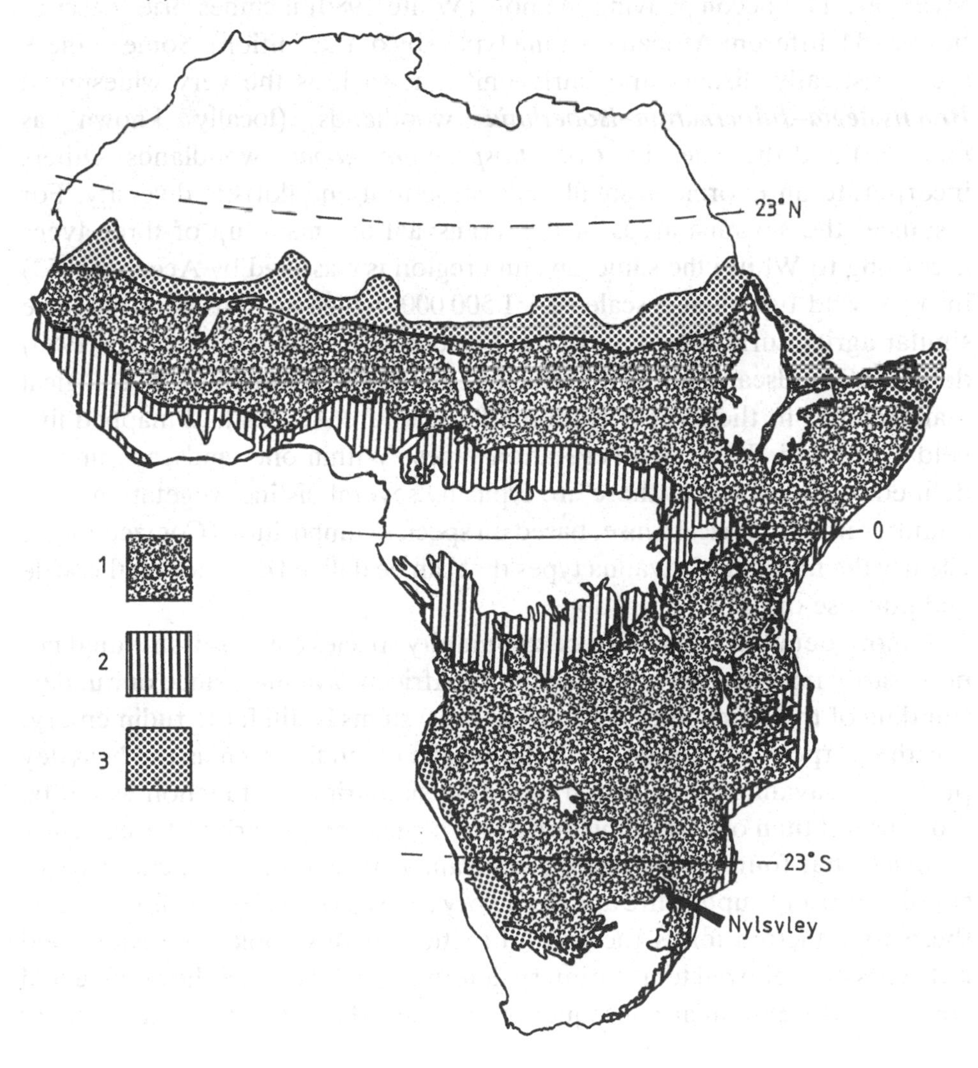

vegetation classes. Formations which he describes as woodland, bushland, thicket, shrubland and wooded grassland fall under our broad savanna definition, along with some dry forests (in particular, the deciduous *Baikiaea plurijuga* forests on Kalahari sand). Nylsvley falls within White's Kalahari Woodland map unit, because of an inaccurately placed boundary rather than a true affinity. On the basis of the vegetation descriptions, it should actually be within the South Zambezian Undifferentiated Woodland and wooded grassland, a map unit which encompasses a wide variety of savannas.

How many different types of savannas are there in Africa? The 1:5 000 000 AETFAT/Unesco map has 27 mapping units which satisfy our savanna definition, out of a total of 90 units. Some of these include only a single vegetation type, and others are a mosaic of types, some of which are not savannas. The accompanying memoir (White 1980) includes brief descriptions of 34 different African savanna types (by our definition). Some of these are floristically distinct and fairly uniform, such as the very widespread *Brachystegia–Julbernardia–Isoberlinia* woodlands (locally known as *miombo*) and the extensive *Colophospermum mopane* woodlands. Others incorporate an enormous small-scale structural and floristic diversity. For instance, the savanna areas of the Transvaal are made up of three types according to White; the same savanna region is classified by Acocks (1953) into 13 'veld types', at a scale of 1:1 500 000, which he considered to have similar agricultural potential. In the Eastern Transvaal, Gertenbach (1983) defined 19 landscape units (at a scale of 1:250 000) for purposes of ecological management in the Kruger National Park where Acocks had mapped five veld types and White two vegetation types. Within one landscape unit as defined by Gertenbach, there are typically several distinct vegetation communities in the classical sense, based on species composition (Coetzee 1983). Clearly the number of savanna types that can be defined depends on the scale and purpose of the investigation.

A more detailed knowledge of community species composition would not necessarily improve the management of African savannas, since our understanding of the functioning of savanna ecosystems is still fairly rudimentary. For the purpose of extrapolating the results of studies such as the Nylsvley project, a savanna classification based on similarities of function would be more useful than one based on floristic or structural similarity. It is currently believed that four factors dominate savanna structure and function: water supply, nutrient supply, fire and herbivory (Frost *et al.* 1986). A discussion of these four factors forms the second section of this book. The water and nutrients are referred to as 'primary determinants', because they define and constrain the potential consequences of the other two. For example, no amount of fire protection will allow a savanna receiving 500 mm of rainfall to become a tall, closed woodland. A given savanna can be positioned in the space defined by water supply on the one axis and nutrient supply on the other; with fire and herbivory as third and fourth axes if appropriate. This approach emphasises the continuity of savannas rather than their separation into distinct classes. The underlying assumption is that savannas lying close together in this 'determinant-space' will be more similar ecologically than those located far apart. In essence, it is a direct gradient analysis technique where the gradients have been selected on the basis of the pooled experience of the scientists collaborating in the intercontinental 'Responses of Savannas to Stress and Disturbance' programme (Frost *et al.* 1986).

Savannas occur in areas with annual rainfall ranging from 350 to 1800 mm (and up to 3000 mm on very infertile sites: Nix 1983). The water availability gradient is therefore very obvious. The distinction between 'wet' and 'dry' savannas is implicit in savanna descriptions from early times; for instance, it largely underlies the Yangambi conference separation of savanna and steppe. However, the degree of aridity is a continuum, which makes any classes based on it unavoidably arbitrary. A useful distinction can be made between those savannas in which grass production is a strongly linear function of annual rainfall ('dry') and those where the relation is weak or absent ('wet'). This transition occurs at about 1000 mm annual rainfall; it is lower on sandy soils, and higher on clays.

Huntley (1982) pointed out that the African arid–moist gradient was associated with a soil fertility gradient. The ancient surfaces of the African shield tend to be wetter and less fertile ('dystrophic') than the surfaces formed since the breakup of Gondwanaland ('eutrophic'). This correlation is not coincidental: weathering and leaching on the older, wetter surfaces is more intense as well as more prolonged than that on the younger surfaces, which have the additional advantage of more base-rich parent materials. African savannas fall more naturally into discrete fertility classes than they do into aridity classes. The reasons for this are explored more fully in Chapters 4 and 12. Since structural features such as biomass, spinescence and leaf size tend to support the distinction between the fertile and infertile savannas, it seems a promising basis for a functional classification of savannas. The distribution, body size and biomass of savanna herbivores is to a large degree predictable on the basis of soil fertility and annual rainfall (Bell 1982; East 1984), lending further support to the idea that nutrient and water availability are fundamental determinants of savanna ecology.

Following the 'functional classification' approach, the Nylsvley study site has representatives of both main African savanna types. The *Burkea africana*-dominated broad-leafed community is an example of the moist, dystrophic savannas, and the spinescent, small-leafed *Acacia tortilis* community represents the dry, nutrient-rich savannas.

Summary

Savannas, defined briefly as tropical ecosystems in which the primary production is contributed both by woody plants and grasses, are the dominant land cover in Africa, and a major world biome. The main determinants of savanna distribution, structure and function are water availability, nutrient availability, fire and herbivory. Savannas have existed in Africa for at least 30 million years, ever since the advent of hot climates with markedly seasonal rainfall made fire a frequent ecological factor. There are two broad classes of savanna in Africa, the broad-leafed and fine-leafed savannas, which tend to occur in nutrient-poor high-rainfall, and nutrient-rich low-rainfall areas, respectively. Nylsvley has representatives of both types.

2

The people of Nylsvley

There is a tendency among ecologists, especially if they habitually work in conserved areas, to regard people and their effect on ecosystems as an unnatural disturbance. In practice, the distinction between 'natural' and 'man-made' ecosystems is blurred and frequently arbitrary. This is particularly true of African ecosystems, which have an ancient association with humans, but only a recent exposure to Western technology. This chapter describes the successive inhabitants of the Nylsvley region, and the impact which they have had on the savannas there.

The first white settlers believed that they had come into a pristine landscape, despite the fact that it was populated to varying degrees, and had been for millennia. Their attitudes persist in a nostalgic perception by the South African public of the 'pre-colonial period' as a time of undisturbed natural harmony and equilibrium. In fact, disturbances play a critical part in maintaining the structure of the vegetation. In the context of savannas, many of these disturbances are partly or wholly of human origin: burning, wood-collecting, grazing, and cultivation to name a few. The historical land-use pattern is important in that it is one of the determinants of the ecosystem as we see it now. The 'wilderness' landscapes of Africa are in part a consequence of centuries of human influence.

Nylsvley offers, within a small area, all the ingredients for successful human habitation: water, grazing, arable soils, fuelwood and rocky hills for refuge in times of unrest. It is therefore not surprising that it has been the site of repeated settlement. Some of the earliest hominid relics (*Australopithicus africanus*, dated to 2–3 My BP) were discovered 60 km to the north, at Makapansgat (Brain 1981). The fossils and the associated faunal remains indicate that these were omnivorous savanna-dwelling ape-men, as much hunted as hunters. Their effect on the environment was probably minimal, and they were extinct by 1 My BP.

The Stone Age people

The ability to make and use tools is the hallmark of early man, and it substantially increased his impact on the environment. In particular, the use of fire for both hunting and domestic purposes would have significantly altered the frequency and season of bush fires. The controlled use of fire appears to be as old as the genus *Homo*, and possibly older (Brain & Sillen 1988). *Homo habilis* and *H. erectus*, the first toolmakers, appear in the fossil record before the demise of the ape-men. In southern Africa, the earliest traces of an indisputably stone tool-based culture are from the 'Cave of Hearths' at Makapansgat (Mason 1988), dated to about 250 000 BP. This is referred to as the Early Stone Age, or Acheulian culture.

Among the earliest remnants of 'modern' man (*H. sapiens*) in the southern African interior is the skeleton found at Tuinplaats, about 20 km to the south of Nylsvley (Inskeep 1979), thought to be from the Middle Stone Age (about the time of the last interglacial, 30 000 BP). It has some anatomical features of *H. erectus*, and was probably ancestral to the modern-day San (bushmen). By this time language and a sophisticated hunter–gatherer culture had evolved, probably quite similar to that of the present San. Stone implements from the Middle and Late Stone Ages are widespread in the Nylsvley region (Fichardt 1957), including some from the banks of the Nyl river, near the study site. They are difficult to date with any precision. In all probability Stone Age hunter–gatherers used Nylsvley up to and beyond its settlement by Bantu-speaking people, and possibly as recently as a few centuries ago.

Anthropological research among the modern San shows that they make extensive use of bush fires, at all times of the year, in order to attract game to the newly-burned patches. The !Kung bushmen of northwestern Botswana (Bicchieri 1972) live in an environment quite similar to Nylsvley. A nomadic band consists of 23–40 people, and uses an area of about 250 km^2, giving a density of about one person per 10 km^2. Although their diet is 60–70% vegetarian, they can have a significant impact on game populations as well. Data from the G/wi bushmen indicate an annual consumption of about 380 kg of meat per person, ranging from giraffe to tortoises. Therefore the offtake of meat during this period amounted to about 0.4 kg ha^{-1} y^{-1}, which is approximately 1% of the herbivore standing crop, and about 4% of the typical offtake from a modern-day cattle ranch.

The Bantu-speaking people

The next big change in the ability of people to modify the southern African savanna environment occurred around 2000 BP, with the appearance of the Bantu-speaking people. These are Central African people with a culture which included pottery, iron-working, domestic animals and crop agriculture. It is clear that these technologies diffused southwards from North, West and Central Africa, but it is not clear if the people themselves moved *en masse*, and if so, from where (Inskeep 1979). Their settlement clusters were relatively large and permanent, giving rise to a population density which was probably between one and five people per km^2. The legacy of these Iron Age people is a major feature of the Nylsvley study site. The style of pottery (Figure 2.1), and ^{14}C dating of charcoal excavated from archaeological sites at Nylsvley show that the settlements belonged to the Tswana culture of the Middle Iron Age (Fordyce 1980).

This was a cattle-centred culture. Cattle, sheep and goats are thought to have been domesticated in Asia Minor starting around 9000 BP. The earliest record of cattle in sub-Saharan Africa is in southern Sudan about 5000 BP (Epstein 1971; Phillipson 1977; Cumming 1982) but they reached southern Africa only around 1500 BP (Inskeep 1979). The delay may in part have been attributable to the wide belt of tsetse fly (*Glossina morsitans*) infested country which covers south Central Africa (Ford 1960, 1971). Tsetse fly is the vector of *nagana*, a fatal trypanosoma disease of cattle. The cattle were small beasts (200–300 kg each) by modern standards, known as *sanga* cattle. Evidence from culturally similar groups during the historical period suggests a stock density of one cow and two or three goats per person on average, which implies a stocking rate of 5–20 kg ha^{-1}. The wild herbivore biomass, inferred from annual rainfall (Coe, Cumming & Phillipson 1976; East 1984) would be about 40 kg ha^{-1} at Nylsvley in the absence of cattle. The high density of predators such as lion, leopard and spotted hyena necessitated the penning of cattle in the middle of the household-village at night.

Figure 2.1. A 14th century pot from site 12 at Nylsvley, typical of early Sotho-Tswana speaking peoples.

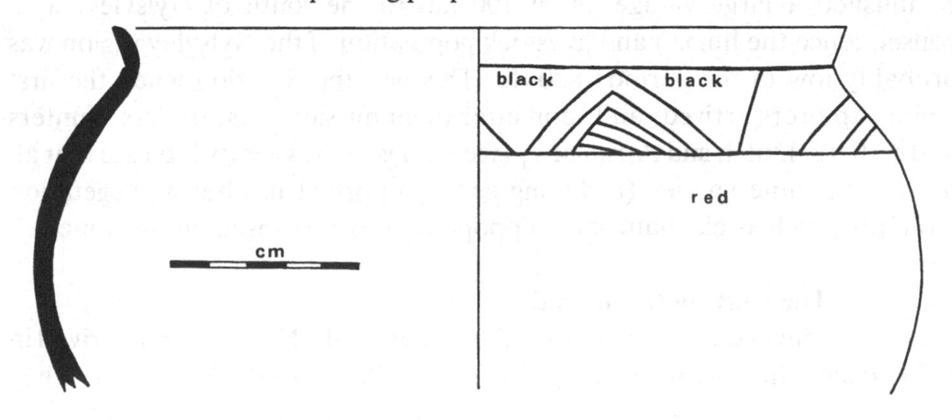

The traditional grain crops, sorghum (*Sorghum bicolor*) and millets (*Pennisetum* sp. and *Eleusine coracana*) originated in the Sahelian savannas, and have been cultivated since about 5000 BP (Phillipson 1977). Indications of a grain-based agricultural economy in southern Africa (granaries, grinding stones and pottery) appear after AD 500. Sorghum remains have been excavated within the Nylsvley study site. Judging from contemporary Tswana practices, the fields were typically small (1 ha per family unit), close to the village and interplanted with a variety of grain and vegetable crops. Each field is cultivated for a few years before being abandoned, using oxen, wooden ploughs and hand-held hoes. The diet was supplemented with wild fruits such as *Sclerocarya birrea*, tubers and leaf vegetables. The cattle were not primarily kept for meat production; meat was provided by goats and the hunting of antelope and birds. Like the San, they used fire to attract game and to provide green grass for their livestock.

Apart from hunting, gathering, grazing and cultivation, these people had a substantial impact on vegetation structure through wood collection for fuel and construction. Each person would have used about 1 ton of wood per year for domestic fires (Gandar 1986; Eberhard 1987). Each family cluster (10–20 people) would have 3–5 huts, each requiring about 1 ton of wood in its construction, and lasting about 20 years. The livestock pens and the family unit were surrounded by a palisade of branches (a *boma*) containing up to 20 tons of timber. Fields were fenced with brushwood, using about 4 tons per 100 m (von dem Bussche 1988). Iron-smelting furnaces, ash and slag-heaps have also been located at Nylsvley. Iron-working would have added substantially to the wood consumption.

In the early nineteenth century, strife associated with the emergence of the Zulu military empire disrupted the settlements of the central Transvaal. The stone-walled enclosures on the hill at Nylsvley probably date from this time of unrest (Mason 1962). A militant offshoot of the Zulus, the Matabele, established a large village about 100 km to the south of Nylsvley; as a consequence the human and livestock population of the Nylsvley region was probably low in the period 1820–40. This was the situation when the first white explorers arrived; and soon after them missionaries, traders, hunters and settlers. Dutch and Afrikaans place names in the vicinity indicate that all the major game species (including some important in altering vegetation structure, such as elephants and hippopotami) were present at this time.

The white settler period

The first major wave of white settlers in the Nylsvley area arrived in 1838, occupying the well-watered fringe of the Waterberg plateau. They were Dutch-speaking pioneers known as *Trekboers*, with an economy based on cattle, sheep, wheat, maize and hunting. After a long trek north from the Cape, with the bible and vague explorers' reports as their only guide, they came upon a strong stream flowing northwards, and called it the *Nyl* (Nile). Where it flowed out onto the Springbok Flats it formed a broad marsh or *vley*; this spot became known as Nylsvley (spelled *Nylsvlei* in the modern Afrikaans form).

Land struggles with the African tribes continued up until the 1880s. This, coupled with malaria, rinderpest (a disease of wild and domestic ungulates), and low rainfall, prevented permanent white settlement at Nylsvley until after the Anglo–Boer war at the beginning of the twentieth century. During this period populations of game animals were drastically reduced by hunting, many becoming locally extinct. Around 1915, the Springbok Flats were settled by white farmers, including the pioneer botanist E. E. Galpin. The farm Nylsvley (3120 ha) was purchased by Mrs A. M. Skirving in 1914. The house which now houses the research station was built in 1917, and the stone buildings which are now the Warden's residence a decade later. The property was managed as a cattle ranch with 100–200 head of cattle for the next three decades.

From 1946 to 1974, the farmer was Mrs Skirving's grandson, George Whitehouse. During this period the cattle numbers averaged 500 head, with a maximum of 700. The cattle were Afrikaner cross-breeds with a mean mass around 270 kg. There was very little wild game in 1914, as a result of decades of uncontrolled hunting and the rinderpest epidemic. Under a strict no-hunting policy, impala (*Aepyceros melampus*) and kudu (*Tragelaphus strepsiceros*) began to return in the 1940s. By 1974 there were several hundred impala on the farm Nylsvley, mostly on the alluvial soils near the river.

George Whitehouse applied a multi-camp rotational grazing system (Grunow 1974). It was necessary to keep the cattle out of what is now the study area between July and January every year to prevent them eating the poisonous plant *Dichapetalum cymosum*. Therefore, 3 weeks after the first rains, the cattle were herded onto the *Combretum* savannas to the west of the Nyl river, where they stayed until mid January. From mid-January to mid-May they grazed in the broad-leafed savanna that was to become the study site. The broad-leafed savanna was divided into three camps each of 185 ha, and a fourth (which included the rocky Maroelakop) of 300 ha. The beef herd was divided into two groups of 250 animals each, which were rotated between the camps. This is the equivalent of a continuous stocking rate of about 40 kg ha^{-1}, or 9 ha per Large Stock Unit. From mid-May to after the spring rains, the cattle grazed on the floodplain and the adjacent *Acacia* communities. The

Table 2.1. *A summary of the history of human occupation of the Nylsvley region and the likely impacts which they had on the savanna environment*

Period	People	Impact on savannas
Early & Middle Stone Age 500 000–10 000 BP	c. *Homo erectus*	Burning, harvesting of plants and animals at low levels
Late Stone Age 10 000 BP–AD 1700 (?)	San (about 0.01 people/ha)	Burning, harvesting of plants and animals at low levels
Middle Iron Age AD 1200–1840	Bantu (about 0.1 people/ha)	Burning, stocking at low levels with cattle, sheep & goats, wood collection
1840–1900	Trekboers	Hunting, light stocking with cattle
1900–1974	European settlers	Cattle stocking at high densities. Extensive cultivation on fertile soils
1974 onwards	State land	Conservation management. Reintroduction of endemic herbivores

major difficulties in maintaining Nylsvley as a cattle ranch were veterinary diseases and the low price of beef. A no-burning policy was followed, but wildfires nevertheless occurred in the study area approximately once every 3 years, usually in late winter. The interval between fires for a given spot in the broad-leafed savanna is thought to have been 5–10 years. The old fields in the study site were cultivated for a period of about 5 years in the early 1960s, but were found to be unprofitable.

The modern period: Nylsvley as a Nature Reserve

Nylsvley was purchased by the state in 1974. It is managed as a nature reserve by the Transvaal Division of Nature Conservation, with the primary objectives of providing a savanna research site, breeding rare antelope, and protecting the bird communities of the Nyl floodplain. Many antelope species thought to have been indigenous to the region have been reintroduced. The populations of large herbivores are controlled by culling. The reserve is burned in blocks of several hundred hectares each, according to an assessment of the degree of grass moribundity in each block. The blocks comprising the study site have been burned every 4–10 years since 1974.

Table 2.1 summarises the land-use history of the Nylsvley region.

Summary

Nylsvley is not a savanna ecosystem free of human influence, and some of its most interesting ecological features are the result of that influence. It has been under almost continuous occupation by humans since at least the Middle Stone Age, more than 20 000 years ago. As the Stone Age hunter–gatherers made way for Iron Age pastoralists, who were in turn displaced by European settlers, so the magnitude of the impact of humans on the Nylsvley ecosystem increased.

3

The climate at Nylsvley

Climate is the ultimate determinant of savanna distribution. The prolonged hot dry season permits the frequent, hot fires which are essential for maintaining the tree–grass mixture.

Nylsvley has a typical savanna climate: a hot wet season and a warm dry season. A brief review of the climatic processes which influence the weather at Nylsvley illustrates some features which are common to savannas in general. The details differ between the northern and southern African savannas because of the greater width of the continent in the north. The influence of the oceans in the southern subcontinent causes the rainfall isohyets there to lie in a north–south direction, rather than the east–west orientation typical of the Sahelian Region. The East African savannas span the equator and therefore have a monsoonal climate, with two wet and two dry seasons per year.

In global terms, savannas are located between the equator and about 30° latitude. In relation to the global atmospheric circulation, this is between an area of divergence (subsiding air) known as the subtropical high-pressure ridge, where the Hadley and Ferrel cells meet, and an area of convergence (rising air) where the north and south Hadley cells meet. Descending air is dry and stable, and therefore not conducive to rainfall; the opposite is true of rising air. The latitudinal location of these zones moves north and south in response to the annual cycle of the earth–sun system. During the winter months the high-pressure ridges lying over the savanna regions intensify and block out the moist tropical air. In summer they shift pole-ward, and allow the ingress of tropical air, resulting in rain.

This general pattern is evident in southern Africa, but is slightly modified by the relationship of the continental land mass to the oceans on either side (Preston-Whyte & Tyson 1988). The southern African circulation is controlled by a pair of subtropical high-pressure cells (anticyclones), one over the Indian Ocean, and one over the Atlantic, and a high-pressure cell over the interior of the subcontinent. To the north are the tropical easterly waves associated with the Inter-Tropical Convergence Zone (ITCZ) (which are fairly stationary in southern Africa, unlike those in the Sahel), and to the south the eastward-moving temperate westerly waves. Both the temperate and tropical waves can produce weather disturbances, and their consequences depend on the position and strength of the subtropical highs.

During the winter the anticyclone over the interior is strong and stable. The descending air associated with it is dry, and the presence of inversions prevents convective storms. Cold fronts associated with westerly waves regularly penetrate deep into the interior. During the summer, the oceanic anticyclones migrate from 21° S to about 31° S, and the ITCZ moves into the southern hemisphere. The continental high is weaker, and lies on the eastern seaboard, allowing low-pressure troughs or cells associated with the moist, tropical easterly waves to penetrate southwards. This brings unstable air to the Nylsvley area, and the possibility of rain for a few days. Alternatively, the subtropical anticyclone may form a ridge on the east coast, which draws in moist, warm air from the Indian Ocean. A westerly wave may act as a trigger for the instability caused by these air movements, especially if it generates a cutoff low over the interior.

The consequence of these processes is that rainfall at Nylsvley occurs mostly as convective storms of high intensity and short duration, which are usually not associated with frontal systems. Rainfall tends to occur on several successive days, followed by a few days or weeks of clear, sunny weather.

Much of the detail which follows in this chapter is drawn from Trevor Harrison's doctoral thesis (Harrison 1984).

Radiation

Most ecosystem processes are ultimately driven by radiant energy from the sun, and savannas are among the sunniest ecosystems in the world. The mostly cloudless conditions in the subtropics result in a below-atmosphere annual radiation load which is generally higher than that received nearer the equator. On average, Nylsvley receives 75% of its potential annual total of 4371 sunshine hours. The total annual solar radiation at canopy level is about 7316 MJ m^{-2}, which is 61% of the radiation received above the atmosphere. The extreme instantaneous irradiance can reach 1430 W m^{-2}, which exceeds the solar constant of 1360 W m^{-2}. This occurs briefly when a cloud reflects extra radiation downwards, just before and after it blocks the sun. The components of this radiation are illustrated in

Figure 3.1. The annual course of solar radiation received at the surface at Nylsvley is shown in Figure 3.2. The annual total Photosynthetic Photon Flux Density (PPFD) (that is, the energy available for primary production, with wavelengths between 300 and 700 nm, expressed in quantum units) is 16 242 mol m^{-2}.

Temperature

The most robust definition of 'the Tropics' is one which rests on the magnitude of annual variation in temperature, rather than on strict latitudinal limits. The seasonal trend of air temperature is summarised in Table 3.1. The mean annual temperature at Nylsvley is 19.0 °C and the mean annual range (the difference between the mean temperatures of the coldest and warmest months) is 10.6 °C, which is less than the mean daily range (17.3 °C in June to 11.9 °C in February).

The absolute recorded temperature range is between 38.5 °C and −3.2 °C. Minimum temperatures below 0 °C can be recorded between May and August, but because the atmosphere is dry in these months, frosts are fairly rare (a severe frost occurs about once every 5–10 years).

The annual mean soil temperature is close to 22 °C at all depths between 0.1 and 1.2 m, but the daily and seasonal variation is highest in the surface horizons. At 0.1 m, the monthly mean values range from 14 °C (June) to 29 °C (December), while at 1.2 m they range from 18 °C (July) to 24 °C (February). The mean daily ranges at 0.1, 0.2, 0.3, 0.6 and 1.2 m are about 13, 4, 1, 0.2

Figure 3.1. Components of the mean annual radiation balance in broad-leafed savanna at Nylsvley.

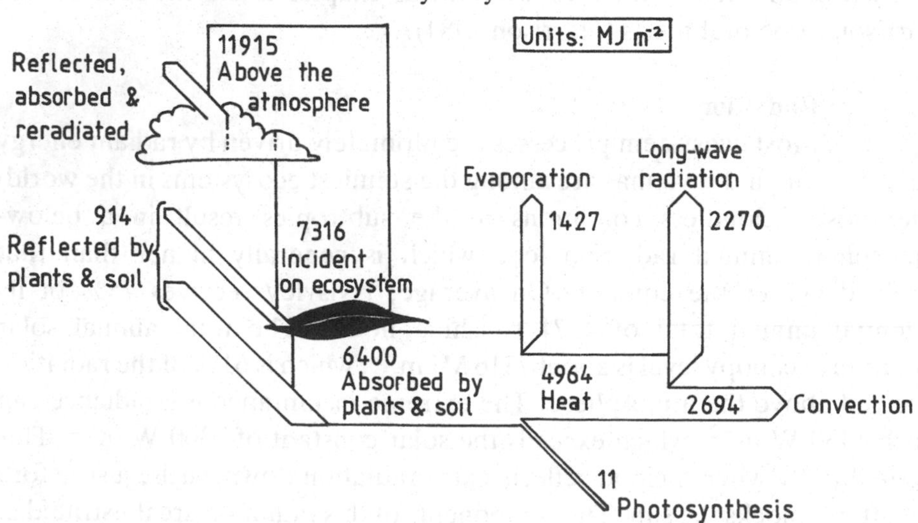

Table 3.1. *Monthly mean values of climatic variables measured at the weather station in the study site over the period July 1975 to June 1982 (Harrison 1984). Most of these values have been interpolated from graphs. The least significant digit may therefore be in error*

	Jan.	Feb.	Mar.	Apr.	May	Jun.	Jul.	Aug.	Sep.	Oct.	Nov.	Dec.
Radiation												
Net solar (MJ m^{-2})	753	624	614	639	524	504	468	530	600	716	609	735
PAR (mol m^{-2})	1632	1385	1363	1418	1163	1118	1038	1177	1332	1590	1352	1632
Air temperature												
Daily maximum (°C)	29.2	28.5	27.4	26.0	24.0	21.0	21.2	24.1	27.4	28.8	29.2	29.2
Daily minimum (°C)	16.8	16.6	15.0	11.6	7.7	4.0	4.0	7.8	11.6	13.2	15.6	16.0
Daily mean (°C)	23.0	22.3	21.0	18.6	15.8	13.0	12.6	16.0	19.6	21.0	22.2	22.4
Wind speed (m s^{-1})	1.45	1.30	1.25	1.25	1.20	1.25	1.25	1.70	1.80	1.90	1.80	1.60
Absolute humidity (g m^{-3})	14	14	13	10	7	6	5	6	8	9	12	13
Rainfall (mm)	150.6	87.2	69.0	27.8	6.0	0.8	0.04	8.6	21.6	53.1	106.9	81.3
Evaporation												
Equilibrium (mm d^{-1})[a]	5.4	5.2	4.8	4.7	4.1	3.8	4.0	5.0	6.2	6.7	6.0	6.1
Class A Pan (mm d^{-1})	7.2	7.0	6.5	6.0	5.1	4.8	5.0	6.6	7.7	8.7	7.7	7.8

[a] '... the asymptotic value approached far enough downstream from the leading edge of an extensive, freely transpiring vegetated region, where the air is close to saturation...' (Thom 1975).

and 0 °C, respectively. The intra-annual range at 0.5 m depth is about 9 °C, which places the soil outside the 'Soil Taxonomy' definition of a tropical (isohypothermic) soil, but at the upper end of the thermic range (Soil Survey Staff 1975).

Humidity

Relative humidity (RH) is strongly dependent on air temperature. For this reason the water content of the air will be expressed in absolute terms, from which the RH can be calculated. The mean monthly atmospheric water content varies from 5.2 g m^{-3} in July (RH = 40%) to 13.5 g m^{-3} in January (58%), with an annual mean of 9.4 g m^{-3} (50%). The atmosphere is therefore relatively dry throughout the year, but is especially dry in winter. The water content of the atmospheric column is mostly controlled by the large-scale movement of air masses, and so does not usually vary greatly on a daily basis, but the RH responds to the diurnal temperature cycle. Close to the vegetation, it also responds to the diurnal rhythm of transpiration. Humidity is close to saturation near dawn, and declines rapidly during the day.

Wind

The mean annual windspeed at 2 m height is 1.45 m s^{-1}, as measured at the weather station, which is in a clearing about 50 m in diameter. Within the savanna vegetation, where the tops of the tree canopies average 6 m, the average above-canopy windspeed measured at a height of

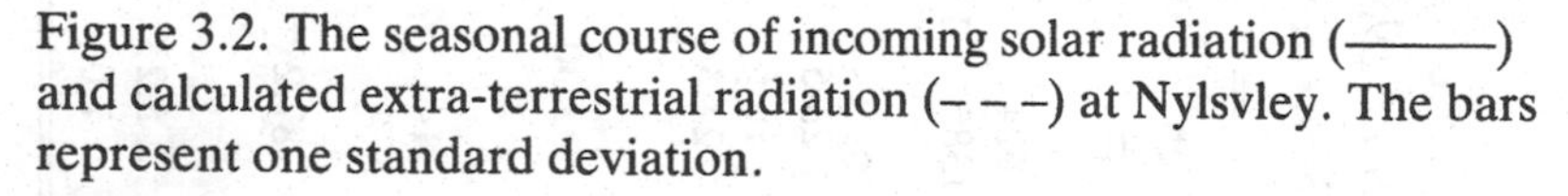

Figure 3.2. The seasonal course of incoming solar radiation (———) and calculated extra-terrestrial radiation (– – –) at Nylsvley. The bars represent one standard deviation.

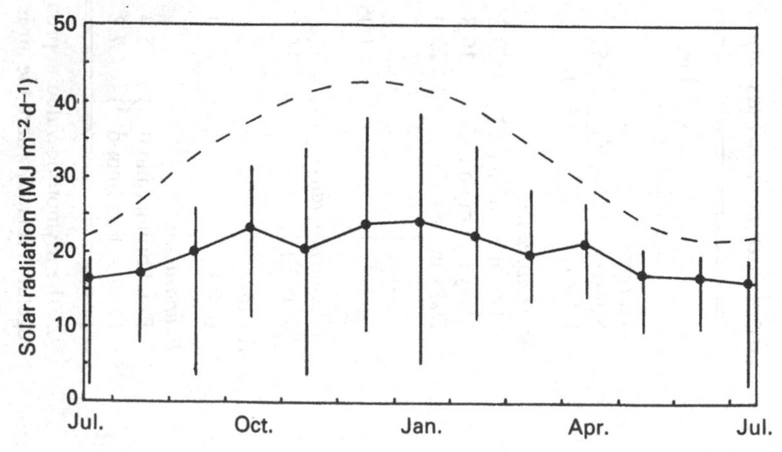

11.5 m is 2.99 m s^{-1}. At a height of 2 m it is reduced to about 1.08 m s^{-1}. From August to January the weather station monthly mean windspeeds are between 1.4 and 1.9 m s^{-1}, and for the rest of the year between 1.1 and 1.3. These calm averages conceal the brief gusts of up to 22 m s^{-1} associated with 'dust devils' and approaching storms, which are ecologically important since they are the final element in a long causal chain which leads to the death of large trees (Chapter 8).

Evaporation

The high radiation and the consequent high temperatures and saturation deficits lead to high potential rates of evaporation. The choice of a reference standard is difficult. The widely used 'short, well-watered green grass' reference is hardly appropriate in a vegetation which is none of those things. The unshielded Class A pan total evaporation for the year averages 2400 mm, with little interannual variation. The seasonal trend of daily Class A evaporation is shown in Table 3.1. The evaporation from a hypothetical extensive water body is often used as a 'reference evaporation' for research purposes (E_0). For Nylsvley, as estimated by the Penman equation (Thom 1975) it averages 1897 mm per annum.

Rainfall

A rainfall gauge has been maintained at the old farmhouse, 5 km west of the study site, since 1916. The annual rainfall totals from this station are illustrated in Figure 3.3. The annual (July to June) rainfall totals are approximately normally distributed, with a mean over the period July 1916 to June 1982 (56 years) of 623 mm and a standard deviation of 134 mm. The high interannual variability is a feature of semi-arid climates, since the coefficient of variation increases as the annual rainfall decreases (Green 1969). There is no evidence for an overall long-term trend in this series, or any other raingauge data from the South African interior (Tyson, 1986). Therefore the changes in vegetation structure which have been observed in the period of record cannot be attributed to a changing climate. What is evident is that there are prolonged periods where the annual rainfall is substantially below and above the mean. In particular, the droughts of 1925–33, 1961–6 and 1982–4 have earned a place in folklore. The most recent drought has been linked to the El Niño phenomenon, an episode of abnormally high sea-surface temperatures in the Pacific Ocean. The mechanism of this linkage is via a global shift in pressure belts, known as the Southern Oscillation.

Smoothing the annual data (with a running mean, for instance) gives the impression that the variation has a cyclic pattern with a period of about 18 years. This cycle has been suggested to be a consistent feature of the summer-rainfall interior regions of southern Africa (Tyson 1986). Proving its existence beyond reasonable doubt is difficult, given the short data record that is available, but tree-ring studies suggest that it has been present for at least 300 years. On a site-by-site basis the predictive value of the proposed cycle is limited, since it accounts for only a small fraction of the interannual variability.

The probability of receiving less than three-quarters of the mean annual rainfall in a given year at Nylsvley is 10%. Plant and animal mortalities in these environments usually begin only after two successive drought years (Walker *et al.* 1987). Assuming that the annual rainfall totals are independent of one another (in other words, ignoring the possibility of a cycle), the chance of receiving less than three-quarters of the mean rainfall two years in a row is only 1%, or once in every 100 years. The presence of a cycle would increase this probability. An instance of two successive years of less than 75% of the mean rainfall has occurred once in the 72-year rainfall record at Nylsvley.

Clearly this is a drought-prone environment. By the same measure, however, there are spells of above-average rainfall which provide establish-

Figure 3.3. Long-term annual rainfall, accumulated from 1 July of one year to 30 June of the next, measured at the Nylsvley farmhouse.

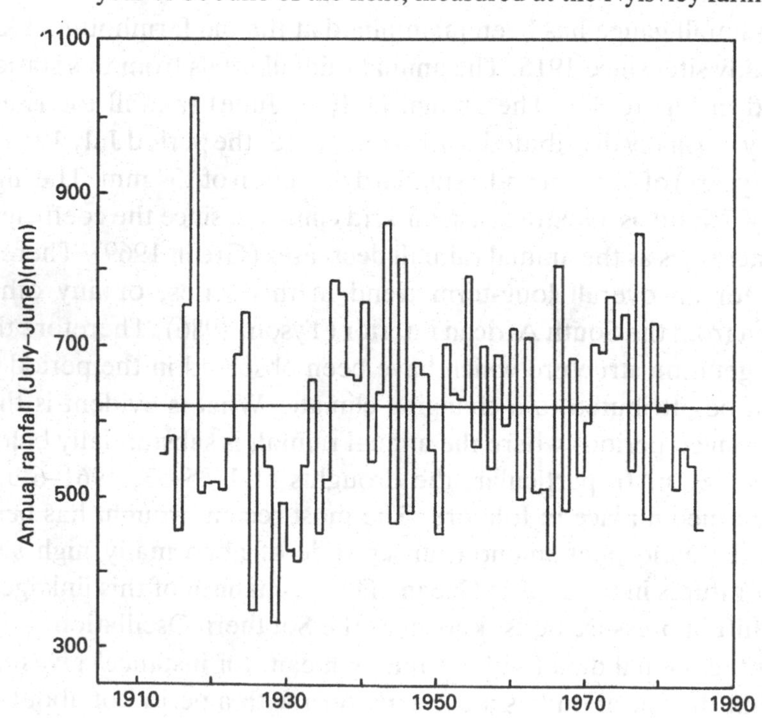

ment opportunities, and permit biomass accumulation and groundwater recharge. The concept of an expected rainfall amount has little biological reality in this situation, particularly for short-lived plants. It is more useful to view their life histories in relation to favourable and unfavourable events and combinations of events. This leads to a stochastic (or 'event-driven') view of ecosystem dynamics, rather than a deterministic view.

The seasonal pattern of rainfall is given in Figure 3.4 and Table 3.1. The main features are:

1. the potential evaporation greatly exceeds the rainfall in all months; and
2. rainfall probabilities are highest from October to April.

The 'rainy season' is substantially longer than the actual number of days on which growth is possible. Water-balance simulation using the long-term rainfall data shows that the summer rainfall is sufficient only to satisfy the reference evaporation from an open water body (E_0) for about 90 days per annum. Evaporation from a savanna with unrestricted water supply will occur at a rate of about 60% of E_0 (Chapter 6); however, a degree of water stress is the norm in the wet season as well as in the dry.

Figure 3.4. Mean monthly rainfall and Class A pan evaporation measured at the study site weather station over the period 1975–90 and 1976–80, respectively.

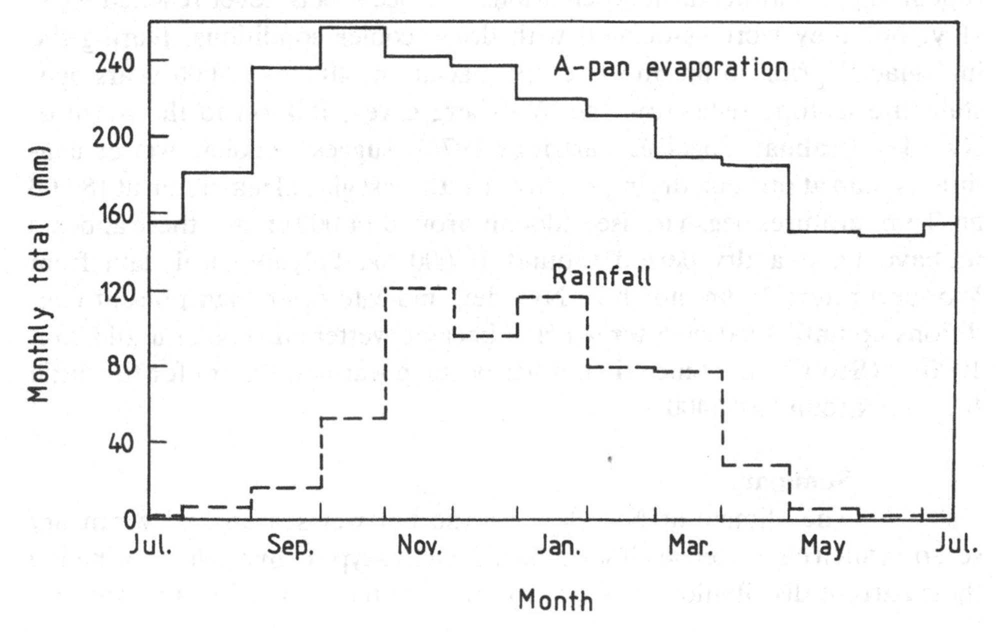

Past climates at Nylsvley

Has the climate at Nylsvley always been conducive to savanna-type vegetation? Tyson (1986) presents a detailed review of past climates in southern Africa: what follows is a summary of important events from a savanna ecology perspective.

At the time that higher plants began to appear in the fossil record (about 400 My BP), the part of Gondwanaland which was to become southern Africa was located over the South Pole, and was extensively glaciated. As Gondwanaland drifted northward, so the climate became more temperate, and coal beds were laid down near Nylsvley about 250 My BP. The climate became progressively drier around 200 My BP, resulting in desert-like conditions. The breakup of Gondwanaland about 120 My ago lead to the formation of the present-day oceans and a series of coolings. Around 25 My BP the interior of southern Africa was probably a mesic woodland or subtropical forest. This was the period of the evolutionary radiation of grasses and the bovid fauna of Africa (Vrba 1985).

The present east–west rainfall gradient in southern Africa was established about 5 My BP. Early hominid and other fossil evidence from Makapansgat, 50 km north of Nylsvley, suggests a savanna climate, but at 3 My BP it was a thicker, wetter bushland than at present (Partridge 1986). A drying trend after 2.5 My BP led to more grasses and grazing animals, of species still found in the Nylsvley region (Vrba 1974, 1975, 1980).

The Quaternary Period, since 1.8 My ago, has been characterised by repeated glacial/interglacial oscillations. The ice sheets never reached Nylsvley, but they were associated with drier, cooler conditions. During the interglacial prior to the most recent glaciation, 40 000–20 000 years ago, stalactite isotope data from the Wolkberg caves, 100 km to the north of Nylsvley (Talma, Vogel & Partridge 1974), suggests cooler, wetter conditions than at present, drying out towards the last glacial maximum at 18 000 BP. Temperatures began to rise suddenly around 14 000 BP, and there appears to have been a dry period around 10 000 BP. Palynological data from Wonderkrater, 25 km north of Nylsvley, indicate drier than present conditions up until 4000 BP, after which it became wetter and cooler until about 1000 BP (Scott 1982). Since then it has become warmer, except for the 'little Ice Age' around AD 1400.

Summary

The climate at Nylsvley has the hot wet season and warm dry season which characterise all savannas. Savanna-type climates have occupied their current distribution in Africa for many millions of years. The present-day vegetation communities at Nylsvley have probably been in place for at least 1000 years, although their component species evolved long before that. With a mean annual rainfall of 630 mm and a mean annual temperature of 19 °C, Nylsvley is towards the dry, cool end of the savanna range. The annual solar radiation at Nylsvley is, like other savannas, among the highest received by any terrestrial ecosystem.

4

Geology, landform and soils

On a global scale, the distribution of savannas is determined by climate. Within this area of broad distribution, however, the occurrence and type of savanna at any given point are mostly determined by soil conditions. The influence of soil is especially clear in African savannas because of the range of weathering states that is represented. In recent landscapes, such as the glaciated regions of the northern hemisphere, all the soils tend to be fertile and of similar texture, since the soil-forming processes have not had time to differentiate them. At the other extreme, the soils of very old or intensively weathered landscapes, such as tropical forests, tend to converge to an infertile and clayey endpoint, irrespective of the original material. In drier stable landscapes, more of the diversity of the parent materials is expressed in the soil. African savannas span a range of rainfall, parent geology and landscape age, and therefore exhibit considerable soil variation. Extremes of fertility and texture occur along with a range of intermediate states.

The soil interacts with plants growing in it through a combination of chemical exchanges and physical effects. Soil chemistry primarily influences plant nutrition, while soil physics plays a crucial role in plant water supply. Since the processes of nutrient mineralisation, transport to the roots and uptake are water-dependent, and dry savannas are by definition water-limited, soil physics is an especially important determinant of dry savanna structure and function. Vegetation boundaries in the field are more often associated with textural changes (a physical attribute) than with chemical changes.

Africa has been one of the most stable continents since the breakup of Gondwanaland. Large areas of the erosion surface formed at that time, called the African surface, are still present. Much of the African surface is underlain by Archean granites of the African shield, a large block of the original crust. This forms the core of the African savannas, with an altitude above 1000 m and annual rainfall in excess of 1000 mm. The soils are deep, infertile, and dominated by 'low activity' clays such as kaolinites and aluminium oxides. These are ferrallitic and ferruginous soils: Oxisols and Ultisols in the USDA Soil Taxonomy nomenclature (Soil Survey Staff 1975); Ferralsols, Acrisols and Nitosols following the FAO classification (FAO/Unesco 1974).

The breakup of Gondwanaland and subsequent continental warping resulted in a drop in river base levels, especially on the east coast of Africa. This caused the major river systems actively to erode back into the African surface, creating newer 'Post-African' erosion surfaces. These surfaces are lower, hotter, drier and younger than the African surface. Their soils are at an intermediate state of development, and are therefore much more strongly related to their parent material. There are strong chemical and textural contrasts vertically in the soil profile, and horizontally down the slope of the land. The catena concept, which describes predictable topographically linked soil–vegetation sequences, was originally developed to account for patterns observed in African savanna landscapes (Milne 1936).

Where the parent material is sandstone, quartzite or recent windborne sand, the soils are sandy and infertile (USDA Entisols, FAO Regosols). Acid lavas (such as felsites and rhyolites) weather to loamier soils which are a little more fertile, but tend to be thin and stony (Lithosols). Granites give rise to soils with a noticeable increase in clay with depth in the profile (USDA Alfisols, FAO Luvisols) and very clear catenas. The upslope positions have sandy, relatively infertile soils, while the downslope soils have clayey, fertile soils. The bottomlands are frequently influenced by excess sodium (USDA Natrustalfs, FAO Solonetz). Shales and mudstones yield loamy, moderately fertile soils (also Alfisols).

The stresses associated with the breakup of Gondwanaland also resulted in the outpouring of lavas from deep within the Earth's mantle. The deeper the origin of the lava, the higher its content of the metallic elements needed for plant growth, and the more 'basic' it is said to be. The 'acidic' rocks which make up most of the crust are dominated by lighter elements such as silicon, which is not a plant nutrient. Thus the African surface is intrinsically less fertile than the areas of Post-African lava.

Basic lavas such as basalts, gabbros and dolerites weather to produce soils that are fertile. They have high initial mineral contents and are dominated by 'high activity clays' such as the smectite group, the best known of which is montmorillonite. However, they tend to be arid soils, because of the high clay content, and exhibit shrink–swell characteristics (USDA and FAO

Vertisols). Soils on ultrabasic intrusions, such as the serpentines of the Great Dyke in Zimbabwe, frequently have such high metal contents as to be toxic to trees, and support a grassland.

The soils deposited by rivers (USDA Fluvents, FAO Fluvisols) are silty and mostly fertile, but small in extent. The most fertile soils of all are on lava and ash deposits resulting from recent volcanic activity (FAO Andosols). They occur in restricted areas in East Africa, associated with the rift valley. Nylsvley occurs at the intersection of three major geological systems (Figure 4.1): the Bushveld Igneous Complex, the Waterberg system and the Karoo system. The Bushveld Igneous Complex formed a series of vast intrusions in the central Transvaal around 1950 My BP (Truswell 1970). It was overlaid in the northwest by sediments of the Waterberg system immediately thereafter, and in the northeast by the upper members (Stormberg Series sandstones and basalts) of the Karoo system about 200 My BP. The main bulk of the Waterberg plateau lies to the west of Nylsvley, and is thought to be a remnant of the African erosion surface (Cole 1986).

To the east of Nylsvley, the Stormberg Series sandstones of the Karoo system are chemically quite similar to Waterberg sandstones. The Stormberg

Figure 4.1. Geology and stratigraphy of the Nylsvley region (simplified from the South African Geological Survey 1:250 000 Sheet 2428 (Nylstroom).

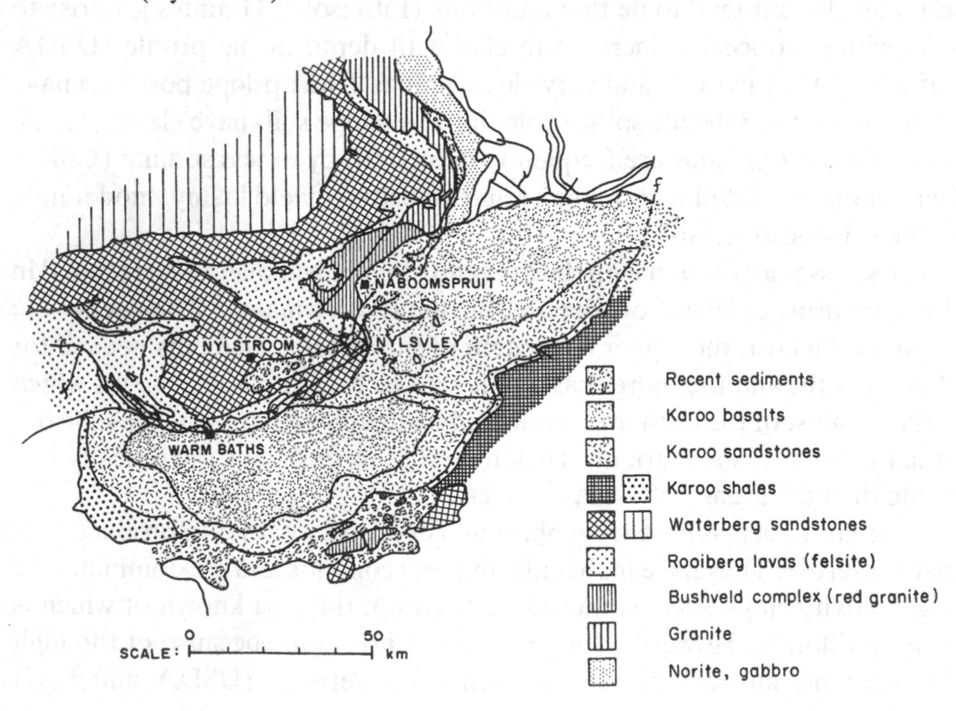

sandstones, and the overlying basalts, eroded rapidly to form a level Post-African plain known as the Springbok Flats, which is the northern portion of the Bushveld Basin. There has been considerable warping, faulting and uplift along the interface of the Waterberg plateau and the Springbok Flats, which has exposed the top layers of the Bushveld Igneous Complex, the Rooiberg felsites.

The soils of the entire Nylsvley Nature Reserve were mapped at a scale of 1:15 000 by Harmse (1977). A much-simplified map is presented as Figure 4.2. The soil properties have been summarised by Frost (1987). The study site itself is located on a remnant of Waterberg sandstone forming the eastern lip of the broad, shallow trough (syncline) down which the Nyl river flows, before it spreads out onto the Springbok Flats. The topography is gently sloping, with a few rocky outcrops. The soils are characteristically sandy, infertile, and 1–2 m deep (Figure 4.4 below). Rooiberg felsites form the western side of the trough, abutting onto the Waterberg sandstones along a fault line immediately to the west of the study site. The floor of the trough is filled with sediments. Adjacent to the Nyl river are deep alluvial soils, strongly influenced in places by sodium left by evaporating floodwaters. Between them and the study site is a tongue of deep, black clay, known locally as *turfveld* (Galpin 1926). It consists of the erosion products of Karoo basalts located 20 km to the southeast, and is therefore dominated by smectite. It exhibits the classical 'melon-hole' gilgai microrelief typical of Vertisols, and is virtually free of woody plants. The felsites on the western trough margin yield a shallow (less than 0.3 m) stony soil rich in iron. Hard plinthite (better known as 'laterite' or *ouklip*, an Afrikaans word meaning 'old stone') occurs along the slope break between the alluvial sediments and the trough slope. Figure 4.3 illustrates the spatial relationships between the soils and vegetation of the Nylsvley reserve.

The soils of the study site, excluding the rocky hills and the lower fringe, belong to the Hutton and Clovelly Forms of the two-level South African soil classification system (MacVicar *et al.* 1977). These forms both have an Orthic A horizon (ie, structureless and low in organic matter) overlying an Apedal B horizon (i.e. one without structure). In the case of the Clovelly Form the B horizon is yellow-brown, while the B horizon of the Hutton Form is reddish. This tends to indicate that the Hutton soils are better drained, but the chemical difference between the soils is insignificant. Bioassays using radish plants (Whittaker, Morris & Goodman 1984) show no significant fertility difference between them. There is also very little variation in soil texture, since all the soils are derived from Waterberg sandstone. Although there is a slight increase in clay content with depth in the profile, it is insufficient to qualify as an argillic horizon. Under the USDA soil taxonomy, both these forms are Ustoxic Quartzipsamments. Soil depth varies greatly at a scale of a few metres, from 0.3 to 2.5 m (Figure 4.4), but is not related to soil type. The very shallow soils are Mispah Form in the South African system, and Lithosols in the USDA and FAO systems. The Mispah soils are chemically and physically very similar to the deeper soils of the study site.

Figure 4.2. Major soil groups of the Nylsvley Nature Reserve (simplified from Harmse 1977).

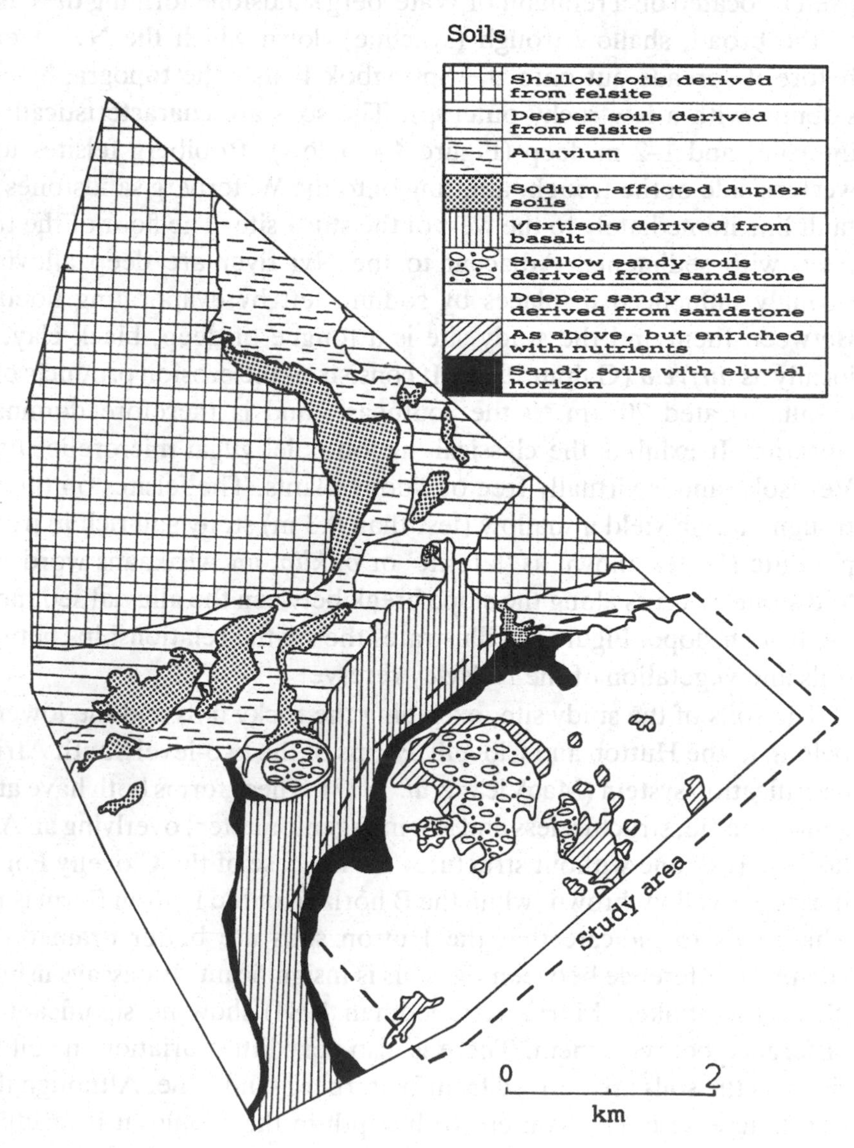

Figure 4.3. Soil and vegetation transect in a NW–SE direction across the Nylsvley Nature Reserve. The Savanna Biome Research Area encompasses most of the broadleafed *Burkea* savanna.

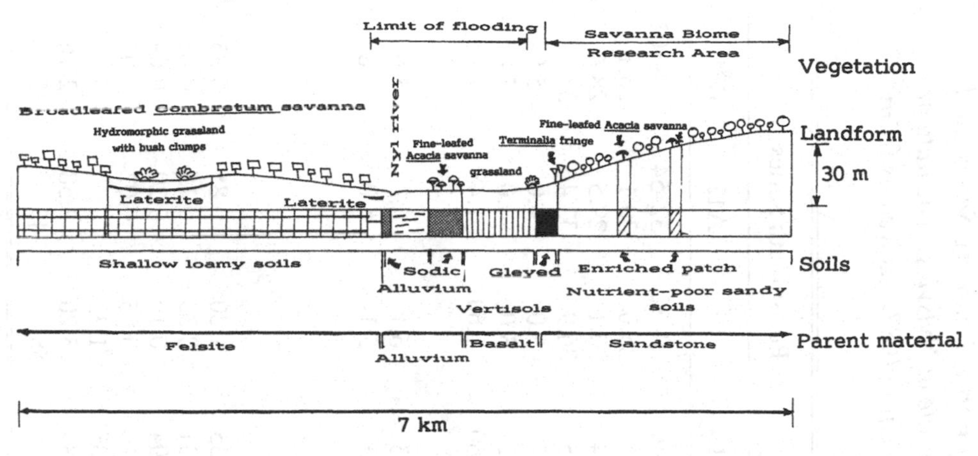

Figure 4.4. Soil depth variations within a 200 × 200 m area of the study site.

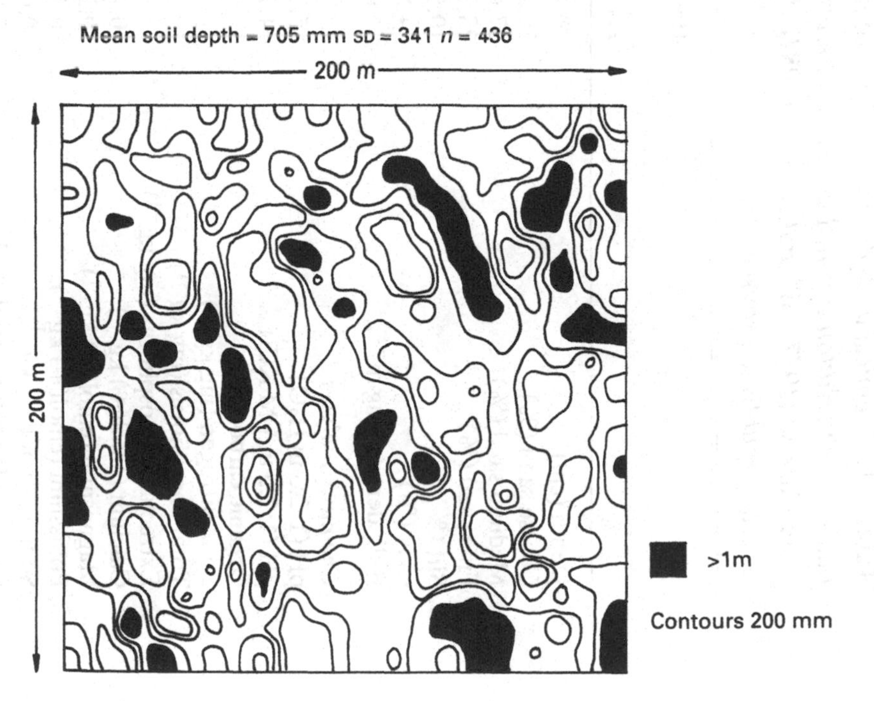

Table 4.1. *Chemical and physical analyses of nutrient-poor (Hutton Form, Bontberg Series) and nutrient-rich (Hutton Form, Portsmouth Series) soils from the study site at Nylsvley. The data are from Harmse (1977), Whittaker* et al. *(1984), Frost (1987) and Meredith (1987), as well as from several unpublished sources*

	Bontberg Series			Portsmouth Series		
	A1	B21	B21	A1	A12	B21
Depth (cm)	0–25	25–57	57–92	0–32	32–54	54–102
Coarse sand (%)	16.9	16.6	10.1	25.0	30.1	26.1
Medium sand (%)	37.6	40.3	40.4	30.1	29.5	28.6
Fine sand (%)	28.7	25.6	29.5	36.9	32.4	33.3
Silt (%)	10.5	10.1	12.7	5.0	4.0	5.0
Clay (%)	4.5	12.4	10.2	3.0	4.0	7.0
Bulk density (mg m^{-3})	1.60	1.60	1.60	1.60	1.48	1.48
Chemical analyses						
pH (1:2.5 in H_2O)	5.2	5.2	5.3	6.1	6.2	5.6
pH (1:2.5 in 1 M KCl)	4.1	4.2	4.3	5.1	5.0	4.8
Organic carbon (%)						
Total nitrogen (mg kg^{-1})						
Total phosphorus (mg kg^{-1})						
Extractable phosphorus (mg kg^{-1})						
Calcium (cmol(+) kg^{-1})	0.45	0.40	0.35	1.30	0.85	0.85
Magnesium (cmol(+) kg^{-1})	0.23	0.34	0.31	0.33	0.15	0.25
Potassium (cmol(+) kg^{-1})	0.10	0.05	0.04	0.19	0.20	0.09
Sodium (cmol(+) kg^{-1})	0.08	0.09	0.10	0.05	0.02	0.05
Extractable acidity (cmol(+) kg^{-1})	2.08	1.83	2.05	1.33		
Cation exchange capacity (cmol(+) kg^{-1})	2.94	2.71	2.85	3.20	2.58	2.76
Base saturation (%)	29	32	28	58	47	45

The real soil contrasts within the study site are at the second level of the South African classification, the soil series. The series within Hutton and Clovelly Forms are differentiated on the basis of the clay percentage (above and below 6%), the grade of sand (medium or coarse) and the degree of base saturation. As it happens, the Nylsvley soils are right on the dividing line for the first two criteria, so they tend to be arbitrarily split in several series. On the third criterion, however, there is a clear separation between nutrient-poor (mesotrophic: sum of bases > 5 but < 15 cmol(+) kg^{-1} clay) series and nutrient-rich series (eutrophic: sum of bases > 15 cmol(+) kg^{-1} clay). The origin of this distinction will be discussed in Chapter 12; suffice it to say that it results in dramatic ecological differences. Typical analytical data are given in Table 4.1.

There is a discernable vegetation sequence from the ridgetop down to the *vlei* margin, but it seems to be more associated with water status than soil fertility (Yeaton, Frost & Frost 1986). Because of their low clay content, the study site soils have a low water-holding capacity. Therefore, despite the low rainfall, significant amounts of water do percolate to the boundary with the underlying rock, which is abrupt. The unweathered sandstone is poorly permeable, so the water moves downhill. During the rainy season a perched water table is present within a metre of the soil surface near the bottom of the catena, and the subsoil there exhibits mottling and gleying (Longlands Form, Waldene Series). Immediately above this is a band of leached, white sand, indicating a seasonal seepline (Fernwood Form, Maputa Series). As is typically the case on seeplines in southern African savannas, this strip supports an almost pure stand of *Terminalia sericea*. This landscape pattern, with broad-leafed savannas on the nutrient-poor uplands, and a grass-covered, marshy, fertile bottomland (*dambo*) is endlessly repeated in south Central Africa. In many respects, Nylsvley is the southernmost representative of this typical landscape.

The sandy, stone-free soils of the study site have greatly facilitated research work, especially on belowground processes. However, they have led to suggestions that the study area is atypical of savannas in general. While it is true that many South African savannas have substantially more clay than the Nylsvley study area soil, there is a significant extent of sandy-soiled savannas in South Africa. Within the Nylsvley region, Waterberg sandstones occupy about 40 000 km^2. The widespread Karoo sandstones yield an almost identical soil. The ridgetop positions on granites are functionally very similar. Outside South Africa are vast areas (such as the Kalahari sand sheet) with very similar sandy soils. Furthermore, we will argue that the fundamental soil distinction in savannas is between nutrient-rich and nutrient-poor, and that situation is well represented at Nylsvley, without the usual confounding factor of a simultaneous texture difference.

Summary

The Nylsvley savanna research site has two main soil types: nutrient-poor and nutrient-rich. Both are very sandy, since they are derived from sandstone, and about 1 m deep. They represent the main distinction in African savannas, between the savannas on the acid, infertile soils of the African erosion surface, and the savannas on the more fertile soils of the later erosion surfaces.

5

The Nylsvley biota

Savannas are not generally regarded as especially diverse ecosystems, but in many instances they have a species richness well above the global average. The Nylsvley Nature Reserve includes, in an area of 3120 ha, over 600 plant species, 325 bird species, 67 mammal species, a large but unknown number of insect species (including at least 194 butterflies, hawk moths and emperor moths, 60 grasshoppers, 21 termites, and 78 dung-associated beetles), 18 amphibian species and 54 reptile species (Table 5.1). The biodiversity found in this small savanna reserve compares favourably with that of entire regions in Northern temperate biomes. The species diversity at Nylsvley is attributable to three factors: the diversity of habitats represented there; the climatic, geomorphological and biogeographical history of the region; and the long period of evolutionary development, uninterrupted by catastrophic events such as glaciations.

Nylsvley lies at the intersection of three rather different geological formations. This, coupled with the local land-form, leads to five distinct soil groups. Within each of these are variations of soil depth, water regime and fertility. Seven discrete plant communities occupy the broad soil groups, with 12 community variations and 4 subvariations (Coetzee *et al.* 1976; Figure 5.1 is a slightly modified map of their distribution). Superimposed on this soil–vegetation mosaic are a number of disturbances such as fire, grazing and burrowing animals and windfalls, which lead to temporary microenvironments with specific attributes. Three further 'secondary' plant communities are recognised on old fields and settlement sites.

The climate at Nylsvley has a high degree of short- and medium-term unpredictability, which permits coexistence of closely related species. Climate variations during the evolutionary past can act as a stimulus to speciation, by repeatedly isolating and re-mixing the gene pool. If they are

Table 5.1. *The main sources for information on species occurring at Nylsvley*

Taxonomic group	Source
Vascular plants	Coetzee *et al.* (1976)
Mammals	Jacobsen (1977)
Birds	Tarboton (1977)
Butterflies and moths	Grei (1990)
Dung-associated beetles	Endrödy-Younga (1982)
Grasshoppers	Gandar (1983)
Termites	Ferrar (1982a)
Amphibians and reptiles	Jacobsen (1977)

too extreme, however, they can lead to mass extinctions. Africa was not subjected to extensive glaciation during the Pleistocene Epoch, but did experience repeated wet and dry, warm and cool periods. Therefore the present biota is based on an old and diverse genetic stock. The changing form of the landscape, brought about by erosion induced by the breakup of Gondwanaland and subsequent continental tilting, has also led to the isolation of biota on the remnants of older surfaces, and the evolution and invasion of new species on the more recent surfaces. The incision of the Limpopo and Zambezi river valleys may have isolated southern Africa to a degree from speciation occurring in the vegetation of Central Africa (Cole 1986).

The plant communities of the Nylsvley Nature Reserve

The Savanna Biome Programme study site occupies the south-eastern quarter of the reserve. Excluding the rocky hill (Maroelakop) and the *vlei* margin, where no experimental work was done, this area of about 800 ha has two main soil variations (Chapter 4), supporting two different plant communities. On the infertile sands is an *Eragrostis pallens–Burkea africana* savanna. The community names used by Coetzee *et al.* (1976) do not necessarily reflect the dominant species, but rather the species most consistently associated with the type. This is a savanna dominated by broad-leafed (> 10 mm) trees and shrubs. The woody layer is about 6 m tall, with a projected aerial cover of 30–40%. The herbaceous layer consists mostly of tufted, poorly palatable grasses up to 0.75 m tall. The grass aerial cover in savannas is usually said to be 'continuous', but this is a misperception. Closer inspection shows that it is often only about 50%, even in high-rainfall years.

This community is considered by Cole (1986) to be an impoverished

Figure 5.1. Vegetation map of the Nylsvley Nature Reserve (simplified from Coetzee *et al.* 1976).

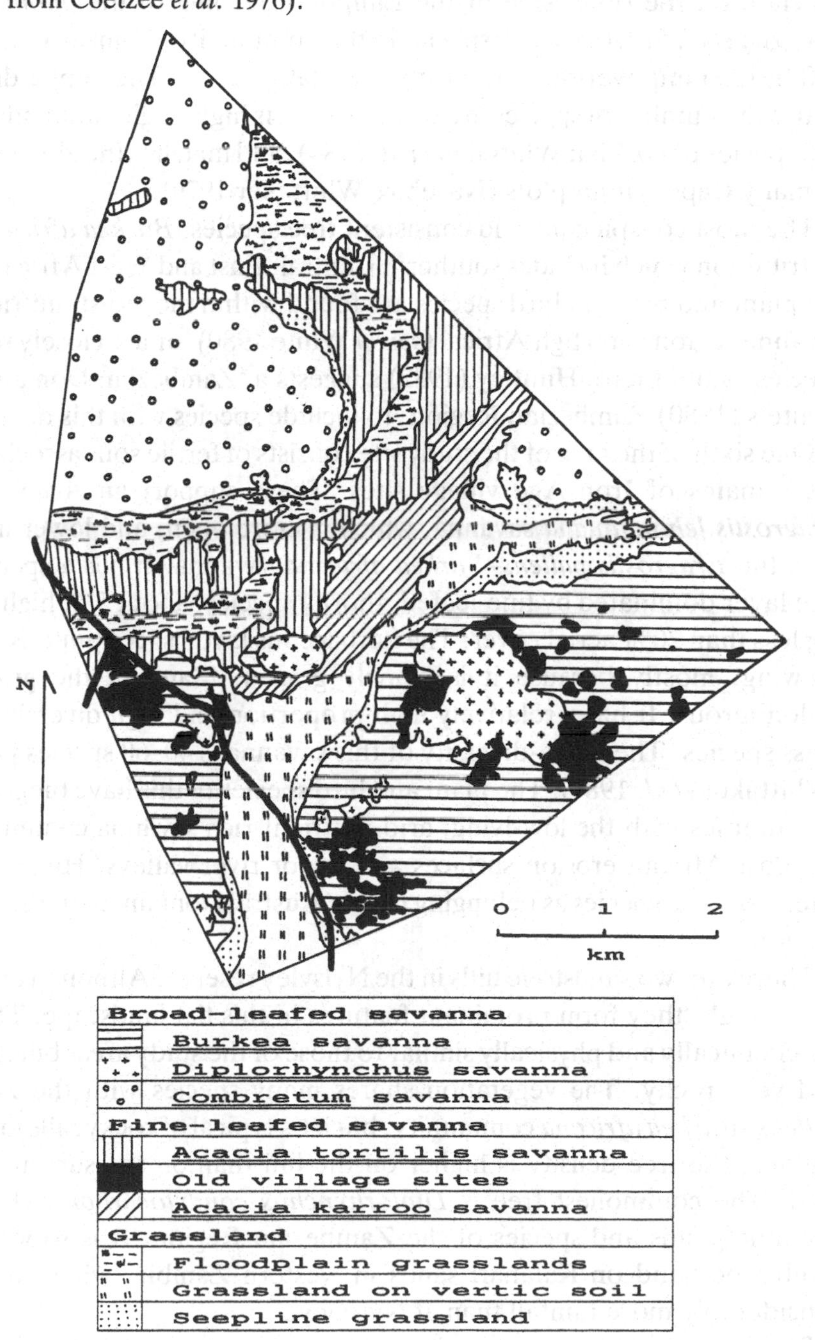

remnant of the Gondwanan flora, the main bulk of which continued to speciate on the other side of the Limpopo, leading to the *miombo* flora. Coetzee *et al.* (1976) list 319 species in this community (from an area of about 600 ha), so impoverishment must be a relative term. The 'alpha-diversity', that is the number of species in a defined area, is high by global standards (80–100 species per 0.1 ha: Whittaker *et al.* 1984), and matches the alpha-diversity of many Cape *fynbos* plots (Naveh & Whittaker 1979).

The most conspicuous and consistent tree species, *Burkea africana*, has a distribution which includes southern, Central, East and West Africa. Most of the plant and resident bird species are found within the moist, nutrient-poor savanna regions of 'High Africa' (*sensu* White 1980), or are closely related to species found there. Huntley (1982) suggests a 'Zambezian Domain' within White's (1980) Zambezian Region, to include species with this distribution.

One sixth of the area of the study site consists of fertile soils associated with the remains of Iron Age village sites. These support an *Acacia tortilis–Eragrostis lehmanniana* savanna, which is more open, and lower in stature than the *Eragrostis pallens–Burkea africana* savanna. It has a species-poor tree layer dominated by fine-leafed, thorny species, about 5 m high, providing less than 20% aerial cover. The herbaceous layer is 'continuous' but low-growing, mostly because it is heavily grazed. Many of the grasses are stoloniferous. It has a relatively high proportion and high diversity of non-grass species. The alpha-diversity of this savanna is 40–60 species per 0.1 ha (Whittaker *et al.* 1984). The plant and bird species of this have biogeographical affinities with the low-lying, arid, nutrient-rich savanna communities of the Post-African erosion surfaces and major river valleys. Huntley (1982) refers to these species as belonging to the 'Austral Domain' of the Zambezian Region.

There are two sandstone hills in the Nylsvley reserve. Although only about 100 m high, they form prominent features in this flat landscape. Their soils are chemically and physically similar to those of the study area, but shallower and very rocky. The vegetation shares many species with the *Eragrostis pallens–Burkea africana* community, but as is typical of rocky hills in savanna regions, the tree density is higher on the hill than on the surrounding flat land. The commonest tree is *Diplorhynchus condylocarpon*, which is an endemic genus and species of the Zambezian Region. It is widespread in scrub woodland on Kalahari sands in western Zambia, where it receives considerably more rainfall than at Nylsvley.

Fine-leafed savanna communities also occur on the alluvial flats surrounding the Nyl river. Where the soil is vertic but seldom flooded, they are dominated by *Acacia karroo*, which is actively invading the *Aristida bipartida–Setaria woodii* grasslands which occupy the more frequently flooded vertic soils. Where the soil is sodic, *Acacia tortilis* dominates the tree layer and *Euclea undulata* the shrub layer. *Sporobolis ioclados* and *S. nitens* are consistently in the grass layer of the sodic soils.

The felsites in the north of the Reserve support a broad-leafed savanna dominated by *Combretum apiculatum*, a species associated with iron-rich soils (Theron 1973). This community shares some species with the *Eragrostis pallens–Burkea africana* savannas, but the species proportions are different, and the vegetation is lower and more open. This is thought to be due to the shallowness of the soil, which also leads to periodic saturation, particularly at the bottom of the slope. Where the felsite slopes meet the floodplain alluvium there is a hydromorphic grassland, often overlying laterite. Within the grassland are bush clumps, usually associated with termitaria. The flora of the bush clumps frequently includes forest species such as *Mimusops zeyheri*, and have the multi-layered, interlocking crown structure of forests. In moist positions, the felsite communities also include elements normally only found at higher elevations in the Waterberg, such as *Faurea saligna*, a primitive member of the Proteaceae. The occurrence of Proteaceae on high-lying, infertile outcrops throughout the Central and southern African savannas is interpreted by Cole (1986) as evidence for a common, proteaceous flora during the Tertiary.

Similar bush clump–hydromorphic grassland mosaics are found at the bottom of the slope occupied by the study site as well. The catenal sequence is *Eragrostis pallens–Burkea africana* savanna at the top of the slope, with a gradual transition to an almost pure belt of *Terminalia sericea* near the bottom. Below this is a band of hydromorphic grassland and bush clumps, ending in the vertic clays of the *vlei* (which do not owe their origin to this catena). The catena is also reflected in the changing proportions of grass species downslope. The uplands are dominated by *Eragrostis pallens* and *Digitaria eriantha*, while the bottomlands are dominated by *Diheteropogon amplectens* and *Elionurus muticus* (Yeaton *et al.* 1986).

A regional vegetation perspective is provided by Galpin (1926), for the Springbok Flats, Acocks (1953) for the whole of South Africa, and Werger & Coetzee (1978) for southern Africa. The Springbok Flats is an extensive area on Karoo basalts and sandstones east of the reserve, supporting communities analogous to the non-flooded vertic soils and the sandy study site soils at Nylsvley. Galpin, who was one of the first farmers to settle in the area after the Anglo–Boer war, states that the grasslands on vertic soils have become progressively invaded by *Acacia karroo* since settlement by white farmers. He attributed this encroachment to changes in the fire regime. The communities on sandy soils have apparently been stable over this period.

The Acocks (1953) classification is at a very broad scale (Figure 5.2). Nylsvley is mapped as falling within his Mixed Bushveld veld type, which he admits to being a 'Daedalian maze'. The *Eragrostis pallens–Burkea africana* savanna of the study site corresponds to the subdivision Mixed *Terminalia–Dichapetalum* Veld of this veld type, a distinction proposed by Irvine (1941). The felsite communities correspond to the Sourish Mixed Bushveld and the

Figure 5.2. Vegetation patterns in the Nylsvley region (after Acocks 1953).

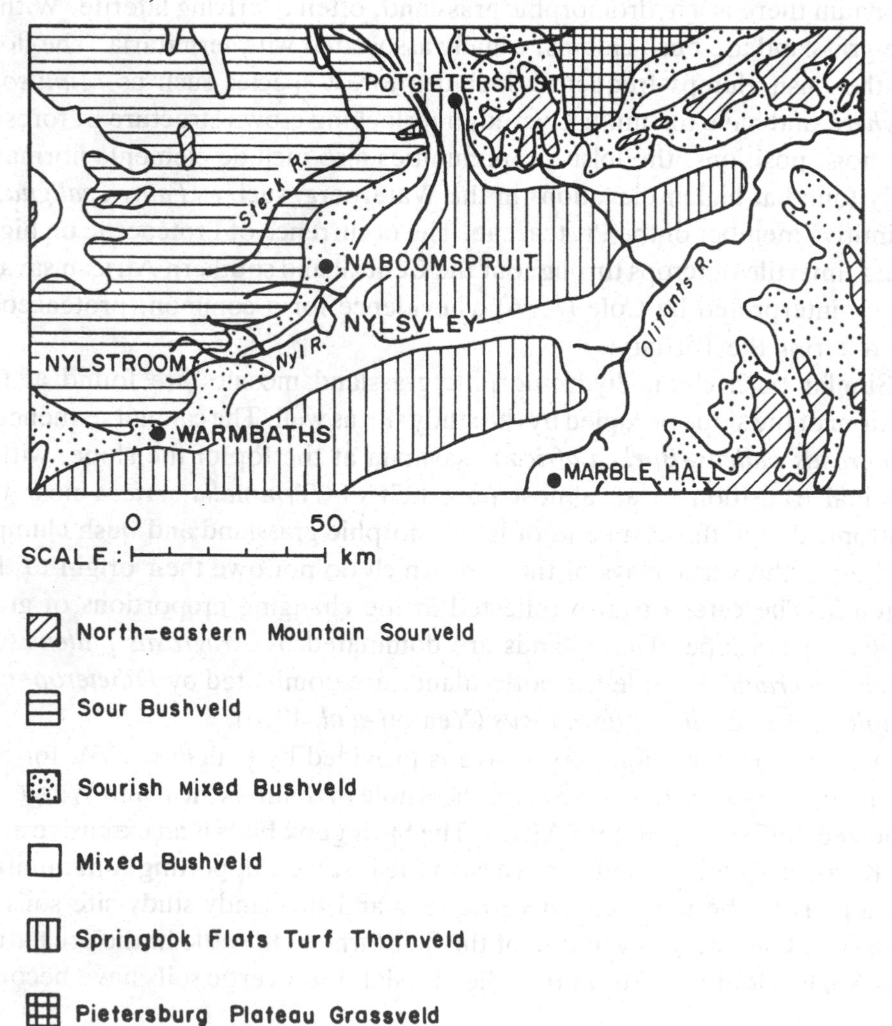

Sour Bushveld veld types, the main bulk of which lie in the Waterberg. Werger & Coetzee (1978) call this type 'Broad Orthophyll Plains Bushveld'. The vertic soils represent Springbok Flats Turf Thornveld, the main extent of which is to the south and east.

The fauna

To the limited extent to which they have been studied, the zoological data support the biogeographical affinities inferred from botanical studies. Although birds and larger mammals are more mobile than plants, and are therefore able to range over several plant communities, they nevertheless tend to be found within a particular set of communities. Their choice reflects the dominant vegetation within their wider distribution.

Of the 325 bird species recorded at Nylsvley by Tarboton (1977), 197 are resident species, 64 are migrants and 64 are sporadic visitors and vagrant species. Many of these bird species are waterfowl associated with the Nyl river wetlands. The savannas of the study site include 174 bird species (Tarboton 1979). Fifteen of these species were exclusively found in the broad-leafed savanna, and 12 species exclusively in the fine-leafed savanna patches. A further 24 species were largely confined to one or other type (Tarboton 1980). The Africa-wide distribution of these species shows the same pattern: those with broad-leafed affinities are members of the mesic east and central *miombo* avifauna (Hall & Moreau 1970), while those that select the fine-leafed savannas are drawn from the xeric southwestern savanna region. The species-pairs *Anthoscopus minutus/A. caroli* (penduline tits) and *Melaenornis mariquensis/M. pallidus* (Marico and Pallid Flycatchers) show this particularly well. The distribution of the two flycatchers, relative to savanna types, is shown in Figure 5.3. Overall, the broad-leafed savanna had a slightly higher avian diversity than the fine-leafed savanna, although bird densities and biomasses showed the opposite trend.

The typical ungulates of the arid savannas (impala, kudu and giraffe) show a marked preference for the fine-leafed patches in the study site, while the broad-leafed communities are selected by species such as grey duiker, reedbuck and roan antelope, typical of the moist savannas. The 67 mammal species recorded for the reserve include 8 Insectivora (shrews), 7 Chiroptera (bats), 3 Primates, 3 Lagomorpha (hares), 20 Rodentia (mice and rats), 18 Carnivora (dogs, cats, mongooses), 13 Artiodactyla (pigs, antelope and equids) and 1 antbear species (Jacobsen 1977; J. Coetzee, Nylsvley Warden). The ungulates (Artiodactyla), most species of which have been reintroduced since 1974, comprise the bulk of the mammal biomass. A detailed ungulate species list is given below in Table 9.1.

Detailed studies of the study site biota

Vegetation

Coetzee *et al.* (1976) describe three variations of the basic *Eragrostis pallens–Burkea africana* savanna within the study site, the first of which has in turn three subvariations (Figure 5.4). The phytosociologists are obviously more sensitive to compositional subtleties than are practitioners of multivariate analysis, who have subsequently failed to find much explainable pattern in these variations (Whittaker *et al.* 1984; Theron, Morris & van Rooyen 1984). The variation within the broad type is not very great: only 0.7 'half changes' on the first ordination axis of a 1 km long transect. In simple

Figure 5.3. The distribution of two conspecific bird species. The Marico Flycatcher (*Melaenornis mariquensis*) is confined to fine-leafed savanna (stippled), while its close relative the Pallid Flycatcher (*M. pallidus*) is found in broad-leafed savanna (after Huntley 1982).

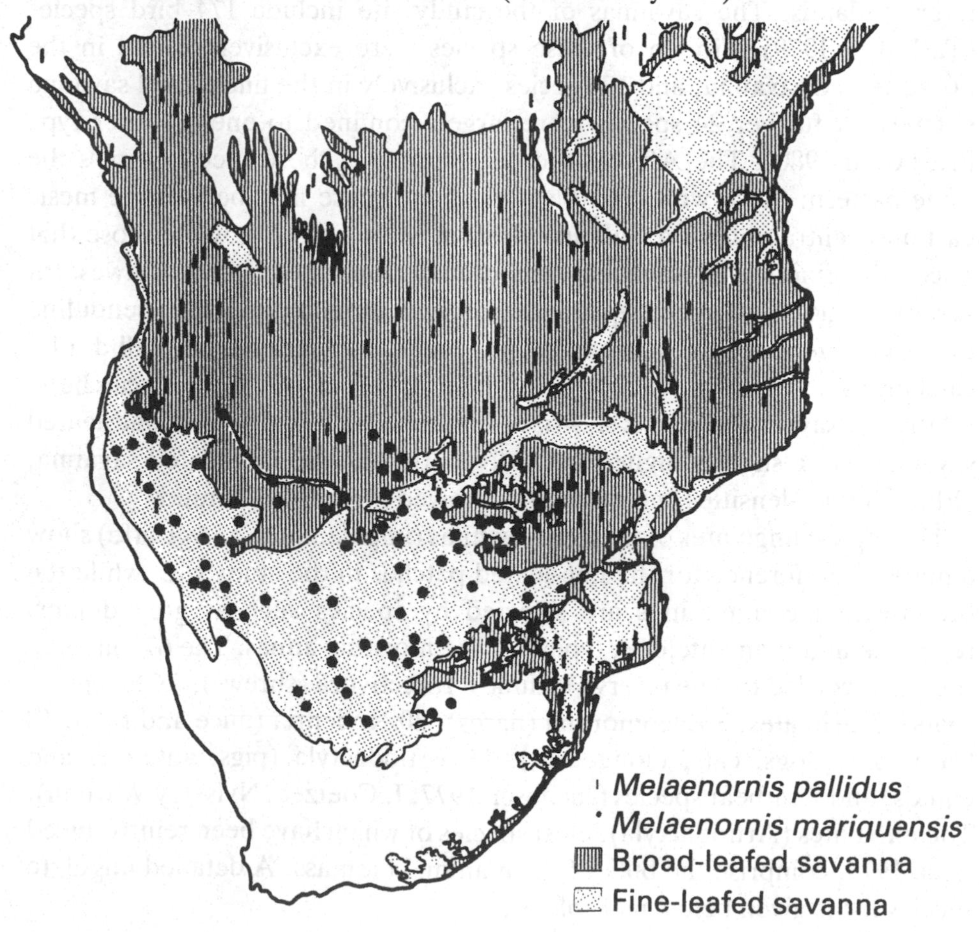

Figure 5.4. Part of the detailed phytosociological classification of the vegetation of Nylsvley proposed by Coetzee *et al.* 1976. The communities outlined in bold are those which formed the focus of the research at Nylsvley. The inset table shows the relationships of the communities to each other and to the soil types.

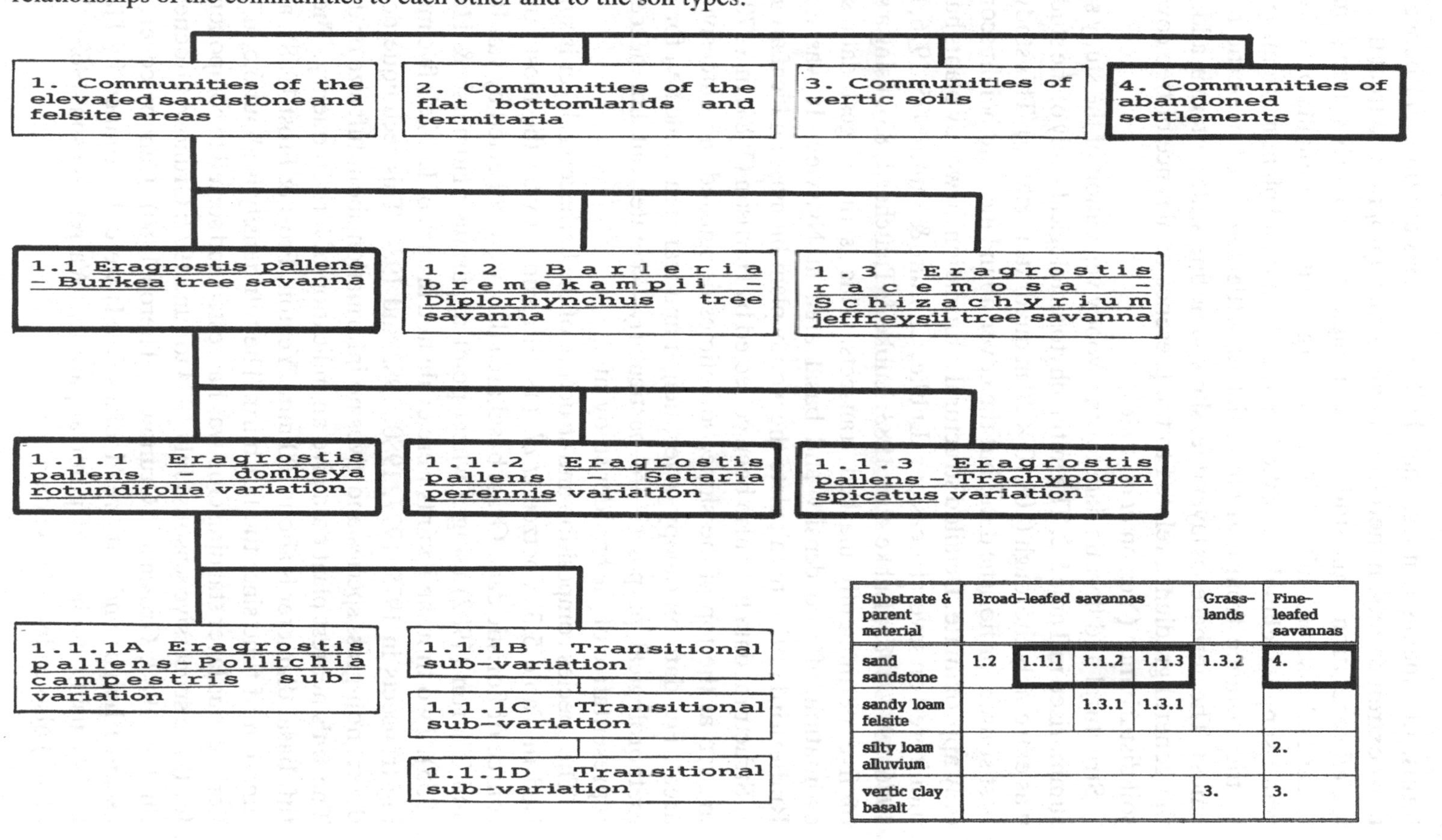

Substrate & parent material	Broad-leafed savannas				Grass-lands	Fine-leafed savannas
sand sandstone	1.2	1.1.1	1.1.2	1.1.3	1.3.2	4.
sandy loam felsite			1.3.1	1.3.1		
silty loam alluvium						2.
vertic clay basalt					3.	3.

terms, this means that less than half the species are different between the most extreme plots. It therefore seems more appropriate to treat the broad-leafed vegetation of the study site as a single, continuously varying unit than as several discrete units. Within the large unit are gradients of soil depth, water supply (related to position on the catena) and plant nutrients (related to the distance from nutrient-enriched settlement sites or vertic bottom-lands). The herbaceous layer does show some fine-scale pattern in addition to the general gradients, related to the presence or absence of tree cover and soil disturbance (Yeaton *et al.* 1986).

Structural and floristic data for the woody vegetation of the study site are summarised in Table 5.2. The initial phytosociological study of the study area was extremely thorough (108 10 × 20 m quadrats in 800 ha). This study is the best source for floristic data, but the cover-abundance and height scores are only approximate. Detailed structural information on woody plants has been monitored on several occasions (Lubke, Clinning & Smith 1975, 1976; Lubke 1976, 1977, 1985; Lubke *et al.* 1983; Lubke & Thatcher 1983), using a system of five permanently marked transects. This is the largest data set for estimation of tree density and basal area at Nylsvley. It was used by Rutherford (1979) to calculate the woody plant biomass and leaf area.

Structural data have also been collected by Harrison (1984) in a 270 × 270 m area at the top of the slope, sometimes designated the 'intensive study area'. His data on canopy cover, height and leaf area, collected for micro-meteorological purposes, are particularly accurate, but not neccessarily representative of the broader study site.

The species composition and structure of the herbaceous layer is summarised in Table 5.3. Coetzee *et al.* (1976) again provide the most complete source of floristic data. Grass basal area data were collected by van Rooyen and Theron (1982) using a wheel-point apparatus (Tidmarsh & Havenga 1955). Two thousand points were evaluated in each of Lubke's five monitoring transects in 1975, 1977, 1980, 1982 and 1984. This technique does not detect non-grass species and gives no information about tuft size or density. Theron's nearest plant data give an indication of forb frequency. The mean tuft basal diameter is about 100 mm (Yeaton, Frost & Frost 1988), and is a function of time since the last burn. The tuft density is about 6.5 m^{-2}. The herbaceous layer standing crop was harvested and sorted to component level for three successive seasons (July 1974 to June 1977: Grunow, Groeneveld & du Toit 1980, Grossman, Grunow & Theron 1980). Canopy cover, inflorescence height and leaf area indices (LAI) were determined by Harrison (1984) using an inclined point apparatus (Warren Wilson 1963). Pendle (1982) also collected LAI data.

Table 5.2. *Summarised floristic and structural data for the woody plants of (a) the broad-leafed savanna; and (b) the fine-leafed savanna in the Nylsvley study site*

(a) Broad-leafed savanna

Species	Density[a] (plants ha^{-1})	Basal Area[b] ($m^2 ha^{-1}$)	Biomass[c] (kg ha^{-1})	Canopy cover[d] (%)	Leaf Area[e] ($m^2 ha^{-1}$)
Burkea africana	411	2.32	8687	13.1	2854
Ochna pulchra	4553	0.50	2136	4.7	2266
Terminalia sericea	111	0.98	1734	11.3	977
Dombeya rotundifolia	14	0.25		1.6	148
Combretum molle	41		353	0.5	107
Combretum zeyheri	19		691	0.5	263
Strychnos pungens	225		312		76
Strychnos cocculoides	6		448		59
Grewia flavescens	195		256		325
Lannea discolor				0.3	
Securidaga longipedunculata	13		207		34
Vitex rehmannii	105		815		
Other	108	0.20	634		131
Total	5801	4.25	16273	32.0	7240

(b) Fine-leafed savanna

Species	Basal area[e] ($m^2 ha^{-1}$)	Canopy cover[e] (%)
Acacia tortilis	2.63	18.6
Acacia nilotica	1.00	7.1
Other	0.04	
Total	3.67	25.7

[a]Lubke *et al.* (1983). 5 belts 30–50 m wide, total area 8.32 ha.
[b]Whittaker *et al.* (1984). 3 belts 10 m wide, total area 1.2 ha.
[c]Rutherford (1979). Used the same belt transects as [a].
[d]Whittaker *et al.* (1984). Line transect 1000 m long.
[e]Whittaker *et al.* (1984). Belt and line transect 10 m wide and 300 m long.

Faunal studies

Several groups of organisms occurring in the study site have been studied in considerable depth in the course of the Nylsvley programme. Many of these will be dealt with in detail in subsequent chapters, in relation

Table 5.3. *The rooted basal cover and canopy cover of herbaceous species within (a) the broad-leafed and (b) fine-leafed savannas at Nylsvley. Both the basal and the aerial cover are highly variable in time and space, in response to rainfall, time since burning, catenal position, and tree cover. The basal cover averaged over five measuring periods for the period 1976 to 1984 was 4.89%*

(a) Broad-leafed savanna

Species	Basal cover[a] (%)	Aerial cover[b] (%)
Eragrostis pallens	1.59	6.5
Digitaria eriantha	1.46	1.9
Elionurus muticus	0.15	3.5
Diheteropogon amplectens	0.40	1.3
Setaria perennis	0.19	9.4
Panicum maximum	0.11	0.2
Heteropogon contortus	0.04	0.1
Other (15 spp.)	1.55	6.0
Grass total	5.49	28.9
Forb total		4.2
Total	5.49	33.1

(b) Fine-leafed savanna

Species	Aerial cover[b] (%)
Eragrostis lehmanniana	29.5
Panicum maximum	4.8
Cenchrus ciliaris	3.3
Schmidtia pappophoroides	3.6
Aristida congesta var. *congesta*	1.3
Other grasses (2 spp.)	0.9
Grass total	43.4
Forb total (22 spp.)	14.1
Herbaceous layer total	57.5

[a]Theron *et al.* (1984). 10 000 points with wheel-point apparatus in 1976.
[b]Whittaker *et al.* (1984). Visual estimation in 200 1 m^2 quadrats in 1979.

to specific ecosystem processes such as herbivory or decomposition. Here follows a list of the main sources:

termites by Ferrar (1982a,b,d);
dung-associated insects by Endrödy-Younga (1982);
Lepidoptera by Scholtz (1976, 1982) and Grei (1990);
social Hymenoptera by Kirsten (1978);
grasshoppers by Gandar (1982a,b, 1983);
acarid mites by Theron (1974) and Olivier (1976);
ticks and other parasites of ungulates by Horak (1978a,b, 1980a,b, 1981, 1982, 1983) and Londt, Horak & de Villiers (1979);
birds by Tarboton (1977, 1980) and Dean (1987);
reptiles by Jacobsen (1977);
small mammals by Körn (1986a,b, 1987a,b) and Körn & Braak (1987);
ungulates by Zimmermann (1978, 1980a,b), Monro (1979, 1980) and Cooper & Owen-Smith (1985, 1986), Owen-Smith (1982, 1985a,b), Owen-Smith & Cooper (1985, 1987a,b, 1989) and Owen-Smith, Cooper & Novellie (1983).

Summary

Nylsvley Nature Reserve has a high plant and animal biodiversity, both relative to other savanna regions and other biomes. The Savanna Biome Study Area, which forms one quarter of the reserve, has two main vegetation types: a broad-leafed savanna dominated by *Burkea africana* in the tree layer and *Eragrostis pallens* in the grass layer; and a fine-leafed savanna with *Acacia tortilis* as the dominant tree, and *Eragrostis lehmanniana* as the characteristic grass. The bird, insect and mammal communities tend to reflect the vegetation pattern. The broad-leafed savanna is floristically, structurally and functionally representative of the nutrient-poor savannas of Africa, while the fine-leafed savanna is a representative of the nutrient-rich savannas.

Part II

The key determinants: water, nutrients, fire and herbivory

6

Water

The strong association between savanna vegetation and climates with a hot, wet summer and a warm, dry winter provides the first clue that water availability is a key factor of savanna ecology. The dominance of water availability as a determinant of savanna structure and function is particularly strong at the dry end of the savanna spectrum. The pattern of water supply in relation to the water requirements of the plants influences both the physical vegetation structure of dry savannas and their ecological composition. A traveller passing down the aridity gradient from a moist savanna, receiving perhaps 1000 mm rainfall per annum, into a desert shrubland or grassland receiving 300 mm per annum will be struck by the progressive decrease in the height and density of the trees, and the consequent change in the proportion of trees to grasses. A similar change can be noted when passing from a sandy to a clayey soil under the same climate and is due, in part, to the different hydrological characteristics of the two soils. The obviousness of the importance of water in savannas can sometimes be a hindrance to understanding their ecology, since it conceals the importance of other more subtle factors.

Water availability determines savanna function by controlling the duration of the period for which processes such as primary production and nutrient mineralisation can occur. Walter (1939) first noted the linear relationship between annual rainfall and grass production in the dry savannas of Namibia, and similar relationships have now been documented for savannas in many parts of the world (Houerou & Hoste 1977; Rutherford 1980; Dye & Spear 1982).

The movement of water in a terrestrial hydrological cycle is controlled by many interacting factors, including climatic variables, soil conditions and plant characteristics. Since most of these factors are difficult to measure on a continuous basis, and they interact in complex ways with other variables, the most practical way to explore their importance is within the framework of a simulation model. A hydrological model forms the core of this chapter. The model is based on our best current quantitative understanding of the hydrological system at Nylsvley. Analysing the situations in which the model fails to match observed data enables us to advance our understanding of the system.

Soil hydrological characteristics

The marked difference between savannas under the same climate but on different soils has two main causes: first, the effect of soil particle size distribution ('texture') on the availability of water; and secondly, the effect of clay chemistry on soil fertility. Some attempts to classify savanna types have used mean annual rainfall as one classification axis, and soil clay content as the other (Walker & Noy-Meir 1982; Johnson & Tothill 1985). There are two problems with this approach: first, the physical and chemical effects of clays become confounded; and secondly, there are extensive savanna soils (Oxisols) which are clayey according to their particle size distribution, but behave hydrologically like sands, due to the aggregation of the clay particles. To avoid these problems, the soil water retention curve and the saturated hydraulic conductivity should be regarded as the fundamental soil physical parameters, rather than the clay content.

A variety of fitted functions can be used to describe the relationship between soil water content and the soil matric potential, known as the water retention curve. The parameters of two commonly used forms are given in Table 6.1, and the curves are illustrated in Figure 6.1. The base data from which they were calculated are given by Henning (1980) and S. J. McKean (unpublished data). Another useful relationship, that between water content and hydraulic conductivity, can be theoretically derived from the retention curve and the saturated hydraulic conductivity.

The concept of 'field capacity', being the soil water content after excess water has drained from the profile, is physically ill-defined, but empirically useful. Several researchers have estimated the field capacity of the soils of the broad-leafed savanna at Nylsvley to be about 12% volumetric water content (Moore 1980; Knoop & Walker 1984; Meredith 1987; Baines 1989). This corresponds to a water potential of −10 kPa, and a pore space approximately one third filled with water. At this water content, the hydraulic conductivity is 0.47 mm d^{-1}, which puts an upper limit on the amount of water that can drain from the profile by unsaturated flow. Since redistribution of water within the profile occurs within hours after a storm in this soil, and drainage is very slow at water contents below field capacity, a simple 'bucket' type water

Table 6.1. *Soil physical parameters for the A and B horizons of Hutton soils under broad-leafed savanna at Nylsvley*

Horizon	Porosity ($m^3\ m^{-3}$)	*Residual water* content ($m^3\ m^{-3}$)	*van Genuchten*[a]		Campbell[b]		K_{sat} ($\times 10^{-6}$ m s^{-1})
			a	*m*	AEV	*b*	
A	0.471	0.008	1.373	1.442	−1.137	1.710	12.38
B	0.397	0.008	1.453	1.516			12.87

[a]Parameters of the water retention curve as defined by van Genuchten (1980).
[b]Parameters of the water retention curve as defined by Campbell (1985).

balance model is adequate to describe soil water distribution. After a storm, each soil layer, starting from the top of the profile, is filled to field capacity, until all the rainfall has been used. Water is then extracted from each layer by evaporation over the period until the next storm.

The particle size distribution varies little within the study area, and only slightly within the profile. There are, however, substantial and significant differences in the water retention curve of the topsoil and subsoil, which are mostly attributable to the effect of soil organic matter. The same applies to differences between subcanopy and between-canopy soils, and soils from the fine-leafed and broad-leafed savannas. The soils of the fine-leafed savanna patches have twice as much organic matter as the broad-leafed savanna soils, and therefore, despite an almost identical particle size distribution, have a water retention curve more 'clayey' and 'arid' than that of the broad-leafed savanna. The field capacity, for instance, is about 17% by volume. This partly accounts for the dominance of the fine-leafed species, typical of clay soils and arid climates.

Figure 6.1. Water retention curves for A and B horizon soils from the broad-leafed savanna at Nylsvley.

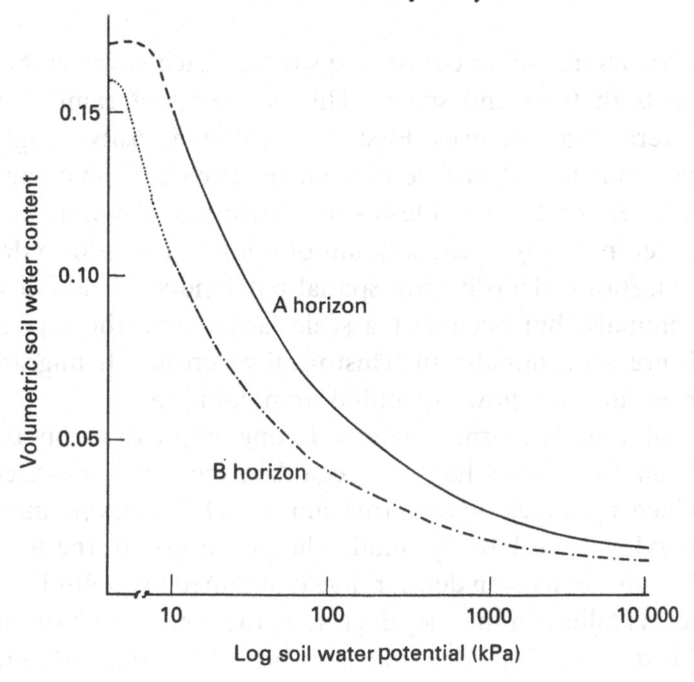

Water balance

The principle of water balance is that the change in the water content of any ecosystem must equal the inputs minus the losses from the system over a given period of time. For a savanna ecosystem, this means that

$$dS = (P_r + P_d) - (E_i + E_s + E_p + D) + (F_o + F_s)$$

where dS is the change in the profile stored water; P_r is rainfall; P_d is dewfall; E_i is evaporation of water intercepted on the canopy and litter surface; E_s is evaporation from the soil surface; E_p is transpiration by plants; D is drainage to below the rooting zone; F_o is net overland flow (run off–run on); F_s is net subsurface lateral flow.

Because of the prolonged winter drought in dry savannas, very little water is stored between hydrological years (July to June at Nylsvley), and rainfall approximately balances losses at a yearly timescale. The water balance equation can be considerably simplified at Nylsvley because several of the components are absent or negligible. For example, snowfall has never been observed, and while dew is quite common, the amount that it contributes to the water budget is very small. Mist is rare, leaving rainfall as the dominant form of precipitation. Drainage and lateral flow are both small, but have important ecological consequences. The main components of the water balance are discussed in detail below.

Rainfall

Rainfall at Nylsvley occurs mainly as convective storms. Each storm is thus relatively discrete in both time and space. This is typical of rainfall in savannas, and has several consequences. First, the rainfall intensity is high, making savanna soils prone to soil-surface phenomena such as capping and erosion by rain-splash. Secondly, the life-history attributes of plants and animals must be attuned not only to the amount of rainfall, but also to dry periods of varying durations. Thirdly, the spatial patchiness of rainfall is important to large animals, but occurs at a scale larger than the typical management unit. Therefore, animals which historically were able to migrate in search of greener pastures are now prevented from doing so.

A dramatic tropical thunderstorm leaves a lasting impression on researcher and ecosystem alike (see Chapter 8 regarding the herbivore–fire–storm interaction which regulates mature tree mortality). However, most rainfall events at Nylsvley are relatively small. The proportion of the total rainfall occurring as storms of a given depth range is obtained by multiplying the proportion of storms falling in that depth class by the mean depth of the class (Figure 6.2). Most rain at Nylsvley falls in events of less than 20 mm.

The small mean storm size results in high proportional interception losses, despite the relatively sparse canopy.

Given a field capacity of 12%, a 20 mm storm will result in wetting of dry soil to 170 mm. This water will evaporate within about 5 days. The plants are forced to place their roots very close to the soil surface if they are to compete effectively with the atmosphere and each other for the fleetingly available water.

Runoff

The sandy soil and gentle topography of Nylsvley result in virtually no runoff at the scale of a plant community. There is some evidence of very localised surface ponding and overland flow during high-intensity rainfall, but there are no gullies or watercourses within the study site.

Drainage

Deep drainage is notoriously difficult to estimate. Strictly speaking, it is water draining below the reach of the deepest roots, to add to the ground-water store. Since the effective rooting depth is unknown, we will take deep

Figure 6.2. The probability of rainfall on a rainy day at Nylsvley exceeding a given amount. The 70-year daily rainfall record from the farmhouse was used to calculate the distributions.

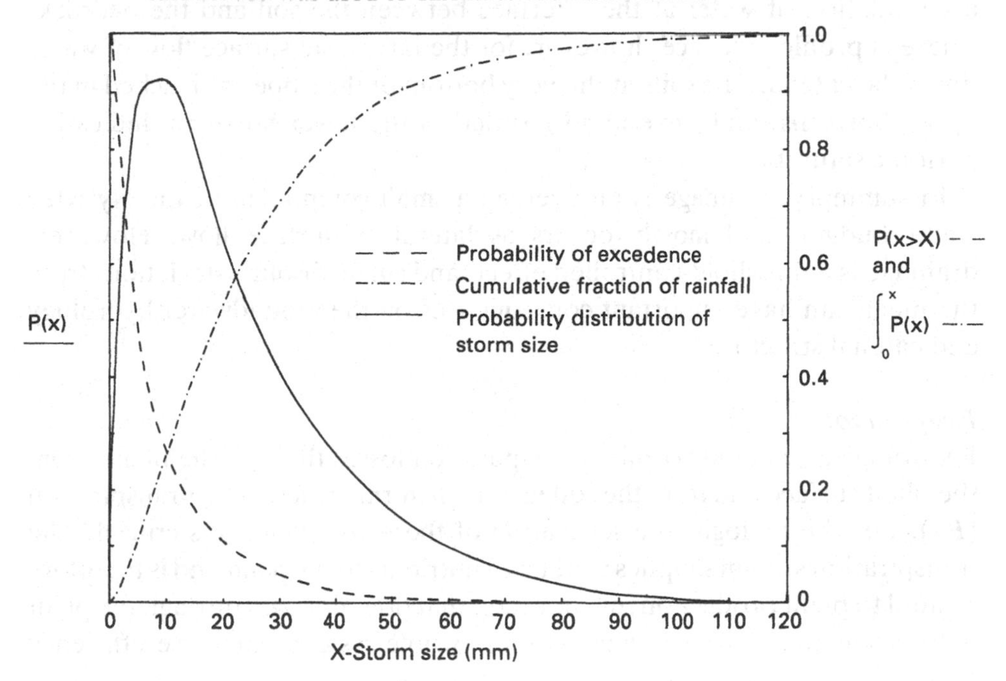

drainage to mean drainage into the rock underlying the soil. In dry regions groundwater recharge is typically only a very small portion of the average rainfall (Martin 1961; Verhagen 1991). Recharge occurs in rare pulses, when a closely-following succession of large rainfall events exceeds the soil profile water-holding capacity and the evaporative demand. There is circumstantial evidence to indicate that this is also the case at Nylsvley. The Waterberg sandstone underlying the Nylsvley study site is a poor aquifer, since it is neither porous nor extensively faulted. Phreatophytes (plants using deep water sources) are not a significant feature of the Nylsvley flora. The water table in the several boreholes in the area is 20 m below the soil surface, and is just as likely to be recharged from the Nyl river 2 km to the north as from deep drainage.

The profile storage capacity at Nylsvley is on average 120 mm of water. A simulation model of soil water movement, using the 70 years of rainfall observations from Nylsvley and the hydrological data for the study site soils, predicts the upper limit of the long-term average annual drainage to be about 6 mm, or 1% of the annual rainfall. In most years the groundwater recharge is zero, while in abnormally wet years it is substantial. It is in the exceptionally wet and dry years that savanna structure is determined, since it is then that woody plant recruitment and mortality are maximised. This is particularly true for landscape features such as seeplines.

There is no evidence in the Nylsvley soil profile for the prolonged accumulation of water at the interface between the soil and the bedrock. There is profile evidence, however, for the lateral subsurface flow of water down the catena. The soils at the very bottom of the slope are leached in the upper horizons, and gleyed and mottled in the lower horizons, indicating periodic saturation.

In summary, drainage is on average a small component of the Nylsvley water budget, and mostly occurs as lateral subsurface flow. However, drainage is a threshold-controlled effect, and small absolute deviations from the mean can have important consequences on the groundwater hydrology and catenal structure.

Evaporation

Hydrologists generally combine evaporation losses through the plant, from the plant surface and from the soil in a single term, called evapotranspiration (E_T). For the biologist the separation of these components is crucial. The transpiration stream supplies water and nutrients to the plant and is therefore related to plant production; evaporation that does not pass through the plant is unproductive. This explains why the whole-system water use efficiency (g DM produced per kg water used) is of the order of 0.4, whereas when it is measured on a single leaf it is around 20.

Moore (1980) calculated the total evaporation rate from 5 × 5 m plots of savanna vegetation, sealed off from the surrounding soil by plastic barriers to a depth of 1.5 m. By measuring the decline in soil water content over time, using a calibrated 'neutron probe' water meter at five depths and two locations in each plot, he found E_T to range from 1.6 to 5.3 mm d^{-1} (Figure 6.3). The evapotranspiration rate depended on the climatic conditions, the leaf area in the plot and, to a lesser extent, the dominant species. The average, over 4 'habitat types' (i.e. dominant species), 3 replicates and 7 measuring periods, was 3.6 mm d^{-1}. These total evaporation rates refer only to periods when water was available in the profile: the annual average includes many dry periods, and is therefore lower. The contribution by each pathway to the total evaporation can be gauged by incorporating the results of a number of individual-pathway studies into a hydrological model driven by the observed Nylsvley weather. This analysis is described later.

Interception

A portion of the incoming rainfall is intercepted by plant parts and litter, and

Figure 6.3. Mean evapotranspiration measured by Moore (1980) in four subhabitats of the broad-leafed savanna. Bars represent the standard deviation of three replicates.

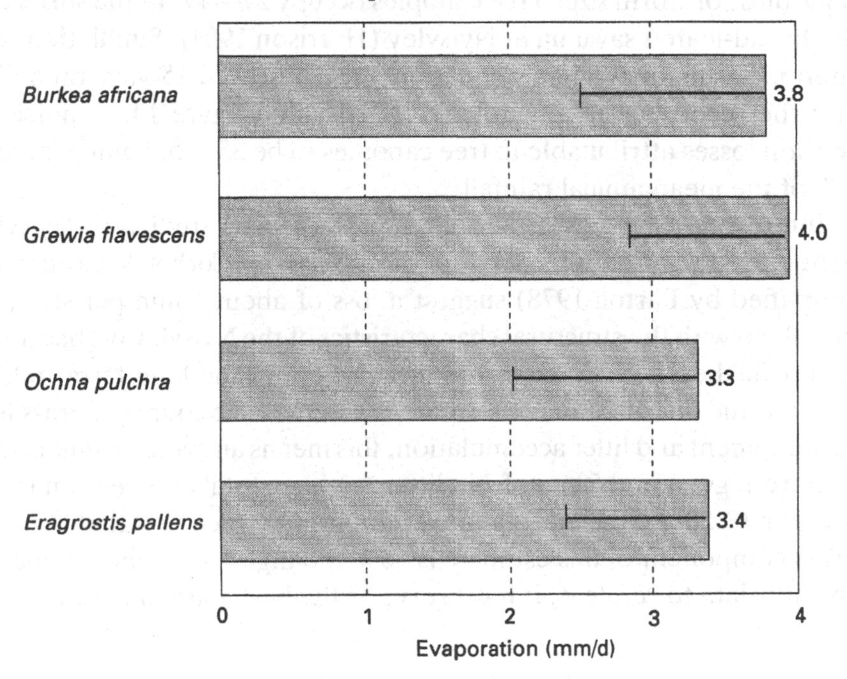

evaporates before it reaches the soil. This is known as the interception loss. The rainfall in savanna regions is usually in the form of discrete, short-duration storms in the late afternoon or evening. Therefore little intercepted water evaporates during the storm, and the interception loss can be treated as an approximately fixed amount per storm, equal to the canopy and litter surface storage capacity. De Villiers (1977, 1981) measured the interception losses under a *Burkea africana* canopy and an *Ochna pulchra* canopy during 34 storms, and indicated that the interception loss per storm was predicted by the following equations:

$$I = 0.07P + 1.16 \qquad r^2 = 0.25 \ (\textit{Burkea africana})$$

$$I = 0.05P + 1.06 \qquad r^2 = 0.17 \ (\textit{Ochna pulchra})$$

where I = interception loss (mm); P = rainfall depth (mm).

The low correlation coefficients are mostly due to storm intensity not being included in the predictive equation. High-intensity storms intercept less water for two reasons. First, a proportion of the evaporation of intercepted water occurs during the storm. Therefore long-duration (low-intensity) storms have greater interception losses. Secondly, the canopy storage volume is reduced by strong winds during intense storms.

The scatter of the data is such that for practical purposes a fixed loss of 2 mm per storm can be assumed for these broad-leafed canopies at Nylsvley, regardless of storm size. Tree canopies occupy 27–44% of the surface area of the broad-leafed savanna at Nylsvley (Harrison 1984). Simulations of interception, using de Villiers's equations, the observed 15-year rainfall record and the recorded leaf phenology at Nylsvley (Figure 14.4) indicate interception losses attributable to tree canopies to be 35 + 5.5 mm per annum, or 6% of the mean annual rainfall.

Interception by grasses and litter has not been studied at Nylsvley, but generalised relationships derived for grasslands (Corbet & Crouse 1968, as simplified by Parton 1978) suggest a loss of about 1 mm per storm from a grassland with the structural characteristics of the Nylsvley herbaceous layer, and a further 1–2 mm from the litter layer. In the long term, taking into account the distribution of storm sizes and the time course of grass leaf area development and litter accumulation, this means an average annual loss of 24 mm from grass and 50 mm from litter, giving a total interception loss of 109 mm, or 18.5% of the mean annual rainfall. Our feeling is that the litter and grass component of this estimate is 50% too high, but in the absence of site-specific data to refute it, it must serve as the best approximation.

Evaporation from the soil

Where the plant canopy is relatively complete, evaporation from the soil is small relative to transpiration. In dry savannas, however, the canopy cover is sparse, especially early in the season. Several studies indicate that evaporation from the soil accounts for about half of the evaporative loss from arid ecosystems (Hide 1954, 1958).

Pendle (1982) used microlysimeters at Nylsvley to estimate the rate of evaporation from the soil surface when bare and covered with litter. By comparing the losses from 20 cm and 40 cm deep lysimeters he showed that evaporation is effectively confined to the top 20 cm of the soil. The evaporation rate was influenced by the evaporative demand (the reference evaporation, E_0, calculated from Penman's (1948) equation), the litter cover, and the number of days since the last rain. The maximum rate of 3 mm d^{-1} for bare soil (about 0.6 E_0) agrees with results obtained from simulations using an hourly-timestep model based on Campbell (1985). Even assuming an infinite evaporative demand, the maximum rate at which this sandy soil can supply water for evaporation from its surface is 3 mm d^{-1}. Since the average daily E_0 is 6 mm, the evaporation process is almost immediately limited by the soil rather than the atmosphere, and the evaporation rate declines very rapidly with time after a storm (Figure 6.4*a*). A consequence of the rapid soil limitation of evaporation is that water loss due to evaporation from the soil surface is effectively restricted to the surface 200 mm (Figure 6.4*b*). Thus it is possible for moisture to persist in the subsoil throughout the dry season, if the plants are dormant.

A complete litter cover (300 g m^{-2}) reduces the rate of evaporation by about 60%, but shading (for instance, by a tree canopy) has hardly any effect (Figure 6.5). On an unshaded soil surface such as occurs between tree canopies, the source of energy for evaporation is direct solar radiation. Standing grass cover and litter simultaneously shade the soil surface and increase the canopy resistance; therefore they decrease the evaporation rate. The direct radiant energy beneath the tree canopy is only one quarter of that between canopies, but sufficient energy is provided by advection of hot air from the between-canopy areas to allow evaporation to continue at close to the maximum sunlit rate. Therefore the effect of litter beneath canopies is mainly to increase the surface resistance to water vapour diffusion.

Combining all these controlling influences in a daily water balance model indicates that evaporation from the soil accounts for about half of the water loss from this savanna. This is lower than for systems with a more clayey soil (Scholes 1987). The model suggests that in this environment the mulching

Figure 6.4. Simulated evaporation from a bare soil surface at Nylsvley. (*a*) trend over time since wetting event; (*b*) depth distribution of water content with time.

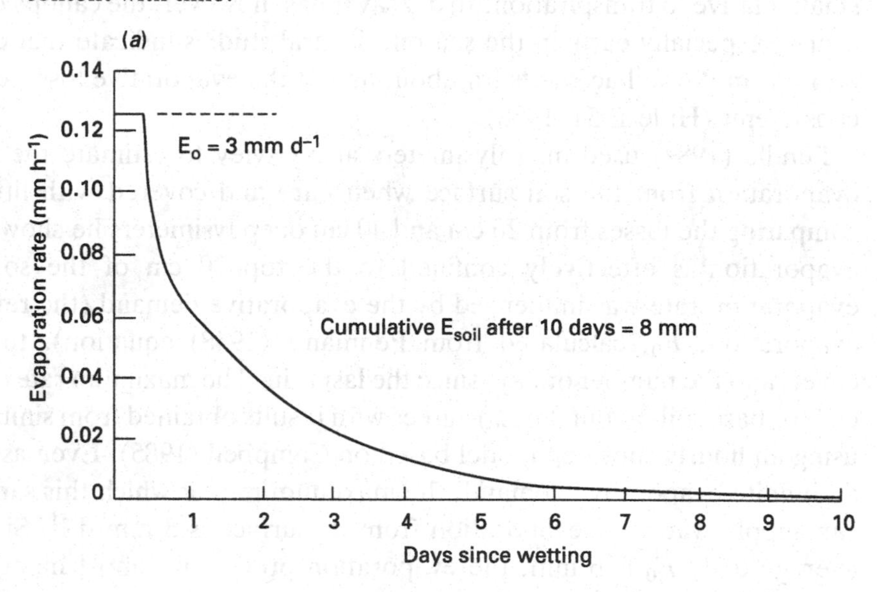

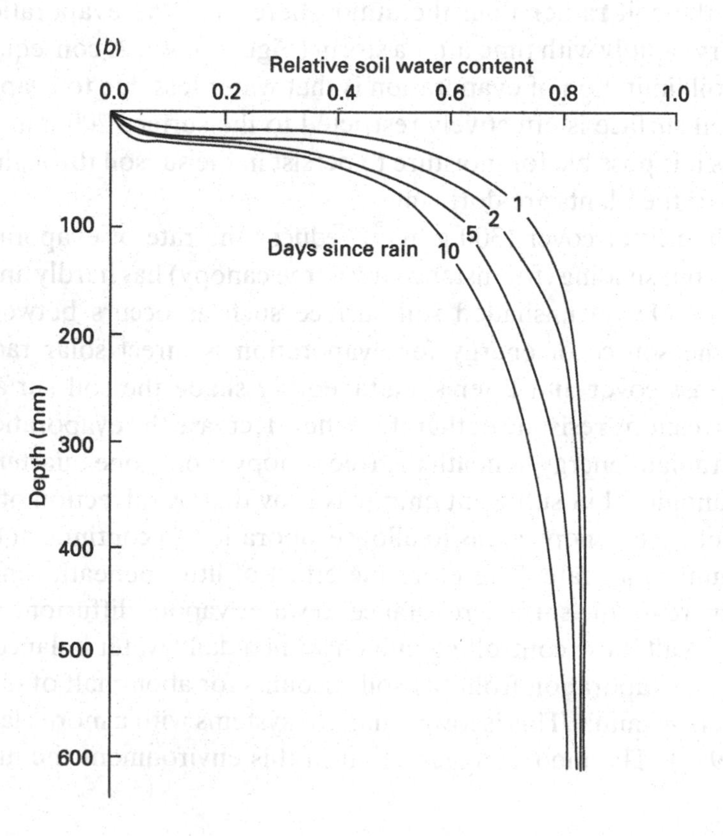

Table 6.2. *Transpiration rates recorded for species growing in the broad-leafed savanna at Nylsvley. The values are averages of data collected shortly after good rainfall, i.e. when soil water supply is at its least limiting*

	E_{max}	
Species	$g\ H_2O\ g\ DM^{-1}\ d^{-1}$	$mg\ H_2O\ m^{-2}\ s^{-1}$
Woody plants		
Burkea africana	22.1[a]	330.2[b]
Ochna pulchra	26.4[a]	268.1[b]
Terminalia sericea	15.3[a]	40.0[c]
Grewia flavescens		234.4[b]
		19.0[c]
Grasses		
Cenchrus ciliaris		495.1[b]
Diherteropogon amplectens	12.6[d]	
Digitaria eriantha	14.6[d]	405.6[b]
Elionuris muticus	14.6[d]	
Eragrostis pallens	11.0[d]	529.5[b]
Eragrostis lehmanniana		625.4[b]

[a]Pendle (1982). Cut shoot weight loss method, 2-hourly measurements integrated over the period of a day.
[b]Kim Olbrich (unpublished data). Average of mid-morning, midday and mid-afternoon readings on 2 days using a LiCor 6000 photosynthesis unit.
[c]Ferrar (1980). Plants in a growth chamber, measured with an Infra-Red Gas Analyser.
[d]Baines (1989). Weighed lysimeters 600 mm deep and 300 mm diameter.

effect of slowly decaying litters does not balance the increased interception loss which they cause.

Transpiration

Pendle (1982) applied several methods to get estimates of transpiration rates by trees and grasses. Most of his data are from the cut-shoot weight-loss technique, in which the rate of mass loss by a recently excised shoot is monitored. Baines (1989) measured the transpiration rates of grasses in the field using a LiCor 6000 photosynthesis system with a 0.25 litre cuvette, and in 0.6 m deep pots by repeated weighing. Their data for unstressed plants are summarised in Table 6.2. A reduction in the transpiration rate occurs after various degrees of soil drying for different species. The soil water content and

Table 6.3. *Permanent wilting points and the soil water content below which transpiration is reduced. Data for trees from Pendle (1982) and for grasses from Baines (1989)*

	Transpiration reduced below		
Species	Soil water content ($m^3\ m^{-3}$)	Xylem pressure potential (kPa)	Permanent wilting point (MPa)
Woody plants			
Burkea africana	0.044	−71.3	−3.1
Ochna pulchra	0.060	−34.0	−3.2
Terminalia sericea	0.070	−24.1	−1.9
Grasses			
Digitaria eriantha	0.094	−12.7	−2.9
Eragrostis pallens	0.064	−29.4	−3.9
Diheteropogon amplectens	0.064	−9.3	−5.5
Elionurus muticus	0.109	−29.4	−2.9

potential at which this occurs has been inferred from their data using the water retention curves, and is presented in Table 6.3, along with the water potential below which transpiration ceases ('permanent wilting point').

Note that wilting points in these wild plants are substantially more negative than the −1.5 MPa commonly applied to temperate crop plants, but that the

Figure 6.5. Effect of litter type and amount on evaporation from the soil surface in shaded and unshaded positions.

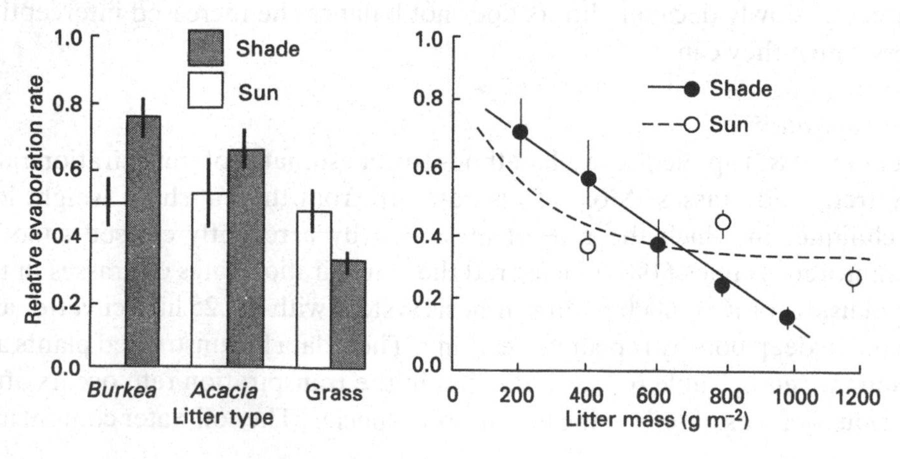

difference in soil water content between −1.5 and −3.0 MPa in the Nylsvley soil is less than 0.01 $m^3 m^{-3}$.

The 'extraction functions' relating transpiration rate to soil water potential are fairly complex, since they are influenced by the soil characteristics, the root distribution, plant physiology and the evaporative demand (Molz 1981). They include such difficult-to-quantify factors as the proportion of the root length which is active in water uptake. Therefore, for modelling purposes at Nylsvley, a fairly simple function was used, which nevertheless captures the main features of the system. Transpiration by species i was calculated as

$$E_i = \Sigma e_{ij} L_{ij}$$

where L_{ij} = factor accounting for the length of active root of species i in layer j; e_{ij} = potential transpirational loss due to species i from soil layer j, calculated as

$$\begin{aligned} e_{ij} &= 1 \quad \text{if } \theta_j > \theta_{cr}; \\ &= (\theta_j - \theta_{wp})/(\theta_{cr} - \theta_{wp}) \quad \text{if } \theta_{wp} > \theta_j > \theta_{cr}; \\ &= 0 \quad \text{if } \theta_j < \theta_{wp} \end{aligned}$$

where θ_j = volumetric water content of layer j; θ_{cr} = water content below which transpiration is reduced; θ_{wp} = water content at permanent wilting point for species i.

If the sum of transpiration by trees and grass and evaporation from the soil exceeds the maximum recorded rate (E_{max}), each component is proportionally reduced. The factors L_{ij} were adjusted until the simulated rate of water extraction matched the data collected by Moore (1980), illustrated in Figure 6.6. The depth distribution of L_{ij} was matched to the measured root distribution (see below, Figure 14.3).

Simulations based on these assumptions indicate that trees extract slightly more water from the Nylsvley system than grasses, over the period of a year. The higher maximum transpiration rate of the grasses is balanced by their lower leaf area duration. This conclusion is not very sensitive to the values chosen for L_{ij}, provided that they are in the range which results in realistic total evaporation rates. This is because the bulk of transpiration occurs while the soil water supply is only slightly limiting. The issue of which growth form uses the water more efficiently is addressed later in this chapter.

A 'best-guess' water balance budget for the broad-leafed savanna at Nylsvley is detailed in Table 6.4. The largest uncertainty is in the partitioning of evaporation between the soil, grasses and trees. Very few detailed water budgets from dry savannas are available for comparison. The key features

Table 6.4. *Best estimate water budget for the broad-leafed savanna at Nylsvley, as predicted by a simulation model based on the data and relationships presented in this chapter and the observed daily climate records from the study site over the period July 1975 to June 1990*

	Mean (mm)			Standard deviation (mm)	Fraction of input (%)
Inputs					
Rainfall	585.8			138.4	100.0
Outputs					
Runoff	0.0			0.0	0.0
Deep drainage	4.8			9.7	0.8
Evaporation	581.0			136.0	99.2
Interception		90.3		15.3	15.4
By trees			34.6	5.9	5.9
By grass			16.2	3.5	2.8
By litter			39.5	6.2	6.7
From soil		275.4		60.1	47.0
Transpiration		215.3		62.6	36.7
By trees			125.7	38.0	21.4
By grass			89.6	25.3	15.3

are the relatively small amount of water available for groundwater recharge, the large amount of water that is lost 'unproductively' through interception and evaporation from the soil, and the unequal division of transpiration between trees and grasses.

Plant water relations

All the species indigenous to the savannas at Nylsvley are to some degree tolerant of water stress, particularly if they are compared with typical crop species. There are several broad mechanisms for drought tolerance or avoidance, and within each a considerable range of variation. Each combination of adaptations yields benefits under a particular set of circumstances, but carries an evolutionary cost under other conditions. Therefore it should not be surprising that species exposed to the same environment can have very different physiognomy, physiology, phenology and life history and be equally successful in the long term. Nor should it come as a surprise that many species share a similar set of features. Plant adaptations tend to come as package deals: a high stomatal resistance, for instance, precludes a high photosynthetic rate. This is the basis for allocating species to 'functional groups'.

Structural and phenological adaptations

The dominant genera of the fertile, arid savannas have fine leaves, while those of the infertile, moist savannas have broad leaves. The most likely reason for this difference is related to water physiology, if not in the present, then in the evolutionary past. The broad-leafed savanna plants are thought to have evolved under more mesic conditions than the fine-leafed savannas. At Nylsvley the two types receive the same amount of rainfall, but more water is available for plant use in the broad-leafed than the fine-leafed savanna, owing to differences in the pathway of water loss. The higher water-holding capacity of the soil in the fine-leafed savanna results in the water being held nearer to the soil surface, and leads to more water being lost by evaporation from the soil.

Since leaves absorb radiant energy, they tend to heat up. If the leaf temperature exceeds the optimum for net photosynthesis, productivity is depressed, and at higher temperatures leaf damage can occur. Ferrar (1980) showed that the temperature optimum for photosynthesis in several savanna

Figure 6.6. The annual pattern of profile soil water content as measured by Moore (1980) and simulated by the model described in this chapter.

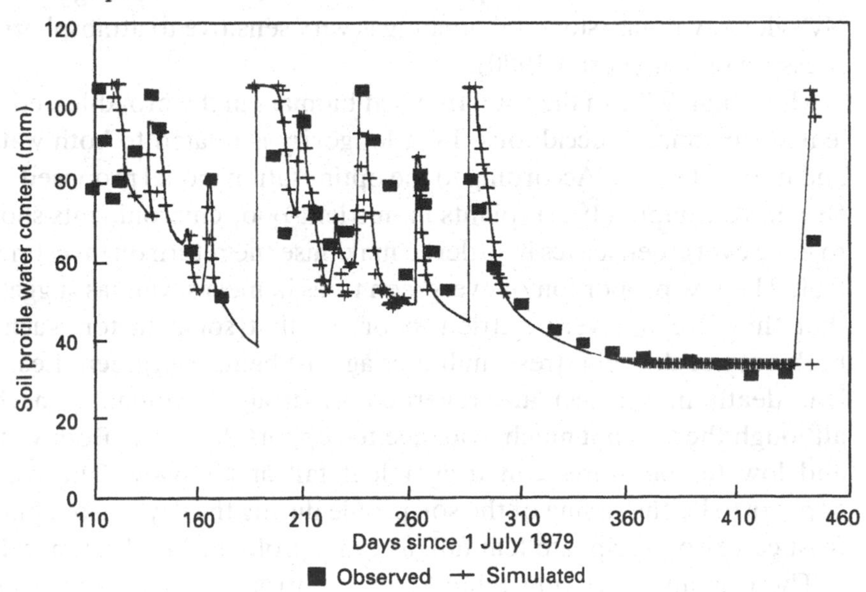

tree species at Nylsvley was around 30 °C, which is close to the mean maximum summertime air temperature. One of the consequences of transpiration is cooling of the leaf. Smaller leaves, such as those on *Acacia tortilis*, assist in keeping the leaf temperature close to the air temperature when there is insufficient water for transpiration. This permits water conservation by stomatal closure without the risk of overheating the leaves. Bate, Furniss & Pendle (1982) explored the effect of leaf size on leaf temperature using a simulation model with parameters matched to the Nylsvley situation. They concluded that even the relatively large leaves of *Burkea africana* lost more heat by convection than by transpiration (even when the water supply was not limited), but that small leaves did stay cooler on average.

A further question relates to the prevalence of compound leaves, particularly in the fine-leafed savannas. Jarvis & McNaughton (1986) offer arguments that suggest that the clustering of tiny leaflets into a compound leaf, typical in the African *Acacia* species, weakens the linkage between atmospheric vapour pressure deficit and transpiration rate, and should therefore increase the capacity of the plant to control its water loss.

The tree leaves within the broad-leafed savannas are mostly relatively thick and sclerophyllous. Table 6.5 describes the leaf characteristics of common species at Nylsvley. The tree leaves are frequently hairy, and have sunken stomata and a well-developed cuticle. In *Ochna pulchra*, which produces leaves well before the first rains, the cuticle completely covers the stomatal openings up until the time that the leaf is fully expanded (Dyer 1980; Ludlow 1987). In the few species which have been investigated from the Nylsvley savannas, stomatal opening is very sensitive to atmospheric vapour pressure deficit (Ferrar 1980).

More than 97% of the total tree leaf biomass in the broad-leafed and fine-leafed savannas is deciduous. Leaf longevity is related to both water stress and nutrient stress. According to the optimisation model proposed by Coley, Bryant & Chapin (1985), plants in nutrient-poor environments should tend to have evergreen leaves in order to maximise the return on investment in the leaf. The low proportion of evergreen trees in most savannas suggests either that they are not very nutrient-poor, or that some factor, such as high herbivory or drought stress, militates against being evergreen. Leaf fall (and leaf death in grasses) are regarded as drought-avoidance mechanisms, although there is not much evidence to support this view. Both water stress and low temperatures can trigger leaf fall at Nylsvley. The usual cause appears to be the drying of the soil profile during the dry season, but an early frost can also precipitate leaf fall, as can a prolonged midsummer drought.

There is an overriding temperature control on leafing-out even in this

Table 6.5. *Leaf characteristics in the broad-leafed savanna at Nylsvley. The peak total leaf area index is 0.78 m^2 m^{-2} for the woody plants, and about 0.5 m^2 m^{-2} for the grasses*

Species	Fraction of total LA (%)	Specific leaf area (m^2 kg^{-1})	Typical dimension (mm)	Type
Woody plants				
Burkea africana	36.5	7.1	30 × 20	Compound, deciduous, smooth
Ochna pulchra	29.0	7.8	80 × 50	Simple, deciduous, waxy
Terminalia sericea	12.5	6.1	80 × 25	Simple, deciduous, hairy
Vitex rehmannii	7.5	7.2	90 × 90	Palmate, deciduous, hairy
Grewia flavescens	4.2		40 × 20	Simple, deciduous, scabrid
Combretum zeyheri	3.4	7.3	60 × 30	Simple, deciduous, hairy
Dombeya rotundifolia	1.9	6.9	80 × 80	Simple, deciduous, scabrid
Combretum molle	1.4		70 × 50	Simple, deciduous, hairy
Strychnos pungens	1.0	5.3	30 × 25	Simple, evergreen, waxy
Strychnos cocculoides	0.8		30 × 25	Simple, evergreen, smooth
Securinega longipedunculata	0.4	7.3	30 × 10	Simple, evergreen, smooth
Other trees	1.7			
Grasses				
Eragrostis pallens	70	3.8	200 × 9	Sword-like, scabrid
Digitaria eriantha	20	10.9	150 × 8	Sword-like, hairy
Other grasses	10	7.4		

subtropical savanna. Midwinter rainfall does not result in immediate bud-break in trees, although carried-over water in the soil can lead to bud-break occurring earlier than usual in spring. Irrigation of the Nylsvley grass layer in midwinter initiates growth when performed in September or August, but not in July (Baines 1989).

The belowground to aboveground biomass ratio in savannas is high by comparison to the typical value in grasslands or forests. This has been variously attributed to water stress, nutrient stress, fire, herbivory or a combination of all these factors. When the confounding effect of heavy structural organs (trunks, branches and coarse roots) is removed by expressing the ratio in terms of assimilatory organs alone (metres of fine-root length per m^2 of leaf area), the value is still high (about 1000, compared to 500 in temperate grasslands). The ratio for the fine-leafed (arid, nutrient-rich) savanna is even higher, suggesting that carbon allocation to the growth of fine roots is predominantly an adaptation in response to aridity. Baines (1989) found an increase in the root to shoot ratio to be the only consistent consequence of prolonged water stress in potted grass tufts.

Prior exposure to water stress resulted in physiological changes in savanna trees from Nylsvley (Ferrar 1980). In all cases, the pre-stressed plants maintained higher photosynthetic rates than the control plants, at a given leaf water potential. The difference was substantial at low water potentials. Dark respiration decreased with decreasing water potential in both pre-stressed and control plants, but in the control plants the ratio of dark respiration to photosynthesis increased with water stress, while it remained constant in the pre-stressed plants. The maximum transpiration rates of pre-stressed plants were lower than those in control plants when water was freely available, resulting in a lower transpiration to photosynthesis ratio. The photosynthetic rates returned to pre-stress levels after rewatering. The consequence of all these changes is to enhance the carbon balance while under water stress.

Photosynthetic stems are common among the woody plants at Nylsvley, and photosynthetic culms among the grasses (Ferrar 1980; Baines 1989). They allow a small amount of photosynthesis (5–10% of the rate exhibited by leaves) to continue after leaf drop.

Minimum water potentials and maximum stomatal conductances

It is generally assumed that water held in the soil at potentials more negative than −1.6 to 2.0 MPa is unavailable to plants, since they wilt permanently at about that potential. Most of the savanna plants are able to continue to transpire at lower water potentials than this arbitrary limit (Table 6.6), even as low as −5.5 MPa for one grass with narrow, rolled leaves. This is

Table 6.6. *The lowest measured wilting point xylem pressure potentials (WPXPP) and of leaf osmotic potential, and the mean maximum stomatal conductance for common species from Nylsvley. Plant water potential data from Pendle (1982) for the trees, and Baines (1989) for the grasses. The stomatal conductance data are from Blackmore (1992)*

Species	WPXPP (MPa)	Osmotic potential (MPa)	Stomatal conductance (cm s^{-1})
Woody plants			
Burkea africana	−3.1	−1.99	0.300
Ochna pulchra	−3.2	−1.82	0.232
Terminalia sericea	−1.9	−1.36	0.243
Dombeya rotundifolia		−2.09	
Acacia tortilis			0.826
Acacia nilotica			0.770
Grasses			
Eragrostis pallens	−3.9		0.325
Digitaria eriantha	−2.9		
Elionurus muticus	−2.9		
Diheteropogon amplectens	−5.5		
Cenchrus ciliaris			0.603
Panicum maximum			0.476

not so great an advantage as it may first seem, since for sandy soils such as those at Nylsvley, very little additional water becomes available at these low potentials. The main advantage is probably in maintaining a low level of physiological activity in the plant, thus allowing a rapid exploitation of growth opportunity following the next rainstorm. *Terminalia sericea*, as always, is the odd one out. Its unspectacular wilting point may be associated with its frequent location just above seepline hydromorphic grasslands. The low water potentials in the other species are permitted by low leaf osmotic potentials and fairly rigid cell walls. There is a tendency for the osmotic potential to decline as the leaf ages (Pendle 1982). In general the permanent wilting potential of grasses is lower than that of trees; in practice grasses are observed to wilt much sooner than trees. This implies either that the trees have access to water unavailable to grasses, or that the trees are able to restrict their water loss more efficiently.

Conversely, when water is abundant, the stomatal conductance of savanna grasses tends to be higher than that of savanna trees (Blackmore 1992), and the conductance of fine-leafed trees is higher than that of broad-leafed trees (Table 6.6). The grasses with the highest potential productivity (*Panicum maximum* and *Cenchrus ciliaris*) have the highest stomatal conductance, and tend to grow in the shade. The dominant grass of the open patches in the broad-leafed savanna, *Eragrostis pallens*, has the lowest stomatal conductance.

Water use efficiency

In determining the success of a species in a particular environment, three things count: the ability to survive periods of adversity; the ability to secure a share of limiting resources; and the ability to convert that resource efficiently into growth. In the case of water, all of these attributes are important, but their relative importance depends on how dry the environment is. Survival of dry periods is crucial in arid systems, competitiveness in water-limited systems, and efficiency in relatively wet systems.

Water use efficiency, in the physiological sense (mg CO_2/g H_2O), is commonly measured almost instantaneously, using an Infra-Red Gas Analysis system applied to a single leaf. In the ecological sense, water use efficiency is measured on whole vegetation patches, over the course of several seasons. The slope of the well-known linear relationship between grass production and rainfall in savannas is a measure of water use efficiency (g aboveground DM m^{-2} per mm rainfall). Efficiency measured according to the physiological definition is several times higher than that measured by the ecological definition. This phenomenon is illustrated in Table 6.7, and has two causes. First, plants measured under laboratory conditions are seldom exposed to the levels of radiant flux encountered in the field. Secondly, instantaneous measures are based on net photosynthesis and actual transpiration. Over the period of a growing season in the field, only a portion (often less than half) of the carbon assimilated is incorporated into primary production, while the rest is lost as respiration or root exudates. Not all the water leaving an ecosystem passes through the plant: at Nylsvley just over a third of the rainfall is transpired. Therefore field-measured efficiencies should be less than half those obtainable in the laboratory, and whole system efficiency should be a sixth or less of the laboratory values. This is borne out by the data in Table 6.7, which is expressed as units of water used per unit of carbon fixed, i.e. the inverse of water use efficiency as defined above.

At a physiological and whole-plant level, savanna grasses are more efficient water-to-biomass converters than are savanna trees, but not always,

Table 6.7. *The amount of water used per unit of carbon fixed for various species and growth-forms, measured at different time and space scales. Instantaneous values are obtained using infra-red gas analyser systems, on a single leaf for a few seconds. Whole plant values are calculated by dividing the harvest mass (converted to CO_2 equivalents) of a plant by the water used in reaching that mass. Ecosystem estimates are calculated from the long-term whole system primary production and water-use estimates. For ease of comparison, all are expressed as g H_2O transpired per g CO_2 fixed. To obtain the more conventional units of instantaneous Water Use Efficiency (mg DM per g H_2O) or of whole system production efficiency (g DM m^{-2} per mm rainfall) divide the given value into 682*

Species	Transpiration/ photosynthesis (g H_2O per g CO_2)	Details
Woody plants (instantaneous)		
Grewia flavescens	53	Growth chamber[a]
	360	Field[b]
Terminalia sericea	74	Growth chamber[a]
Burkea africana	263	Field[b]
Ochna pulchra	125	Field[b]
Acacia tortilis	350	Field[b]
Woody plants, long-term	183	Field[e]
Grasses (single leaf, instantaneous)		
Eragrostis pallens	40	Growth chamber[c]
Digitaria eriantha	43	Growth chamber[c]
	336	Field[b]
Cenchrus ciliaris	526	Field[b]
Eragrostis lehmanniana	1470	Field[b]
Whole grass plant, full growing season		
Unstressed	52	Pots in field[d]
Stressed	47	Pots in field[d]
Grass community, full season	127	Field[e]
Trees and grass, long-term	420	Field[e]

[a]Ferrar (1980).
[b]K. A. Olbrich (unpublished data, collected with LiCor 6000).
[c]Cresswell *et al.* (1982).
[d]Baines (1989).
[e]This study, using water use data from Chapter 6 and production data from Chapter 10.

and not by a large margin. This weakens those models of coexistence in savannas which are dependent on superior water use efficiency in grasses (for example, Walker & Noy-Meir 1982). Furthermore, it is not true at a whole-system level. Savannas with their tree component removed have never been found to have higher productivity than the combined tree-and-grass system (Scholes 1987). The reason is that in the absence of trees, much of the rainfall is lost from the system by evaporation or leaching, at times when the grass is unable to use it. The coexistence of a variety of water use patterns gives the vegetation as a whole a higher water use efficiency.

Summary

Savanna structure and productivity is strongly correlated with plant water availability, especially in dry savannas. The availability of water to savanna plants is determined by the interaction of many factors: rainfall amount and its seasonal and storm size distribution; the seasonal pattern of atmospheric evaporative demand and plant phenology; the proportion of rainfall which is lost to interception, runoff and drainage; the hydrological characteristics of the soil and the plant rooting depth; and the water use characteristics of the plant. Only about one third of the rainfall falling at Nylsvley passes through plants as transpiration: the remainder evaporates, mostly from plant, litter and soil surfaces.

Rainfall satisfies only a small fraction of the atmospheric evaporative demand at Nylsvley, even during the wet season. However, water stress in savannas is not continuous, but episodic. During the growing season, periods of water stress alternate with periods of water sufficiency. Water availability acts as a switch on savanna function: for a few days after rain, it is 'on' and processes such as nitrogen mineralisation and carbon assimilation proceed at their maximum rate. As the soil dries out, the switch turns 'off' relatively rapidly, and these processes are almost quiescent until the next rain.

The plants of the Nylsvley savanna, and probably savannas in general, are tolerant of water stress. The savanna as a whole is a conservative water user, since the maximum evapotranspiration rate is only one third to one half of the evaporation rate from an open water body.

7

Nutrients

The core of systems ecology is the notion that interactions amongst organisms, and between organisms and their environment, can be quantified. The interactions themselves are invisible, but one way of quantifying them is to relate them to the flow of some measurable, necessary factor to and from the organisms. Energy transfers have frequently been used to map the interactions within an ecosystem: this approach is used in Chapters 9 and 10. Any element which is taken up by plants and animals can be used in a similar way. In this chapter we concentrate on the movement of two key elements – nitrogen and phosphorus – not as energy carriers, but as building blocks essential for the growth of any organism. It is useful to examine the pathways of nitrogen and phosphorus against the background of the carbon cycle, since for the major biological portion of their cycles they are found in organic form (that is, as part of carbon-based molecules). This property is shared by sulphur, which was not studied in detail at Nylsvley, but is included here for completeness. Some details of other elements, such as calcium, magnesium, potassium and sodium, are also included.

Among the key processes regulating the passage of N, P and S through the ecosystem is the process whereby they are liberated from their carbon bondage, and become available for uptake by organisms. The other essential elements (such as potassium, magnesium and calcium) are less intimately associated with carbon. They can pass between the organic and inorganic phases of the system much more easily, for instance by leaching out of living or dead tissue.

The trademark of systems ecology is a flow diagram consisting of boxes and interconnecting arrows. The boxes represent reservoirs (also called pools) of a particular element, and the arrows show the rate and direction of transfer (flux) of the element between pools. The fluxes are usually inferred from the rate of change in the size of their source or sink pool, measured over a period of time. It is usually necessary to block some of the pathways selectively, in order to achieve this. If the study does not proceed beyond filling in the boxes, the result is a static and not very useful snapshot of the ecosystem. Identifying and quantifying the fluxes is a considerable improvement; but a complete biogeochemical system analysis requires knowledge of the factors which control the rate of the flux, as well.

The pathways of carbon, nitrogen, sulphur and phosphorus are strongly associated for the biological portion of their cycle. In the remainder of their cycles they form an interesting continuum from a very open cycle (carbon) to a tightly closed cycle (phosphorus), with nitrogen and sulphur in between. The carbon cycle has large fluxes to and from the atmospheric carbon pool which, for the purposes of this study, is defined as being outside of the system. Therefore, at the scale of a patch of broad-leafed savanna, the carbon cycle is 'open'. The nitrogen and sulphur cycles also include an atmospheric loop, but it is small relative to the recycling which occurs within the plant–soil system. The phosphorus cycle has a very small atmospheric flux. Phosphorus cycling is virtually entirely restricted to the soil–plant system, and differs from the nitrogen cycle in the prominence of the inorganic soil pools and the non-biological processes of exchange between them. The degree of 'openness' of the cycle has important consequences on the potential for loss of elements from the system, and on the rate of recovery after such leakage.

Nitrogen and phosphorus are essential for the growth and functioning of all organisms. Their concentration in plant and animal tissues must fall within a circumscribed range (Table 7.1), and their availability is a potential constraint on productivity. Since they are required in large quantities relative to other elements, they are frequently classified as 'macronutrients'. Nitrogen and phosphorus are the most frequently limiting nutrients in terrestrial ecosystems, and savannas are no exception.

The initial work on the nitrogen cycle at Nylsvley was performed by Charlotte Gunton and Guy Bate. Information on nitrogen fixation was provided by Nat Grobbelaar, M. W. Rösch, P. C. Zietsman and N. van Rooyen, and on leaching by Frances Meredith. Peter Frost put together the first nutrient budgets. The remaining details of the nitrogen cycle were provided by Mary and Bob Scholes, who also quantified the phosphorus cycle.

The carbon cycle

Three chapters have been devoted entirely to key processes in the carbon cycle: Chapter 9 on herbivory, Chapter 10 on primary production and

Table 7.1. *The mean and range of tissue carbon, nitrogen, phosphorus and sulphur contents of dominant plants of the broad-leafed savanna at Nylsvley. The mean is averaged over the year, while the maximum is usually attained in the early wet season, and the minimum in the late dry season. N and P data from Scholes & Scholes (unpublished); S data from du Preez* et al. *(1983)*

Component and part	Carbon (%)	Nitrogen ($mg\ g^{-1}$) Mean	Nitrogen Min	Nitrogen Max	Phosphorus ($mg\ g^{-1}$) Mean	Phosphorus Min	Phosphorus Max	Sulphur ($g\ kg^{-1}$) Mean
Woody plants								
Burkea africana leaf	45.8	12.10	4.60	19.10	0.68	0.20	1.19	1.1
Terminalia sericea leaf	45.9	8.41	3.91	15.95	0.59	0.11	1.21	1.1
Burkea africana twig	44.9	6.35	3.80	10.49	0.56	0.11	1.08	1.0
Terminalia sericea twig	44.8	4.22	1.48	7.23	0.72	0.17	1.03	0.9
Burkea africana wood	45.0	8.28	4.80	1.302	0.41	0.21	0.74	0.2
Terminalia sericea wood	47.0	3.08	1.40	8.12	0.35	0.22	0.46	0.3
Coarse roots	42.7	7.13	4.79	9.84	0.21	0.10	0.43	1.0
Grasses								
Eragrostis pallens leaf	45.2	5.16	1.20	11.45	0.63	0.16	1.23	1.4
Digitaria eriantha leaf	41.6	8.23	1.69	2.296	0.99	0.22	2.12	0.9
Eragrostis pallens crown	41.8	3.30	1.08	7.22	0.38	0.12	0.89	
Digitaria eriantha crown	40.2	4.24	1.28	9.06	0.62	0.17	1.39	
Digitaria eriantha fine roots	44.8	3.63	1.03	8.53	0.40	0.10	1.18	0.7
Litter								
Fine roots	42.0	10.84	5.32	13.52	0.49	0.20	0.63	
Tree leaf litter		6.40	3.87	8.86	0.63	0.29	1.11	0.9
Dead wood		6.89	1.27	15.82	0.45	0.11	0.95	
Dead grass	42.1	3.78	1.43	5.17	0.37	0.15	0.62	
Flowers & fruit	45.3	6.21	5.05	7.37	0.62	0.48	0.76	

Chapter 11 on decomposition. This section will therefore be brief, and will concentrate mainly on the sizes and locations of the pools, since the above chapters deal with fluxes. The principal pools and fluxes in the carbon cycle of the broad-leafed savanna at Nylsvley are illustrated in Figure 7.1.

The total carbon pool is about 9000 g C m^{-2}, of which two-thirds is in the soil organic matter. This is probably close to the lower limit for savannas, since the sandy soil at Nylsvley has little potential for stabilisation of carbon. The biomass is average for a broad-leafed savanna at the dry end of the spectrum, but low relative to the wet savannas of Central and West Africa.

Figure 7.2 shows the depth distribution of carbon in the soil beneath and between tree canopies. Note that the carbon content beneath the canopies is over 20% higher than in the grassy area, and that this difference is confined to

Figure 7.1. The mean annual carbon cycle in the broad-leafed savanna at Nylsvley. Values in the pools (boxes) are in g C m^{-2}, while the values linked to fluxes (arrows) are in g C m^{-2} y^{-1}.

Broad-leafed savanna carbon cycle

Total ecosystem carbon stock 9357 g/m ^{2}

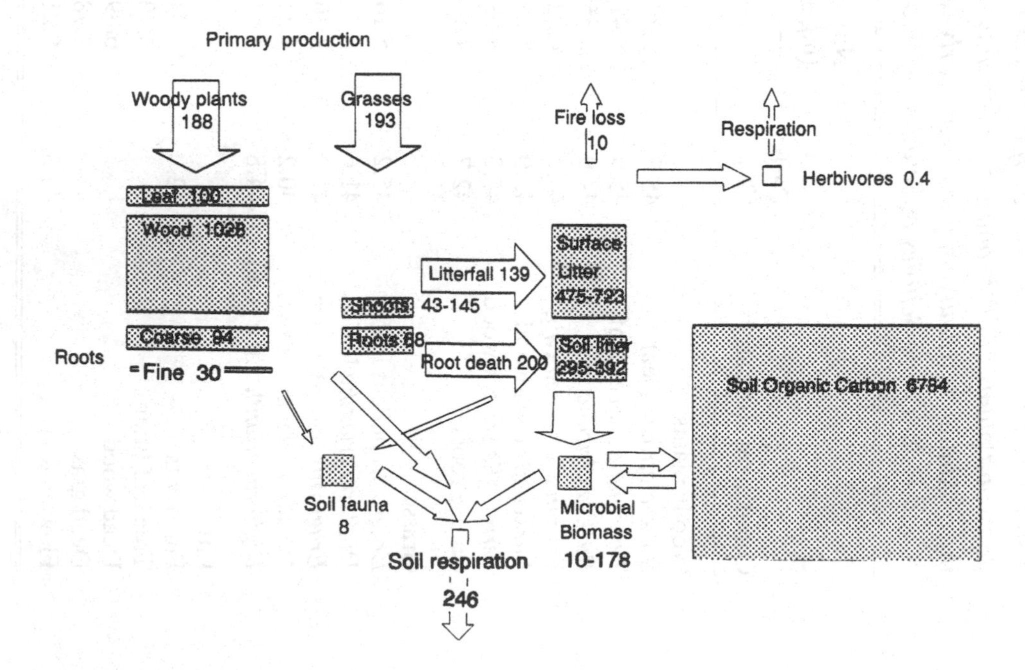

the topsoil. The difference is attributed to the patchy input of tree litter, which accumulates mostly beneath the canopy, and the slow decomposition rate of tree leaf litter relative to grass litter.

The nitrogen cycle

The nitrogen cycle of a terrestrial ecosystem is complicated because of the several forms in which nitrogen can occur and the multiplicity of pathways between them. The nitrogen cycle of the broad-leafed savanna at Nylsvley is diagrammatically illustrated in Figure 7.3. The total ecosystem nitrogen content is about 350 g N m^{-2}.

The largest nitrogen pool in this and most other terrestrial ecosystems is organic nitrogen in the soil. The nitrogen in the biomass represents 8% of the total system nitrogen. The soil nitrogen is spatially variable, both horizontally and with depth (Figure 7.4). The total nitrogen content of the soil beneath a tree canopy (to a depth of 1 m) is 44.6% higher than between trees. A 40% canopy cover was used to calculate the overall soil total nitrogen content.

The soil inorganic nitrogen pools were calculated from 300 samples from the 0–250 mm horizon taken by M. C. Scholes (unpublished data). The nitrate values were consistently larger than the ammonium values, in contrast to the data of Gunton (1981), reported in Bate & Gunton (1982). Wolfson & Cresswell (1984) showed that *Digitaria eriantha*, one of the dominant grasses

Figure 7.2. Depth distribution of organic carbon beneath and between tree canopies in the broad-leafed savanna. Means and 95% confidence intervals for 9 samples in each habitat. – – –, subcanopy habitat; ———, between-canopy habitat.

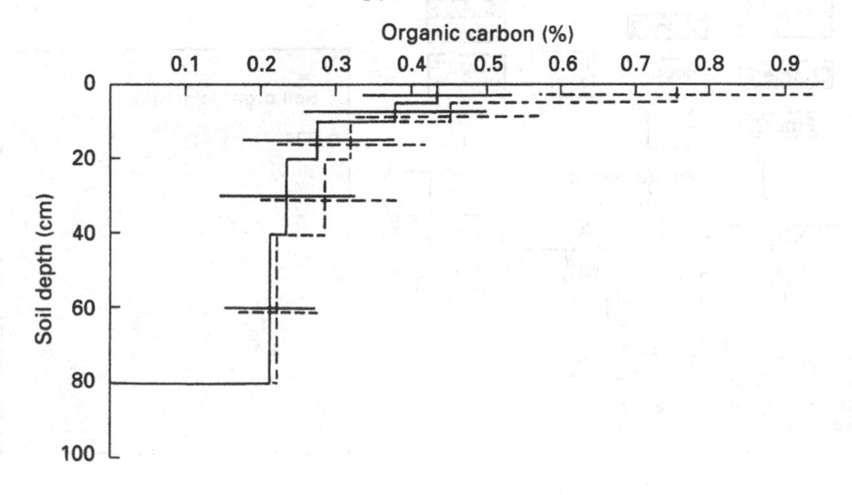

of the broad-leafed savanna, grew best when NO_3^-–N dominated the soil solution. The soil extractable inorganic-N concentrations at Nylsvley are low, so the technique used in their determination is critical. The nitrate values reported in the two studies are quite similar. The high ammonium values reported by Gunton (1981) are thought to be artefacts due to the use of the specific ion electrode, which is close to its detection limit at these concentrations.

The large organic-N and small inorganic-N pool in the soil implies that N-mineralisation is the rate-limiting step in the nitrogen cycle, and the larger nitrate than ammonium pool indicates that production of ammonium, rather than its conversion to nitrate, is the rate-limiting factor.

Nitrogen inputs in precipitation

Nitrogen is received from the atmosphere in both rain and dust (wet and dry deposition). Ionising radiation, in the form of cosmic rays and

Figure 7.3. The nitrogen cycle of broad-leafed savanna at Nylsvley. The size of the boxes is approximately proportional to the pool size in midsummer (g N m^{-2}). Arrows represent annual flux (g N m^{-2} y^{-1}).

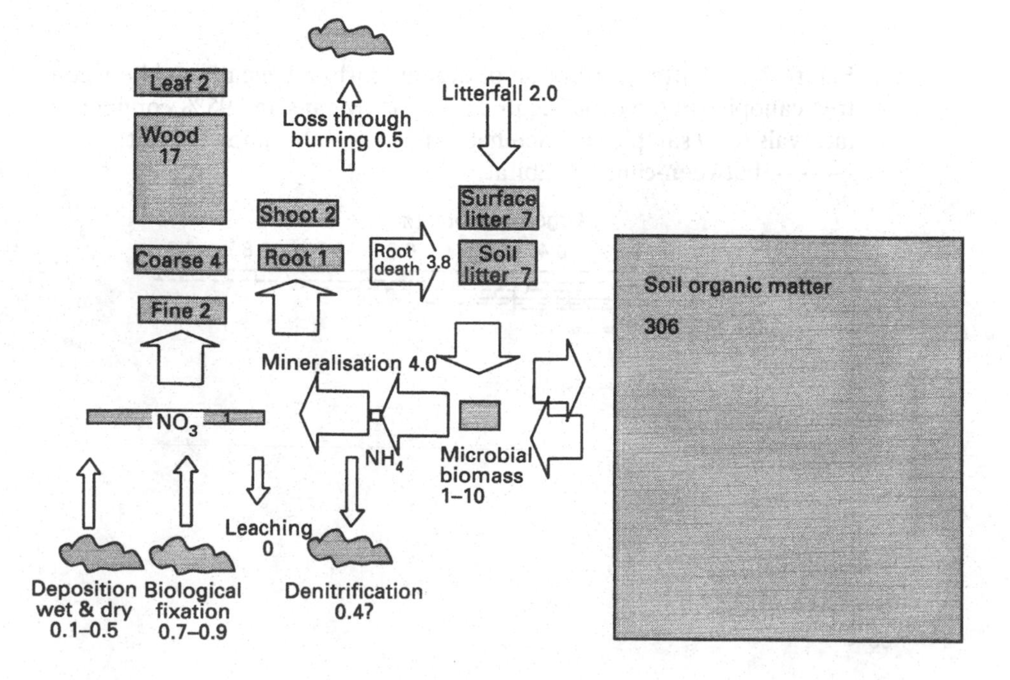

lightning discharges, converts atmospheric dinitrogen (N_2) to ammonia (NH_3), which dissolves in the rain as ammonium (NH_4^+). Oxides of nitrogen, produced by vegetation fires or industrial processes, form nitric acid, which dissociates in the rain to form nitrates (NO_3^-): this is one component of 'acid rain'. Airborne dust contains both organic and inorganic nitrogen.

Gunton (1981) measured the ammonium and nitrate content of rainfall at Nylsvley over a period of 3 years (August 1978 to April 1981). Rain was captured in five collectors with a 150 mm orifice mounted at 1.5 m and protected against contamination by birds and insects. Since the collectors were permanently in the field, they captured a proportion of the dry deposition as well. Figure 7.5 shows the nitrogen deposition per month as a function of the rainfall during the month. The distinctly different relationships for the periods before and after August 1980 suggest a methodological change at this time. Note the reservations expressed earlier regarding the use of the specific ion electrode for ammonium determinations. It is unclear which relationship is more reliable. The early data indicate a mean annual nitrogen deposition (in 630 mm rainfall) of 0.166 g N m^{-2} y^{-1}, while the later data indicate 0.401 g N m^{-2} y^{-1}. In both cases, the interannual variation should be similar to that of the rainfall, which has a coefficient of variation of 25%.

Weinmann (1955) recorded 0.032 to 0.043 g N m^{-2} per 100 mm of rainfall in a savanna in Zimbabwe, 600 km north of Nylsvley. This would translate to

Figure 7.4. Depth distribution of total nitrogen beneath and between tree canopies in the broad-leafed savanna. Means and 95% confidence intervals for 9 samples in each habitat. – – –, subcanopy habitat; ———, between-canopy habitat.

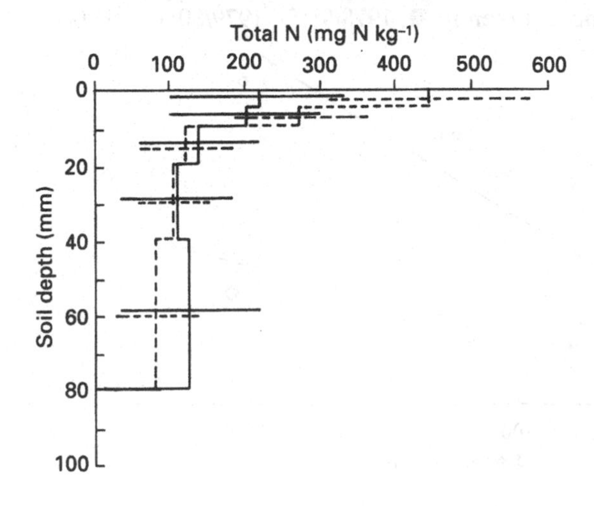

0.192 to 0.258 g N m^{-2} y^{-1} at Nylsvley. Pollution monitoring of the Eastern Transvaal Highveld, 200 km southeast of Nylsvley, indicates a minimum deposition of 0.245 g N m^{-2} y^{-1} (Tyson, Kruger & Louw 1988). These values fall between the two Nylsvley estimates. Nitrogen deposition of 0.25 g N m^{-2} y^{-1} will be assumed for Nylsvley. This is in general agreement with data from the central plains of North America (Reuss & Copley 1971). Nyslvley is upwind of the major industrial centres of the Witwatersrand and the Eastern Transvaal, so these values can be regarded as background levels of nitrogen deposition. The composition of inorganic nitrogen in the rainfall measured at Nylsvley is 21% nitrate-N and 79% ammonium-N. The average for the Eastern Transvaal is 42% (± 2) nitrate-N, reinforcing the suspicion that the ammonium estimates at Nylsvley are too high.

N fixation

Dinitrogen fixation is the process whereby atmospheric N_2 is converted to ammonium. In ecosystems it is mostly performed by bacteria of the *Rhizobium* type, which are located in gall-like nodules on the roots of certain leguminous plants. The relationship between the bacterium and the host is symbiotic, since the host expends a substantial fraction of its carbon assimilation in the maintenance of its captive nitrogen source. Certain free-living organisms, notably soil-surface algae in the Cyanophyta, are also capable of fixing nitrogen at a low rate. Fixation has been reported to occur in the rhizosphere of some tropical grasses, including root sheath-forming grasses such as occur at Nylsvley (Wullstein, Bruening & Bollen 1979).

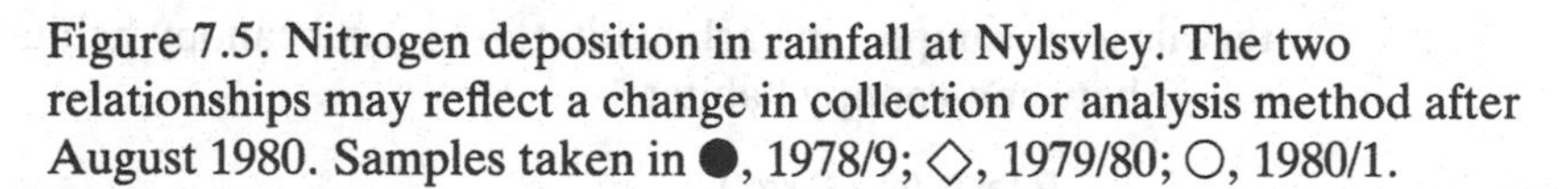
Figure 7.5. Nitrogen deposition in rainfall at Nylsvley. The two relationships may reflect a change in collection or analysis method after August 1980. Samples taken in ●, 1978/9; ◇, 1979/80; ○, 1980/1.

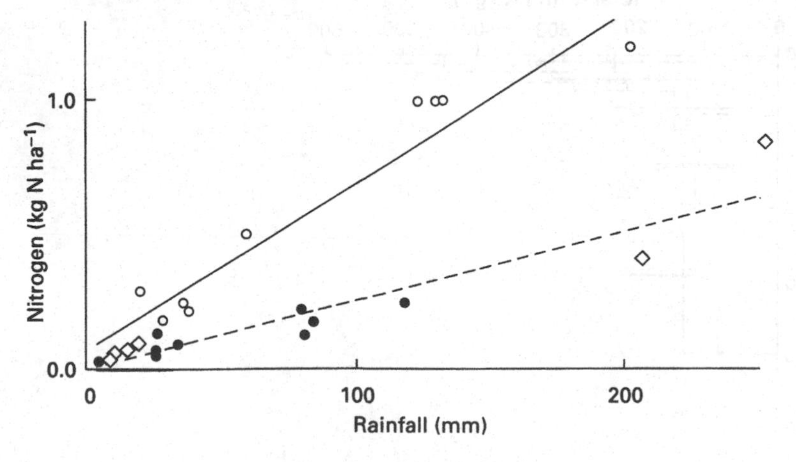

Nitrogen fixation was studied at Nylsvley using the acetylene reduction assay. There are well-known problems with the expression of estimates of fixation obtained in this way in absolute terms, but it nevertheless gives a reliable indication of which species are actively fixing nitrogen, and under what conditions. The main problem lies in the choice of an appropriate factor to convert acetylene reduction measurements to N_2 reduction estimates. The Nylsvley study used a factor of 3, which has been widely applied (Turner & Gibson 1980). The annually integrated flux could be an over- or underestimate, but probably not by more than 50%. Fixation measurements were obtained monthly during the 1975–6 season by Grobbelaar & Rösch (1981) and during the 1980–1 season by Zietsman, Grobbelaar & van Rooyen (1988). Note that the latter paper supersedes the former, since an error was discovered in the gas chromatograph calibration procedure requiring recalculation of the previously published data.

Grobbelaar & Rösch (1981) tested the roots of 37 species of grasses, forbs and woody plants growing in the broad-leafed savanna at Nylsvley for N_2-fixing activity. Zietsman *et al.* extended the survey with an additional 29 non-legume species, and tested aboveground parts as well. Both studies found activity to be associated exclusively with nodulated legumes. Significantly, the dominant tree of this savanna, *Burkea africana*, has never been found to nodulate, despite being a legume. In general, members of the legume subfamily Caesalpinioideae, which dominate the moist, nutrient-poor savannas of Africa, are not nitrogen-fixing. No activity was detected in sieved soil, or (in the latter study) associated with soil-surface algae.

Both studies followed the same sampling design. Forty 56.4 mm diameter cores were taken to a depth of 400 mm at random locations in a 400 × 30 m gridded area. Ten per cent of the soil atmosphere was replaced by commercial acetylene (scrubbed through sulphuric acid) and the rate of reduction of acetylene to ethylene was recorded over a 90-minute period beginning 90 minutes after the addition of acetylene. In the second study, only 62 cores of the 1000 tested showed substantial nitrogenase activity (> 270 nmol C_2H_2 per reduction core h^{-1}). Nodules were discovered in 90% of these. No nodules were found in random samples of low-activity cores. The nodule density in the surface 400 mm is approximately 33 m^{-2}. All legumes known to form active nodules, and growing within 2 m of the 'active' core sites, were recorded. Thirteen per cent of the active cores had no legume within this range. *Elephantorrhiza elephantina* occurred near 74% of the remainder, *Cassia biensis* near 66%, *Tephrosia semiglabra* near 19%, *Indigofera daleoides* near 19%, *T. forbesii* near 13%, *T. longipes* var. *lurida* near 6% and *Rhynchosia monophylla* near 4%. *Indigofera comosa*, *I. filipes* and

T. lupina also occur at Nylsvley and are known nitrogen fixers, but could not be implicated in the field.

Figure 7.6 shows the seasonal pattern of acetylene reductase activity for the 2 years studied. Integrating the area under the curves and converting to N_2-fixation by assuming that three molecules of C_2H_2 reduced are equivalent to one molecule of N_2 reduced gives dinitrogen fixation estimates of 0.63 and 0.85 g N m^{-2} y^{-1} for 1975–6 and 1980–1 respectively. These values are about double the usual range for temperate grasslands (Vlassak, Paul & Harris 1973; Kapustka & Rice 1978), but similar to those quoted for *Acacia holosericea* savannas in Australia by Langkamp, Farnell & Dalling (1982). Apart from the contentious conversion factor, overestimation could be caused by diurnal patterns (all these estimates were taken in the morning, when N_2-fixation tends to be high), and underestimation by damage to the nodules during the coring process.

Figure 7.6. Seasonal pattern of acetylene reductase activity in two sample years in relation to the rainfall pattern.

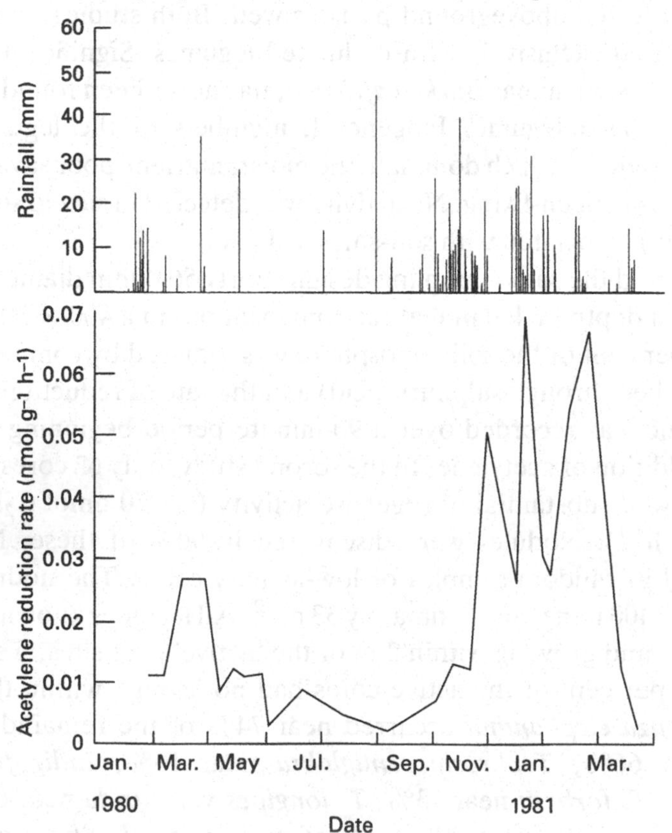

Volatilisation

Volatilisation is the process whereby NH_4^+-N is lost to the atmosphere as ammonia. The principle mechanisms are through the combustion of nitrogen-containing organic residues, direct volatilisation of NH_3 from NH_4^+ in the soil solution under alkaline conditions, and from nitrogenous compounds in animal urine. Most of the nitrogen lost during fires is not lost as NH_3 but as N_2 and N_2O, a process sometimes referred to as pyrodenitrification (Crutzen *et al.* 1979; Crutzen & Andreae 1990; Lobert *et al.* 1990; Cofer *et al.* 1991; Menaut *et al.* 1991). Vegetation burning in African savannas is a significant global source of N_2O, which is both a 'greenhouse gas' and active in the ozone cycle. The proportion of the nitrogen in plant tissues and litter which is lost to the atmosphere during combustion is a function of the fire temperature. Loss of nitrogen is virtually complete above 600 °C, which is easily achieved in an intense savanna fire. However, the fire temperature close to the soil (in the litter layer) is usually much lower. At a fire temperature of 200 °C only 50% of the nitrogen is lost. The present fire ecology policy at Nylsvley is to burn in late winter in the early morning, sufficiently frequently to prevent the accumulation of large amounts of dead grass. All these conditions tend to reduce the fire temperature and volatilisation losses.

The model constructed in Chapter 11 to calculate the pathways of carbon oxidation was used to estimate nitrogen losses during fires. Assuming that a fire consuming 1000 g m^{-2} of fuel results in complete gasification of nitrogen compounds, that half of the litter is burned in a given fire, and a 5-yearly fire regime, it was estimated that about 0.2 g N m^{-2} y^{-1} is lost through fires. This is about 6% of the nitrogen returned annually to the soil by litterfall and root turnover.

The soils of the broad-leafed savanna at Nylsvley have a pH in water of 4.9 (Harmse 1977). Therefore volatilisation directly from the soil solution is unlikely. Volatilisation also occurs from the urine of ungulates, and a small amount of volatilisation can occur directly from plant tissues. These estimates have not been made for Nylsvley.

Mineralisation

Nitrogen in the soil is mostly in the organic form. Microbial attack on the carbon skeleton of the organic molecule frees the nitrogen in the form of NH_4^+: the process is called nitrogen mineralisation. If the C:N ratio of the organic substrate is greater than 16 (as it usually is in fresh litter) then the NH_4^+ is likely to be taken up immediately by the microbes involved in the decomposition process. Thus the decaying organic matter will release no nitrogen for plant uptake; in fact, it may even reduce the available nitrogen in the soil solution. This process is known as microbial immobilisation. Once the C:N of the litter–microbial biomass system has fallen to about 11.5 (the C:N ratio of microbial biomass), NH_4^+ begins to be released from the system. This is known as net nitrogen mineralisation, and is what is measured in the field.

The *in situ* measurement of mineralisation relies on the measurement of the rate of accumulation of inorganic nitrogen (NH_4^+ and NO_3^-) in a column of soil from which root uptake has been prevented. This is achieved by knocking a 50 mm diameter stainless steel cylinder into the soil to a depth of 250 mm, thus isolating the soil inside it without disrupting its structure or temperature unduly (Raison, Connell & Khanna 1987). A duplicate core is analysed immediately for 1 M KCl-extractable NH_4^+ and NO_3^- while the *in situ* core is allowed to incubate in the field. The Nylsvley study incubated over a 2–4 week period, after establishing that the rate of mineralisation of inorganic-N was constant over this duration. Net mineralisation is the difference between the total inorganic nitrogen after and before incubation.

Raison *et al.* (1987) left two tubes in the field, one open and one covered, in order to calculate 'actual' mineralisation and leaching loss. M. C. Scholes & R. J.Scholes (unpublished data) point out that the mineralisation rate is not 'actual', since the water content in the tube soon deviates from the surrounding soil because of the exclusion of roots. They used only one tube (replicated 24 times), which was brought up to field capacity by the addition of deionised water, and covered against rain for the duration of the incubation. This gave an *in situ* mineralisation potential. The actual mineralisation was then calculated by using a model to interpolate between sample dates and integrate over the season on the basis of the soil temperature and water content.

The dependence of the N-mineralisation rate on soil water content was determined in the laboratory (Figure 7.7). The influence of temperature was determined by relating the *in situ* rates to the mean air temperature over the incubation period (Figure 7.8). Singh & Gupta (1977) report the Q_{10} (rate at temperature ($t + 10\,°C$)/rate at temperature t) for soil respiration to be in the range 1.5–3, averaging around 2, which agrees well with this data. Mineralisation in nutrient-poor systems can be constrained by nutrient deficiencies (Purchase 1974), but the addition of available phosphorus to laboratory incubations did not stimulate the mineralisation rate. The mineralisation rate was strongly related to the soil total-N content (Figure 7.9), but there was no evidence of substrate depletion as the season progressed. This suggests that the easily mineralisable nitrogen pool (N_0) must exceed the annual mineralisation.

These three relationships allowed the modification of the water balance model described in Chapter 6 for the purpose of predicting mineralisation rates. More sophisticated models are possible, but in this case a simple model accounts for most of the observed variation. Figure 7.10 shows the simulated time course of nitrogen mineralisation for the sample period. The simulated mean annual net N-mineralisation, based on the 15-year study site weather record, was 3.5 g N m^{-2} y^{-1}. Figure 7.10 illustrates the interplay of temperature and water controls on mineralisation. The overall constraint on the maximum rate is temperature, but at a shorter timescale, mineralisation occurs in discrete pulses when soil water conditions permit.

Nitrification

Nitrification is the process (performed by the bacteria *Nitrosomonas* and *Nitrobacter*) whereby NH_4^+ in the soil solution is converted first to NO_2^- and then to NO_3^-, yielding energy to sustain the bacterium. In many hundreds of *in situ* mineralisation tests performed by Mary Scholes (unpublished data), no significant NH_4^+ was detected, while large amounts of NO_3^- were generated. This means that ammonification (the first step in mineralisation) is the rate-limiting step at Nylsvley. Therefore NH_4^+ does not accumulate, and the rate of nitrification is equal to the rate of mineralisation.

Figure 7.7. Dependence of nitrogen mineralisation rate in soils from the broad-leafed and fine-leafed savanna on the soil water content. These data were obtained in a laboratory incubation in which varying amounts of water were added to initially dry soil.

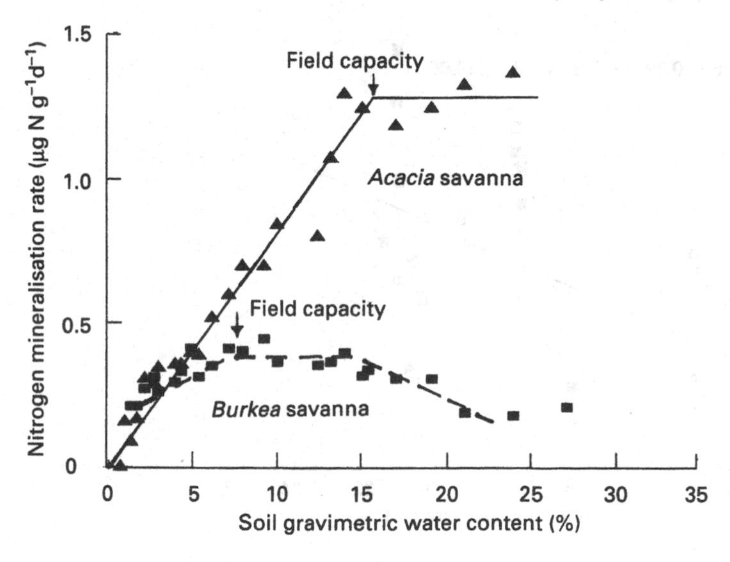

Denitrification

The conversion of NO_3^- to gaseous N_2 or oxides of nitrogen (NO, NO_2, etc.) mostly occurs in anaerobic situations, where nitrate replaces oxygen as a terminal electron acceptor. In many systems it is a major avenue of nitrogen loss. Furthermore, the oxides of nitrogen are important 'greenhouse gases'. The soil need not be completely saturated for denitrification to occur. High oxygen demand in the centre of a compact, moist soil aggregate can lead to a tiny localised anaerobic spot, where denitrification occurs. The soils at Nylsvley are both freely draining and unaggregated, making denitrification by this mechanism unlikely.

A small amount of denitrification accompanies the process of nitrification. Since the nitrification flux is substantial in the broad-leafed savanna, this could be a significant pathway of nitrogen loss. In the absence of field data it has been assumed that denitrification is equal to 5% of the nitrification flux, an amount of 0.175 g N m^{-2} y^{-1}.

The largest denitrification flux occurs as a consequence of the burning of litter and dead grass, and amounts to about 0.2 g N m^{-2} y^{-1}. This 'pyrodenitrification' flux is discussed in the section on volatilisation above.

Immobilisation

Microbial immobilisation is a transient phenomenon in this savanna, owing to the short lifespan of microbes and their sensitivity to water

Figure 7.8. Relationship between mean air temperature and the *in situ* N-mineralisation rate measured over a 4-week period, relative to the rate at 20 °C.

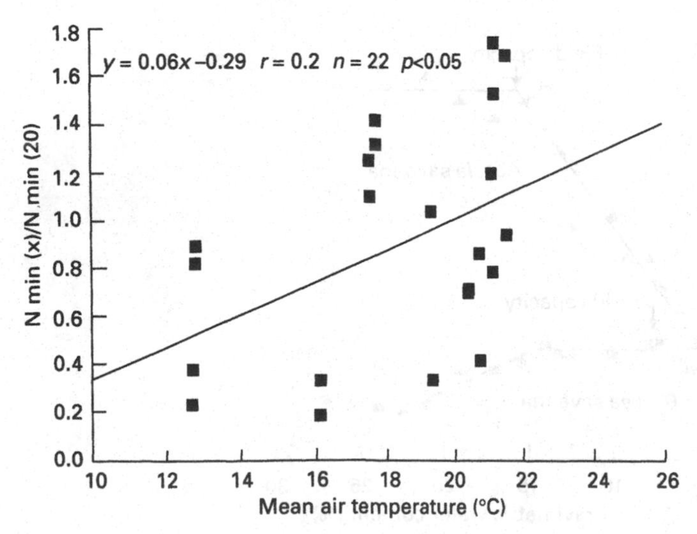

stress. A relatively large amount of nitrogen (10 g m^{-2}) is immobilised in the microbial biomass early in the season, but this declines as the season progresses. The nitrogen does not all appear as net mineralisation, however. Some may be stored in dormant microbial propagules. The early-season microbial N-immobilisation could have an important ecosystem-level consequence by minimising nitrogen losses attributable to leaching, at a time when the plant roots are not fully developed.

Uptake

Plant nitrogen uptake can be calculated in two ways: either by summing the increments of nitrogen in the biomass, or by measuring the nitrogen in the litterfall. The former method is more appropriate for Nylsvley, since the vegetation is not in steady state, but is technically more difficult to achieve due to the considerable retranslocation which occurs within the plant. Multiplying the production components of the savanna by their nitrogen content leads to an uptake estimate of 4.0 g N m^{-2} y^{-1} using the first method, and 5.8 g N m^{-2} y^{-1} using the second. The difference is due partly to nitrogen in the accumulating woody biomass. The uptake estimate is in good agreement with the net mineralisation estimate, which is necessary since inorganic nitrogen was never observed to accumulate in the soil.

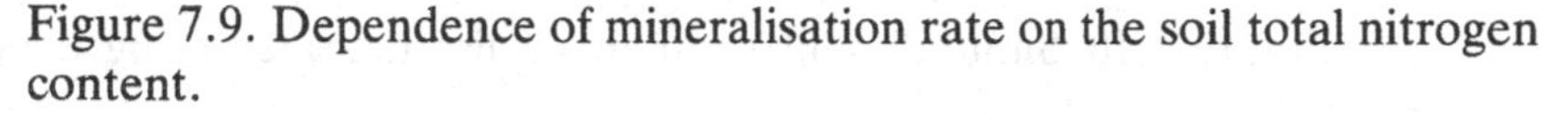

Figure 7.9. Dependence of mineralisation rate on the soil total nitrogen content.

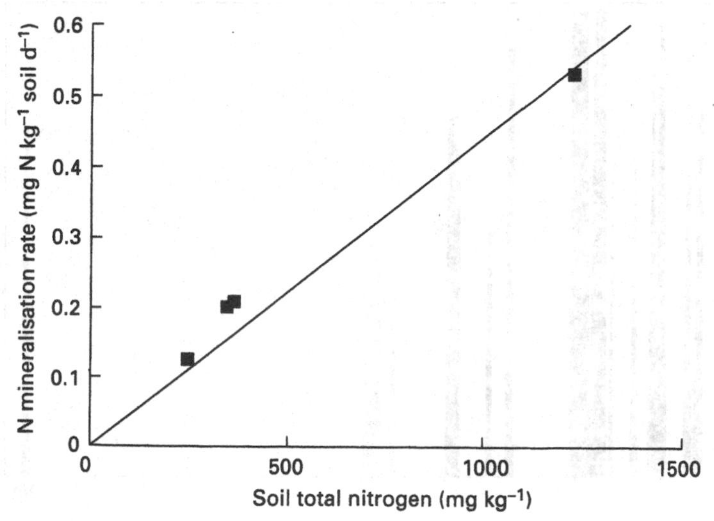

Leaching

Nitrate, since it bears a negative charge, is not held on the cation exchange surfaces of the soil. It is therefore susceptible to loss by leaching. Soils do bear a small anion exchange capacity. A large proportion of the ion exchange capacity of the Nylsvley soil is provided by soil organic matter, which has both negatively and positively charged exchange sites, the ratio of which depends on the soil pH. In acid soils, the anion exchange capacity can be significant. Bate & Gunton (1982) showed that leaching of nitrate from broad-leafed savanna soil cores reconstituted in the laboratory was negligible when the exchangeable NO_3^--N in the soil was less than 10 mg kg^{-1}, but increased rapidly above that level. This suggests that the anion exchange capacity is in the region of 0.07 cmol (−) kg^{-1}, or 4% of the CEC (cation exchange capacity). The average *in situ* soil NO_3^--N content is 1–5 mg kg^{-1}.

The hydrological model presented in Chapter 6 showed that the long-term average amount of water leaching below 1 m depth amounted to 6 mm per annum. If the entire exchangeable NO_3^--N of the soil were in the soil solution (an overestimate, since some is on the anion exchange surfaces), this 6 mm would represent a nitrate loss of 0.16 g N m^{-2} y^{-1}. This is an upper limit on leaching loss from the undisturbed broad-leafed savanna. In fact leaching is an extremely sporadic process. Most years will see no leaching at all. Occasionally the coincidence of several large rainstorms will result in a leaching pulse. More sophisticated water movement, nitrogen mineralisation and solute leaching models (the LEACHM family: Wagenet & Hutson 1987)

Figure 7.10. Simulated nitrogen mineralisation rate during 1988/9.

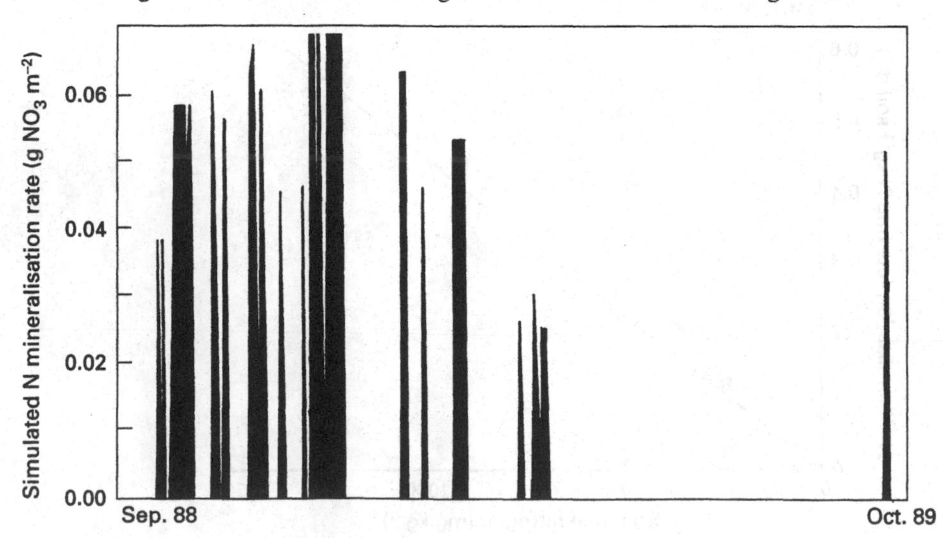

confirm that leaching losses under intact savanna at Nylsvley are likely to be very small.

There is a large potential for nitrate leaching losses following disturbance of the savanna, however. The soil has a low water-holding capacity, and most of the anion exchange capacity is associated with the organic matter. Conversion to a cropland stimulates N-mineralisation and simultaneously reduces the anion exchange capacity. The soil water content remains high during the crop establishment phase, which increases both mineralisation and the probability of leaching. Tree removal (bush clearing) to stimulate grass production also increases the soil moisture duration and therefore leaching losses. Note that nitrogen uptake is dominated by the woody plants.

Meredith (1987) conducted an experiment designed to test whether the disturbances typical of land-use transformations in savannas (burning, mowing, grazing, clearing and cultivation) resulted in nutrient loss or redistribution as a result of leaching. In each treatment, the leachate passing below 400 mm was collected in tray lysimeters and analysed for inorganic nitrogen, major cations and phosphate (Table 7.2). In the undisturbed savanna, nitrate movement below this depth was negligible in relation to the total soil pool, and became significant only in the completely cleared treatments, especially if the soil was disturbed to simulate cultivation.

The phosphorus cycle

Phosphorus is taken up by the plant in several ionic forms. At the soil pH prevailing at Nylsvley, the dominant form is $H_2PO_4^-$. All the ionic forms have a low diffusivity in soil and, coupled with the low concentration of phosphate ions in the soil solution because of their low solubility, this results in the rapid development of a phosphate-depleted zone around the roots. This is one reason why plants are forced continuously to replace their fine root network: new soil must be explored in order to satisfy the plant's phosphorus requirements. Mycorrhizae (symbiotic fungi associated with the root) are important in the phosphorus nutrition of plants growing in low-phosphorus soils (Gerdemann 1968). The thin fungal hyphae are able to explore the soil volume at a lower carbon cost than the thicker roots. The hyphae may have access to phosphorus unavailable to higher plants (Baylis 1972), due to more efficient uptake or perhaps to the secretion of phosphatases or acids which increase the availability of phosphorus in the soil (Tinker 1975).

The depth distribution of total phosphorus in the broad-leafed savanna is illustrated in Figure 7.11. It is noteworthy that, unlike carbon and nitrogen, there is very little decrease in the phosphorus content with depth, nor is there

Table 7.2. *Movement of nutrients below 300 mm depth in undisturbed nutrient-poor broad-leafed savanna, and the nutrient-rich fine-leafed savanna. The values for the increase following clipping (to simulate grazing) and ploughing after removing the vegetation refer to the increase in the concentration of the nutrient in the leachate, not the increase in total amount leached, which also depends on the volume leaching. The leachate concentrations were unaffected by burning the vegetation above the lysimeters. Data from Meredith (1987)*

	Ca^{2+}	K^+	Mg^{2+}	Na^+	NO_3^-	PO_4^{2-}
Input in rain (kg ha^{-1} y^{-1})	7.12	8.51	1.13	6.73	0.28	0.98
Nutrient poor						
Exchangeable 0–30 cm (kg ha^{-1})	128.8	78.8	39.4	ND	5.4	9.5
Leached (kg ha^{-1} y^{-1})	0.9	2.6	0.4	0.6	1.3	ND
Conc. change after clipping (%)	25 to 80	−35 to 5	60 to 180	−60 to 5	10 to 66	
Conc. change after ploughing (%)	120 to 600	−10 to 10	400 to 500	5 to 20	100 to 190	
Nutrient rich						
Exchangeable 0–30 cm (kg ha^{-1})	12285.6	1970.2	2517.7	83.4	9.6	1788.0
Leached (kg ha^{-1} y^{-1})	16.4	6.5	6.1	0.5	3.8	1.0
Conc. change after clipping (%)	−6 to −40	−15 to −45	−10 to −60	5 to 50	10 to 30	40
Conc. change after ploughing (%)	−15 to 15%	15 to 20	−30 to 20	−10 to 10	380 to 670	30

ND = below the threshold for detection.

a large difference between the subcanopy and between-canopy areas. The total soil phosphorus to a depth of 1 m is 346.1 g P m^{-2}, giving a C:P ratio of 11.8, which is extremely low. dDu Preez, Gunton & Bate (1983) calculate the total phosphorus to 1 m in broad-leafed savanna soil, based on extrapolation from the top 225 mm, to be 224.8 g P m^{-2}. These values of total phosphorus are within the normal range for natural ecosystems (Brookes, Powlson & Jenkinson 1982); the abnormally low C:P ratio is therefore due to the low carbon content of the soil.

The distribution of phosphorus in the broad-leafed savanna ecosystem is shown in Figure 7.12. The total amount of phosphorus in the system is 91.2 g P m^{-2} in this figure. The discrepancy between this value and those given above arises because the samples were taken in different places, and analysed in different ways. There is high spatial variability of phosphorus at Nylsvley. Note that less than 1% of the phosphorus is in the vegetation.

Forms of phosphorus

The low solubility of inorganic phosphorus compounds is an important factor in the phosphorus cycle. Consequently the soil phosphorus is conceptually viewed as consisting of a small readily available exchangeable pool, a moderately available labile pool, and a large, unavailable occluded pool, all in equilibrium with one another. The chemical reality is a little more

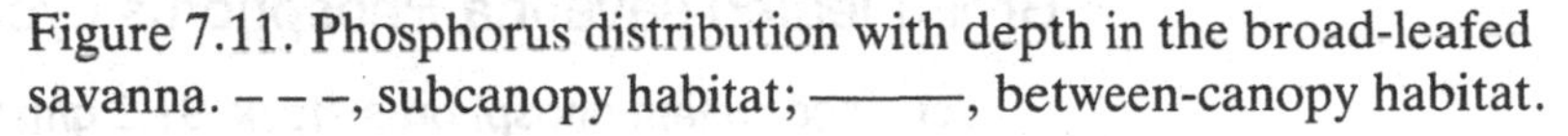

Figure 7.11. Phosphorus distribution with depth in the broad-leafed savanna. – – –, subcanopy habitat; ———, between-canopy habitat.

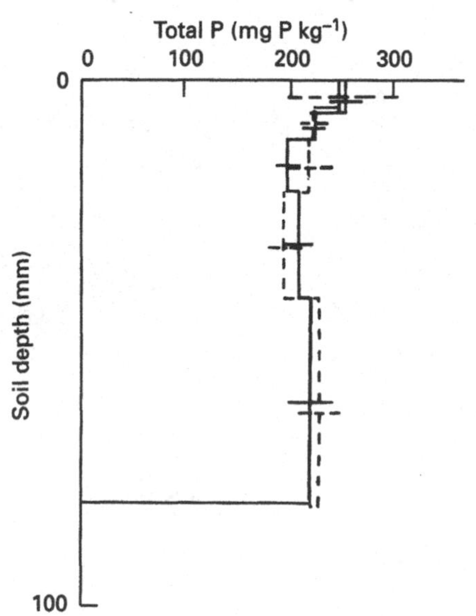

complicated, since phosphorus is present in a large variety of forms in the soil, and these forms do not all fit neatly into the three conceptual boxes. Organic-P is in the soil organic matter and microbial biomass, and is released by microbial oxidation (in the same way that nitrogen is mineralised), rather than by solubilisation. The inorganic-P can be in calcium phosphates (Ca-P) such as the mineral apatite, or in aluminium phosphates (Al-P) or iron phosphates (Fe-P). All of these forms can be occluded by iron and aluminium oxides, which prevent their solubilisation.

The fractionation of the total soil phosphorus within the broad-leafed savanna at Nylsvley is presented in Table 7.3 (Scholes & Scholes 1989). Nearly half is in the organic form. Most of the inorganic phosphorus occurs as phosphates of iron and aluminium, with a lesser amount as calcium phosphate. Very little is occluded, which is consistent with the small amount of iron oxide in the soil. Only 5% of the total soil phosphorus is available to the plants in the short term. The exchangeable phosphorus level of about 5 mg P kg^{-1} would indicate a severe phosphorus deficiency in an agricultural soil; we will argue that they do not represent phosphorus deficiency in the natural

Figure 7.12. The phosphorus cycle in the broad-leafed savanna at Nylsvley. The size of the boxes is approximately proportional to the size of the pools ($g\ P\ m^{-2}$). The arrows represent annual fluxes ($g\ P\ m^{-2}\ y^{-1}$).

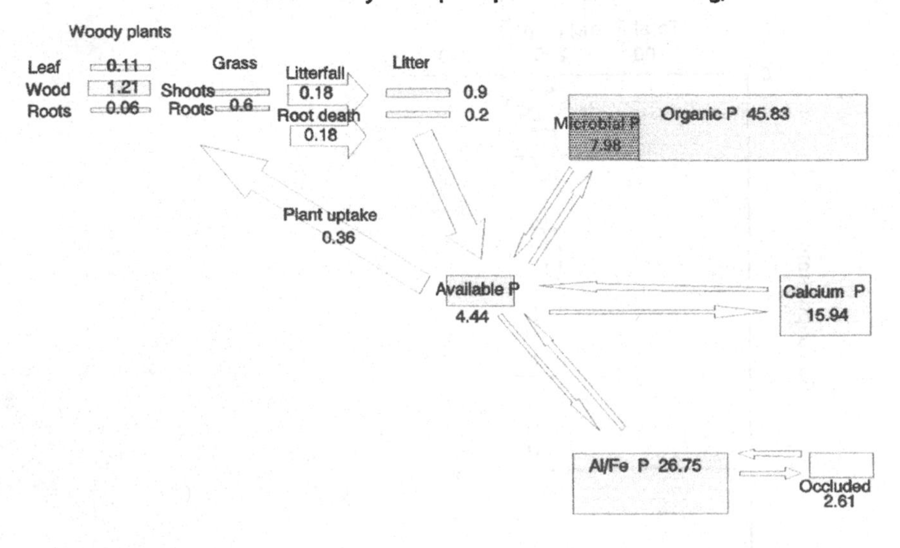

Table 7.3. *Phosphorus fractions in surface soils (0–300 mm) from the broad-leafed savanna at Nylsvley. The fractionation procedure follows Olsen & Sommers (1982)*

Form	mg kg^{-1}	%
Readily available to plants		
Resin-bag extractable	5.1	4.6
Moderately available		
Calcium-bound	18.3	16.7
Organic phosphorus	52.6	47.9
Iron- and Aluminium-bound	30.7	28.0
Occluded		
Reductant phosphorus	0.0	0.0
Within iron oxides	3.0	2.7
Total phosphorus	107.7	100.0

savanna. Given the equal balance between the organic and inorganic labile forms, microbial oxidation and solubilisation are both likely to be important in the maintenance of the available pool.

Tests for available soil phosphorus are notoriously inconsistent. The main reason is that the extractants used by different laboratories are able to remove different proportions of the various fractions. Each extractant is most suitable for a particular soil type, and this has impeded standardisation. In the Nylsvley work, an anion exchange resin was used to extract the phosphate from a soil and water suspension (Sibbensen 1978). This approach closely simulates the behaviour of a root in soil, and has been shown to correlate well with plant response and the available pool determined by radioactive isotopes.

Phosphorus mineralisation

Scholes & Scholes (1989) used the *in situ* mineralisation method developed for nitrogen (Raison *et al.* 1987; described in the nitrogen mineralisation section above) for phosphorus mineralisation as well. The presence of a large inorganic pool of phosphorus in equilibrium with the available pool complicates the interpretation, since it can act as either a source or a sink of phosphorus. Phosphorus mineralisation occurred at a mean rate of about 0.025 mg P kg^{-1} d^{-1} in topsoil from the broad-leafed savanna when the soil was moist. Compare this with a mean potential mineralisation rate for nitrogen of 0.18 mg N kg^{-1} d^{-1} (SE = 0.01, n = 20) during the summer months. The mineralisation rates are in approximate proportion to the total amounts of phosphorus and nitrogen present. From this it can be inferred that the total annual phosphorus mineralisation is about 0.6 g P m^{-2} y^{-1}.

Phosphorus immobilisation: chemical and microbial

Phosphorus can be immobilised by precipitation in weakly soluble phosphates, occlusion in ferric oxides, or sequestration in the microbial biomass. Soils with a high potential for phosphorus immobilisation are said to be 'phosphorus-fixing'. The highly-weathered tropical soils typical of many savannas are strongly P-fixing owing to their high iron oxide content. Scholes & Scholes (1989) showed that the soils of the broad-leafed savanna at Nylsvley did not have a high potential for chemical immobilisation of phosphorus; the sorption potential reached a limit at about 35 mg P kg^{-1}. However, when these soils were incubated in the laboratory with added inorganic phosphorus, 2.6% of the added phosphorus was immobilised per day. Since this soon exceeded the sorption limit, it was assumed that the phosphorus was immobilised by microbes.

A natural gradient of soil total phosphorus exists between the broad-leafed and fine-leafed savannas at Nylsvley. Scholes & Scholes (1989) used *in situ* phosphorus mineralisation measurements along this gradient to show that net mineralisation occurred only when the soil available phosphorus was below 100 mg P kg^{-1}; above this value net immobilisation occurred. The phosphorus content of microbial biomass (and soil organic matter) is quite variable (Brooks *et al.* 1982). At Nylsvley the microbial C:P was assumed to be about 20. The peak microbial carbon in the broad-leafed savanna is equivalent to 193 g C m^{-2}. Therefore about 10 g P m^{-2} is microbially immobilised in the early part of the growing season.

Phosphorus uptake and litterfall

The available soil phosphorus pool is much more variable than the soil nitrate pool (Figure 7.13). It increases during wet periods, especially if they coincide with times when the plants are inactive, and declines during drier periods in the growing season to a consistent minimum value of 0.4 mg P kg^{-1}. This pattern suggests that the phosphorus mineralisation rate exceeds the uptake rate in the short term, and that the excess is gradually removed from the soil solution by uptake, microbial immobilisation, and precipitation.

Mycorrhizal symbionts

Sloff (1982) found four types of mycorrhizal spores in soil from the broad-leafed savanna at Nylsvley. There was 50% vesicular–arbuscular mycorrhizal infection of the grass *Panicum maximum* collected there. However, in an experiment where *P. maximum* and *Eragrostis curvula* were grown in sterilised and unsterilised soil from the broad-leafed savanna, the plants in the sterilised soil consistently outperformed those in the unsterilised soil, at all levels of added phosphate. They also had higher tissue concentrations of phosphorus. No mycorrhizal infections were found on the roots in either treatment. This would suggest that for these grasses at Nylsvley, mycorrhizal associations are not important in phosphorus nutrition.

Phosphorus exchanges with the atmosphere

Phosphorus deposition in dust and rainfall has not been measured at Nylsvley. Typical values from other ecosystems are in the range 0.02–0.1 g P m^{-2} y^{-1} (Brown, Mitchell & Stock 1984). Phosphorus can be lost to the atmosphere by wind erosion of the soil and in particles released by fires. The sandy Nylsvley soil is not highly prone to wind erosion, nor are the wind speeds sufficient to result in significant erosion from a well-vegetated

Figure 7.13. Seasonal trend of soil available phosphorus. Bars represent the 95% confidence interval (n = 4).

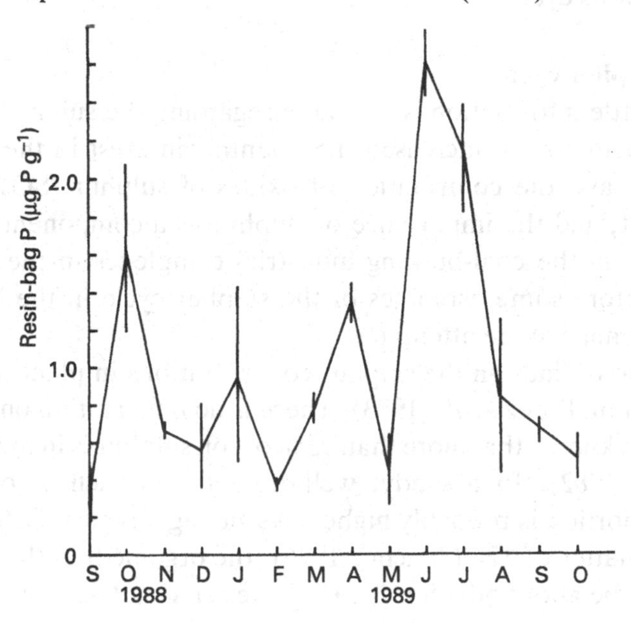

savanna. Loss of a significant proportion of the phosphorus in fuels occurs only when the fire temperature exceeds 600 °C, which is seldom the case under the fire regime currently prevailing at Nylsvley. Therefore the phosphorus exchanges between the atmosphere and the broad-leafed savanna are believed to be small in comparison to the terrestrial pool and the internal recycling.

Is the broad-leafed savanna phosphorus limited?

Despite the low available soil phosphorus pool, which would be considered extremely deficient in the context of an annual crop, there is no convincing evidence that primary production in the broad-leafed savanna is phosphorus-limited. Responses to phosphorus fertilisation were reported by Walker & Knoop (1987) with some inconsistencies. There is some circumstantial evidence that the system is not phosphorus limited: the low C:P ratio, the high phosphorus mineralisation rate relative to the plant requirements, and the occasional accumulation of available phosphorus in the soil. The total phosphorus content of the soil is much bigger than the amount in the vegetation. Coupled with the relative immobility of the phosphate ion in solution, and the absence of a significant atmospheric component to the cycle, the potential for phosphorus depletion through ordinary management practices such as grazing, burning or firewood collection is minimal in this savanna. The sufficiency of phosphorus in this savanna is probably related to the 'closedness' of its cycle.

The sulphur cycle

Very little information is available regarding the sulphur cycle at Nylsvley. Two factors have increased the scientific interest in the sulphur cycle in recent years: the contribution of oxides of sulphur (SO_x) to the greenhouse effect, and the importance of sulphur as a component of 'acid rain' pollution from the coal-burning industrial complexes in the Eastern Transvaal. Therefore some estimates of the sulphur cycle in the Nylsvley broad-leafed savanna will be attempted.

The only source of data on the sulphur concentrations in plant tissues at Nylsvley is that of du Preez *et al.* (1983). There is no information on sulphur in the soil, but it is known that more than 95% of soil sulphur is in the organic form (Tabatabai 1982). In a sandy, well-drained soil such as occurs at Nylsvley the proportion is probably higher. Assuming a typical C:S ratio of the soil organic matter of 126 (Sanchez 1976), the organic-S in the soil to a depth of 1 m will be about 50 g S m^{-2}. Du Preez *et al.* (1983) calculate the sulphur in the dominant tree and grass species, which together make up about half of the biomass, to be 1.7 g S m^{-2}. The mean litter sulphur content is 0.09%, making the annual litterfall sulphur flux (which can be regarded as a measure of sulphur uptake by plants) about 0.9 g S m^{-2} y^{-1}. If the C:N ratio is 19 for soil organic matter in the broad-leafed savanna, and sulphur mineralisation occurs in proportion to nitrogen mineralisation, then the sulphur mineralisation is about 0.6 g m^{-2} y^{-1}. The sulphur inputs in the form of dry and wet deposition of SO_4^{2-} are unknown for Nylsvley, but the typical values for unpolluted sites in the Transvaal are in the order of 0.3 g S m^{-2} y^{-1} as wet deposition, and up to 1.1 g S m^{-2} y^{-1} as dry deposition (Tyson *et al.* 1988). Polluted sites receive about three times this amount. Since the soils are not prone to anaerobiosis, losses as SO_x are likely to be small, and so are losses by leaching of SO_4^{2-}. The major losses are likely to be through gasification during fires. This crude analysis indicates that like nitrogen and phosphorus at Nylsvley, most of the sulphur (about 93%) is in the soil, and most of that is probably in the organic form. There is no reason to suspect that sulphur is a limiting nutrient at Nylsvley, especially given the small difference between the sulphur content of green leaves and litter. The sulphur deposition appears to exceed the plant sulphur requirement.

The basic cations

Calcium, magnesium, potassium and sodium are not generally thought to be important limiting nutrients in dry savannas; however, the situation could be quite different in the wetter, more highly leached savannas. This group of elements, all alkali metals and therefore collectively known as 'bases', are lost from the ecosystem mainly by leaching. Therefore the 'base saturation' (the percentage of the soil cation exchange capacity occupied by bases) decreases with increasing rainfall, and is an index of weathering intensity. Towards the dry extreme of savannas, the soil frequently contains bases in excess of CEC. The excess salts appear in the soil as precipitates: calcium carbonate nodules, calcrete layers, saline or sodic horizons. At the wet end of the savanna range, the highly weathered soils of old erosional surfaces are characteristically 'acid'. The cation exchange capacity is dominated by H^+ and aluminium, with the base saturation typically below 20%. High levels of aluminium in the soil solution are toxic to plants, and this can confine the effective rooting depth to the less-acid topsoil. Addition of calcium to these soils ('liming') usually results in improved plant performance, at least in agricultural systems.

A single-minded focus on plant nutrition conceals the profound impact which the basic cations can have on landscape pattern in the semi-arid savannas. Seasonal climates with an annual rainfall between 400 and 800 mm are neither so wet as to have lost most of their bases, nor so dry that the bases are effectively immobile. Under these circumstances, the soil acts as a landscape-scale chromatography strip, with the cations migrating through it at different rates. It is no accident that the concept of a 'catena' (a predictable, topographically linked sequence of soils and vegetation) was first developed for the semi-arid savannas of Africa (Milne 1936), for it is here that they are most clearly displayed. A crucial factor in determining landscape pattern in dry savannas is the interaction of monovalent cations (particularly Na^+), divalent cations (Ca^{2+} and Mg^{2+}) and clays. Where the divalent cations dominate the soil solution, individual clay particles attract one another despite their negative charge, to form sand-sized aggregates. This promotes free water movement through the soil. Above a critical threshold level of sodium, the clay particles repel one another (deflocculate), which impedes water flow, but promotes clay movement. In conjunction with the differential mobility of sodium and calcium in the soil, this mechanism is responsible for much of the landscape pattern in African dry savannas.

The information in the following section is mostly drawn from the work of Derek du Preez, Guy Bate, Graham von Maltitz, Peter Frost and Frances Meredith.

The cycling and distribution of bases

The biogeochemical cycles of calcium, magnesium, potassium and sodium are similar to one another, but dissimilar to those of nitrogen, phosphorus and sulphur. Unlike the latter elements, the basic cations do not spend most of the cycle bound into organic compounds. Therefore the process of biological mineralisation is much less important, and the process of leaching is much more important. The mean concentrations of calcium, magnesium, potassium and sodium in plant tissues of the broad-leafed savanna are given in Table 7.4.

Over 90% of the ecosystem stock of basic cations is found in the soil. This includes only the exchangeable cations in the soil, not those incorporated in minerals or soil organic matter. Therefore nutrients are not as easily lost from this savanna through vegetation removal as they are, for instance, from tropical forests, where a much higher proportion is in the biomass.

The effects of disturbance and land use

All of the major cations show large concentrations within the fine-leafed savanna patches. The fine-leafed savanna makes up about one sixth of the study site, but contains about half of the calcium, potassium and magnesium, and a quarter of the sodium. The various basic cations show

Table 7.4. *The contents of calcium, magnesium, potassium and sodium in plant parts of the broad-leafed savanna, in midsummer. The data are from Olsvig-Whittaker & Morris (1982), Morris* et al. *(1982), du Preez* et al. *(1983), von Maltitz (1984) and Frost (1985)*

	Content (mg kg^{-1})			
Plant species and part	Ca	Mg	K	Na
Woody plants				
Burkea africana leaves	2860	940	2100	49
Ochna pulchra leaves	6830	1340	3850	36
Combretum molle leaves	12 600	3130	5090	34
Vitex rehmanni leaves	11 680	3740	4450	25
Terminalia sericea leaves	7270	1880	2170	28
Mixed species fruit & seeds	4500	1820	6720	31
Mixed species twigs and bark	15 190	1110	1300	77
Burkea africana twigs		600	3000	
Ochna pulchra twigs		800	4000	
Terminalia sericea twigs		800	3000	
Mixed species branches & trunks		220	820	
Burkea africana branches & trunks		300	500	
Ochna pulchra branches & trunks		300	800	
Terminalia sericea branches & trunks		300	100	
Burkea africana coarse roots		500	2000	
Ochna pulchra coarse roots		500	2000	
Terminalia sericea coarse roots		500	2000	
Grasses				
Eragrostis pallens leaves and culms	5800	300	10 000	
Digitaria eriantha leaves and culms	5800	800	6200	
Panicum maximum leaves and culms	5000	3000	30 000	
Eragrostis pallens roots		200	1800	
Digitaria eriantha roots		400	2400	

differential mobility in the soil. Potassium was five times more prevalent in soil leachate at 25 cm depth than would have been predicted from its proportion on the exchange complex, and sodium twice as prevalent (Meredith 1987). An area of broad-leafed savanna on a nearby farm, which had been protected from grazing for 30 years, had topsoil magnesium and calcium contents nearly twice those on adjacent grazed patches, while 30 years of subsistence maize cultivation led to a 25% increase in topsoil potassium, and a similar decrease in topsoil calcium and magnesium content (Meredith 1987).

Simulated grazing and burning disturbances carried out on small plots fitted with pan lysimeters at 25 cm did not show large leaching losses of basic cations, although they did show losses of nitrate, an anion. Calcium and magnesium increased two- to nine-fold in leachate from cleared and ploughed plots relative to that from undisturbed controls, but even so, the total leachate flux of these elements at 25 cm depth remained of the same order as inputs in the rainfall.

Basic cations can be leached directly out of plant leaves or litter by rainfall. Immersion of fresh leaves in distilled water for one minute resulted in losses of 0.5–2% of the leaf potassium, 4–20% of the calcium and 1–5% of the magnesium (von Maltitz 1984). The rate of leachate loss increased slightly with leaf age, and was markedly higher in woody plants than in grasses, and in plants growing in fertile savannas than in those of infertile savannas.

Monitoring the nutrient content of tree and grass leaves as they aged revealed three patterns of translocation before leaf drop (von Maltitz 1984). Deciduous trees in the infertile savannas remove about half of the potassium and phosphorus from the leaf before it falls, but none of the calcium or magnesium. Deciduous trees in the nutrient-rich savannas, and grasses on both soils, do not retranslocate any of the elements that were monitored. Evergreen trees, which make up a small proportion of the infertile savanna vegetation, show a slow decline in cation content with leaf age, which is attributed to leaching.

Calcium

Although there is no experimental evidence that the plants of the broad-leafed savanna are calcium limited, the calcium availability is sufficiently low to exclude shell-bearing molluscs from the fauna. Land snails are found in the Nylsvley region, but not on the sandy soils of the study site.

The highest plant calcium contents were recorded in *Eragrostis pallens*, the dominant grass of the nutrient-poor savanna, while the lowest were in *Panicum maximum*, a dominant in the nutrient-rich savanna.

Sodium

Sodium is not an essential plant nutrient, except in certain families, such as the Chenopodiaceae. It is an essential element for mammals, however. Mammalian herbivores are therefore frequently sodium-limited, and are attracted to sources such as sodium-rich soils or salt licks. Sodium moves through the soil faster than calcium or magnesium, due to its lower surface charge density. Therefore the bottomlands of landscapes derived from sodium-rich parent materials (such as granites) are frequently 'sodic'. This has a major impact on the soil hydrology, erodability, vegetation pattern, grass productivity and forage palatability.

Sodic sites are a prominent feature of the Nylsvley Nature Reserve, but did not attract much research, since they occur outside of the study site. Their mechanism of formation differs from the usual catenal pattern, described by Dye & Walker (1980). They form on alluvial soils, pedogenically unrelated to the upslope catena. The ionic composition of the waters of the Nyl river is strongly dominated by sodium. This is a feature of rivers in this portion of the subcontinent (King, de Moor & Chutter 1992). When the Nyl periodically floods, this water evaporates, leaving its salt content on the floodplain, resulting in large areas of sodic soils.

The A horizon of sodic soils is highly susceptible to erosion, and typically is present only where protected by vegetation. Most of the area of the sodic site is occupied by the hard, bare surface of the columnar-prismatic B horizon. Much energy has been expended in conservation areas trying to revegetate this surface, which is regarded as a sign of overgrazing and an erosion hazard. However, once it has reached this stage, the surface is extremely stable, particularly at Nylsvley, where it is at the bottom of the landscape. The remnants of Stone Age encampments are frequently found scattered on the surface, suggesting that the exposure of the B horizon occurred long ago.

Despite the paucity of vegetation, mammalian herbivores show a great preference for these sites, which contributes to their erosion susceptibility. The vegetation on the sites is always heavily browsed and grazed. Bailey (1990) found that the vegetation growing on sodic sites had a significantly higher sodium content than vegetation (of the same and other species) from adjacent areas. This could be a major reason for the animal preference.

Summary

The soil fertility of dry savannas (and therefore many aspects of their ecology, including primary and secondary production, structure and species composition, as we will argue in Chapter 12) is mostly related to the availability to plants of nitrogen and phosphorus. This is because the basic cations are not strongly leached in dry climates. Therefore, a standard soil analysis which emphasises basic cations is of little use in prediction of nutrient limitation in savannas. In the case of Nylsvley, nitrogen appears to be a stronger limiting factor than phosphorus, perhaps because its biogeochemical cycle is 'open', that is, including a significant atmospheric portion. There are several avenues of nitrogen loss, of which pyrogenic and biogenic denitrification are the most important. Although the dominant trees at Nylsvley are leguminous, there is no evidence that they support a significant number of symbiotic nitrogen-fixing nodules. The process of mineralisation, whereby nutrients are made available in ionic form by the oxidation of organic compounds, dominates both the nitrogen and phosphorus cycles at Nylsvley, and is controlled by water availability and soil temperature.

At Nylsvley, the concentration of nitrogen and phosphorus in plant tissues tends to be at the lower end of the required range, and therefore fertilisation results in increased growth. In this sense, primary production at Nylsvley is not constrained by the carbon cycle or any of its controlling factors, but by the rate of uptake of nitrogen and phosphorus. The uptake rate is constrained by the rate of release of inorganic nitrogen and phosphorus from dead organic matter into the soil solution through the process of mineralisation. At Nylsvley, and probably in savannas in general, the rate of this process is controlled mainly by the water content of the soil. Therefore, the linkage between rainfall and primary production, so widely observed in savannas, probably operates via the influence of water availability on the nitrogen and phosphorus cycles, which then constrain the carbon cycle.

While there is no evidence to suggest that plant growth at Nylsvley is limited by the supply of basic cations, the distribution of cations in the landscape can have profound effects on landscape pattern in this and other dry savannas.

8

Fire

The role of fire in the maintenance of structure and function in African savannas is probably the oldest issue in savanna ecology, but certain aspects remain contentious (Phillips 1965, 1968; West 1965; Rose-Innes 1972; Gillon 1983). One point of view, noting the ubiquitous occurrence of fires in savannas and the tendency for woody plant density in savannas to increase when fires are excluded, concludes that savannas are fire subclimaxes to woodland or forest. Another view stresses the long history of fire in Africa, with the numerous plant adaptations to surviving fires, and presents fire as a modifier of savanna structure rather than a primary determinant of savanna distribution (Frost *et al.* 1986). Attitudes towards the use of fire as a management tool in savannas have ranged from pyromania to pyrophobia.

Although fire ecology has been an important theme in southern African ecological research (Booysen & Tainton 1984), it was never a major focus of the Savanna Biome Programme. This was more because the topic was receiving attention in other programmes, rather than a belief that fire was unimportant in savanna ecology. However, controlled burning is part of the management system in the Nylsvley Nature Reserve, and several researchers occupied with other projects examined the effect of management fires on the topic of their interest. Mark Gandar synthesised the results of several studies on the fires of September 1978 (Gandar 1982c), and Peter Frost reviewed the effects of fire on savanna organisms in general (Frost 1984, 1985b).

The fire regime at Nylsvley

The fire regime of an ecosystem has four components:

1. frequency, the reciprocal of the mean time between fires;
2. intensity, the rate of energy release ('hotness');

3. season of burning; and
4. type of fire (with or against the wind, on the ground or in the tree canopy).

The fire regime is summarised by the mean values for these factors, but it must be remembered that they are all distributions, showing a natural spread of values. It is just as injurious to biodiversity to impose a rigid, single 'correct' fire regime as not to burn at all.

Fire frequency

The long-term historical fire frequency at Nylsvley is unknown. Natural fire frequencies in moist savannas are thought to be about once every 1–2 years, decreasing to once every 3 or more years in dry savannas (Edwards 1984; Trollope 1984). Nylsvley, at 600 mm per annum rainfall, falls on the threshold between moist and dry savannas, suggesting triennial burning on average, but with considerable variation. When the Nylsvley area was used for grazing cattle by early white settlers (from about 1840 to 1915), it was common practice to burn *suurveld* (fibrous, unpalatable, low-productivity grasslands such as occur at Nylsvley) annually during autumn or winter, to provide green forage at a critical time. Between 1916 and 1974, Nylsvley was a commercial cattle ranch. During this period, fires were not intentionally set in the broad-leafed savanna, and were actively combated when they occurred. Large areas nevertheless burned almost every dry season, due to arson and accidental causes. Any given patch was thought to have burned approximately once every 4–5 years. This frequency was adopted by the reserve managers in order to maintain the vegetation in its present state.

The 800 ha study area is divided into four approximately equal-sized camps, one of which is burned approximately every year, in rotation, just before the first rains (September or October). The grass layer is inspected before burning, and if found not to have accumulated much dry standing grass, the burning is postponed until the next year. The fires known to have occurred within the study site are detailed in Table 8.1.

The increase in the woody plant biomass over the period of the study (Figure 8.1) suggests that the historical fire frequency, or intensity, was higher than at present. Alternative explanations for the increase in woody biomass are that large herbivores, such as elephants, formerly occurred in the area, and would have depressed the woody biomass; or that previous inhabitants harvested trees for fuel and timber. However, the troughs in the overall upward trend of woody biomass coincide with known fires, reinforcing the belief that fires are the major factor controlling woody plant increment.

Table 8.1. *Fires in the Nylsvley study site over the period 1973–90*

Year	Camp 1	Camp 2	Camp 3	Camp 4
1973				
1974				
1975				
1976				
1977				
1978		September		
1979	October			
1980				
1981				October
1982				
1983			September	
1984				
1985	October	October		
1986				
1987				September
1988				
1989			September	
1990	October	October		

Fire temperature profiles

Harrison (1978, reported in Gandar 1982c) measured the temperature profiles during a night-time fire using alloy sticks with known melting points. Fire temperatures at the soil surface depend on the amount of litter which ignited, and ranged from less than 260 °C below *Burkea africana* canopies to over 816 °C in a thick, dry litter bed below *Ochna pulchra*. Many temperatures were above 760 °C. The mean temperature at the base of grass tufts was 460 °C, but the distribution was bimodal, with peaks at 350 and 600 °C. The higher temperatures were measured in grass tufts, which continued to glow after the flames had passed. The temperatures above 1 m height in tree canopies never exceeded 260 °C, the lowest temperature detectable by this method. Within *Grewia flavescens* and *Ochna pulchra* shrubs the temperatures above 1 m exceeded 300 °C in several cases.

Fire intensity

Fire intensity is expressed as the rate of energy release per metre of flame front (kJ m^{-1} s^{-1}). It can be calculated from the rate of spread (m s^{-1}) multiplied by the energy content of the material burned (kJ m^{-2}). The latter can be calculated from the fuel load (g m^{-2}), its energy content (about 20 kJ g^{-1}) and the proportion of the available fuel which actually burned. The fuel load includes litter, but the proportion of litter burned is highly variable. Frost (1985a) estimated that 74% of standing dead grass and grass litter was consumed in a fire at Nylsvley in 1983, while 58% of twig and bark litter and 42% of tree leaf litter was consumed in the same fire. Rutherford (1981) estimated that 20% of the standing dead wood burned in the 1978 fire. Similar values have been recorded in other savannas (Frost & Robertson 1987).

The rates of spread measured in three 100 × 100 m blocks in September 1978 were 1.2, 1.7 and 0.67 m s^{-1}. The fuel load was about 90 g m^{-2} in the first two blocks, and slightly higher in the third, yielding calculated fire intensities of 211, 300 and greater than 1200 kJ m^{-1} s^{-1}. The same trend was observed in the temperature profiles. The first two correspond to 'warm' fires, and the third a 'hot' fire in the terminology of Trollope & Potgieter (1985). The effect of a fire on woody plant mortality is strongly dependent on the relationship between the temperature profile and the plant height. Mortality usually occurs only when the flame height exceeds the canopy

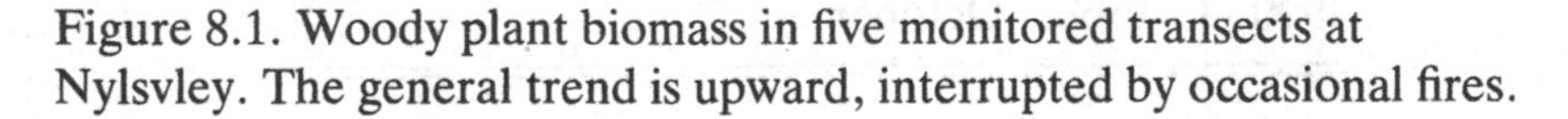
Figure 8.1. Woody plant biomass in five monitored transects at Nylsvley. The general trend is upward, interrupted by occasional fires.

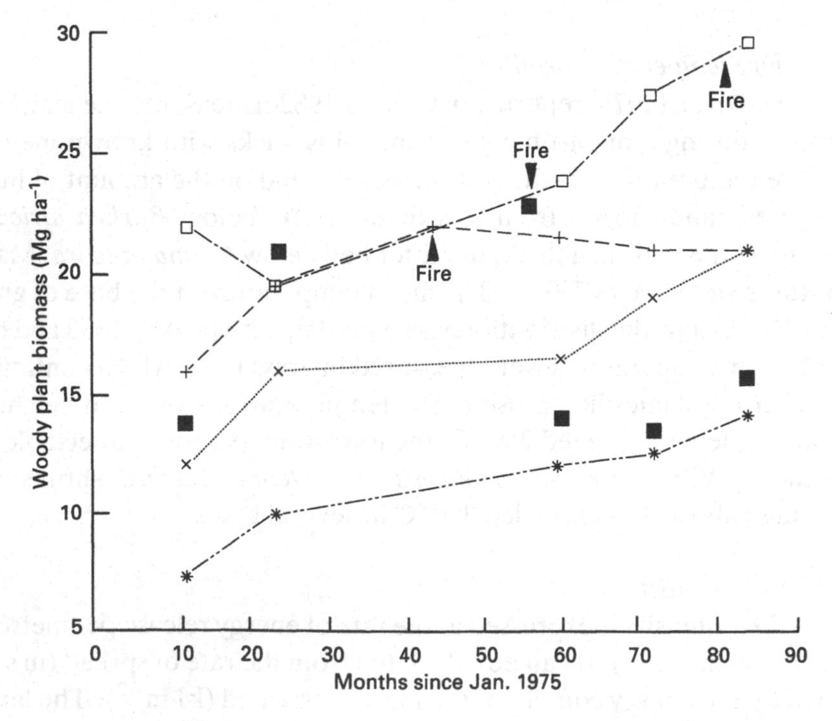

height (Trollope 1984). Since there is a good relationship between flame height and fire intensity, it can be concluded that a hot fire is required to kill the aboveground parts of established woody plants; typically only fires that exceed 2000 kJ s^{-1} m^{-1} are effective for bush control. Young saplings, with their foliage within or just above the grass layer, are susceptible to damage by fires of lower intensity.

A notable feature of this fire experiment was the extreme patchiness of fire intensity at all scales from the individual grass tuft up to the entire burning block.

The post-fire microclimate

The shortwave reflectivity (albedo) decreased after burning as a result of the blackening of the surface, but recovered over a one-and-a-half month period following good rains (Figure 8.2; adapted from Gandar 1982c). Despite the decreased reflectivity, the net downward soil heat flux was not significantly greater on the burned than unburned sites (141 vs 135 W m^{-2}, SD = 14). This was due to the low thermal conductivity of the sandy Nylsvley soils: most of the absorbed radiant energy was returned to the atmosphere as sensible heat. High air temperatures just above the soil surface could have negative implications for the heat balance of seedlings establishing immediately after fire.

The soil water content at 35 and 45 cm depth a month after the burn was

Figure 8.2. Time-course of soil reflectivity (albedo) following a fire at Nylsvley (Gandar 1982c). ●——●, burned savanna; ○——○, unburned savanna.

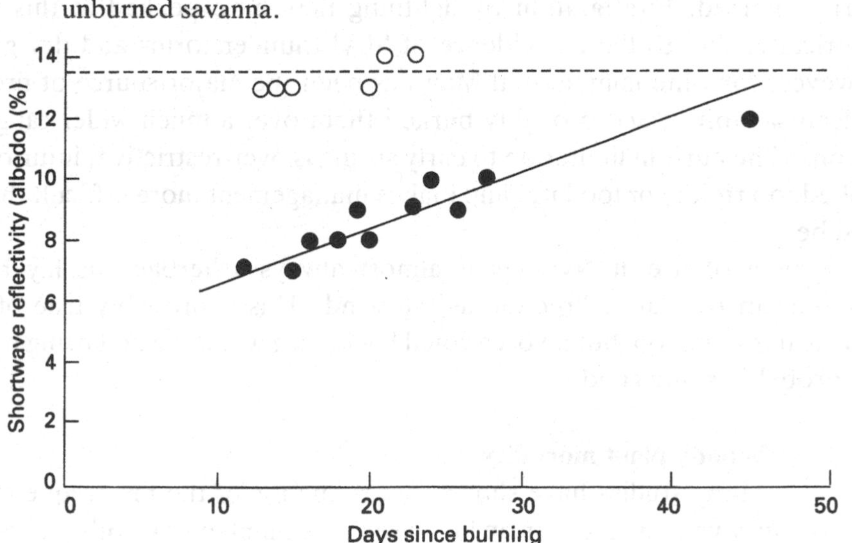

5–50% lower in the burned than unburned patches. Rainfall between the date of the fire and the commencement of soil water measurement amounted to 23.5 mm. There was no difference in the water content of the surface soils: both were dry. Drier soils following fire have been reported in several other studies (West 1965), but are difficult to explain given the sandy nature of the Nylsvley soil. The increase in absorbed radiant energy could have stimulated evaporation from the soil surface, but soil water below 30 cm in a sandy soil should be reasonably well shielded from evaporation. The reduction in transpiration on the burned sites would be expected to lead to a higher, rather than lower, water content in the subsoil in the short term. Gandar (1982c) suggested that the difference could be attributable to decreased infiltration on the burned site, but there is no direct evidence to support this hypothesis. The condensation within the soil of hydrophobic substances present in the smoke is well-known in vegetation containing large amounts of resins and waxes (Cass, Savage & Wallis 1984), but has not been investigated in savannas.

Fire season and type

The historical fire season is not known for Nylsvley, but was probably concentrated in the dry season. It is quite difficult to sustain a fire during a normal wet season, although it is possible in drought years. The prescribed burning season for most savanna areas in South Africa is the early spring, after the first rains. It is argued that burning at this time will cause the most damage to trees, but the least damage to grass, cause the least nutritional stress to grazers, and expose the ground surface to erosion for the shortest period. Fire ignition by lightning may have peaked at this time historically, due to the coincidence of 'dry' thunderstorms and dry grass. However, for time immemorial Man has been the major source of fires in African savannas, and probably burned them over a much wider range of seasons. The current limitation to early spring is over-restrictive, injurious if applied too rigidly or too late, and makes management more difficult than it need be.

The type of fire at Nylsvley is almost always a herbaceous layer fire travelling in the same direction as the wind. This is probably true of the historical regime too, but an occasional back burn would do no lasting harm, and probably some good.

Woody plant mortality

Many studies have shown that reduction of the fire frequency or intensity in savannas results in an increase in the number of woody plants and their height, canopy volume and biomass (Trapnell 1959; Moore 1960; Trollope 1980). Increasing fire frequency while maintaining fire intensity is less easily achieved. Nevertheless, some studies have shown reductions in woody plant density, canopy volume and biomass by increasing fire frequency or intensity (Rains 1963; Kennan 1972). The woody plant basal area, stem density and biomass at Nylsvley has been increasing over the period of record (1974–86), with periodic setbacks coinciding with fires (Figure 8.1).

Rutherford (1981) monitored the effect of moderately intense ('fast') and less intense ('slow') burn on woody plant survival. The overall mortality on a per plant basis was low: 2% in the fast and 5% in the slow burn, compared with 1% in unburned savanna over the same period. The somewhat counterintuitive result of higher mortality in the less intense fire is explained by differences in the temperature profile between the fast and slow burns. The heat is concentrated closer to the ground in slow burns and back burns, leading to higher mortality in small plants. Mortality among small individuals (less than 100 mm tall) was 9% in the fast and 13% in the slow burn. The multi-stemmed shrubs (typified by *Grewia flavescens*) burned vigorously, but no mortality was recorded in the slow burn. In the fast burn, 38% of the individuals were killed, but these were entirely in the less than 250 mm height class.

The canopy top-kill data show the opposite trend: top-kill was greater in the fast burn (43.2%) than the slow burn (23.5%). In both cases, over 90% of the canopies killed were below 2 m in height; no plants taller than 5 m experienced canopy death. Six months after the fire, the leaf biomass of dominant species in the burned plots was less than that of the same species in the unburned plot. The degree of reduction depended slightly on the species and strongly on the size of the individual. *Ochna pulchra* plants shorter than 1 m showed a 90% reduction in leaf biomass, but individuals between 2.5 and 5 m tall showed only a 26% reduction. *Terminalia sericea* showed a 90% leaf biomass reduction in plants shorter than 2.6 m. Stems of *Burkea africana* smaller than 80 mm in diameter supported less leaf biomass in the season after burning than in a no-burn season.

The reduction in leaf mass in the main part of the canopy is partly offset by the proliferation of new basal shoots, which support proportionately more leaf than older, unburned shoots. Most of the species in the savanna at Nylsvley exhibit this response to fire. Although the fire killed most of the pre-existing basal shoots, the number of replacement shoots more than compensated for the loss: the ratio of pre- to post-fire basal shoots in *O. pulchra* was 1.88, and 1.30 in *B. africana*. The mass of new basal shoots produced was inversely proportional to the degree of canopy die-back, which was in turn related to the fire intensity. The amount of woody plant leaf material between ground level and 250 mm increased by about 10% after the fire.

The structure of the vegetation was also altered by the fire. The mean plant height in the unburned control increased by 4%, but that on the fast and slow burn plots decreased. The decrease was greater for small plants than for large, and greater for the fast burn than the slow burn. Reductions in canopy volume followed the same pattern.

In summary, woody plant leaf biomass, height and canopy volume is reduced during the season after the fire, particularly at heights between 0.25 and 3 m above the ground. A leaf area reduction would be expected to be associated with the decrease in radial increment, but no data are available to confirm this.

If mature trees are not susceptible to fire mortality, and the woody biomass is increasing at Nylsvley, why is the savanna there not a closed woodland? As the woody biomass increases in savannas, herbaceous production is suppressed, making intense fires less probable. Under this scenario, surely some savanna patches, sometime in their long history, would have escaped fire control, and irreversibly become woodlands ? One suggestion is that in the pre-twentieth century period, elephants would have kept the woodlands open. However, the main effect of elephants on woodlands is to change their structure, rather than their density.

Yeaton (1988) proposes an alternative mechanism to account for tree mortality at Nylsvley. He observed a large number of felled mature trees following a windstorm. Closer examination showed all the toppled trees to have snapped off close to the ground, having been hollowed out on the one side by fire. The hollowness both weakened the tree and imbalanced it, since the loss of conducting tissue on one side encouraged the tree to grow asymmetrically. Surveys showed that virtually all mature individuals of *Burkea africana* and *Dombeya rotundifolia* had basal fire scars. When the stumps of fallen trees were sectioned, it could be seen that the trees had survived an average of four fires since the formation of the scar, which had grown progressively bigger and deeper with each fire. The initial scars were formed when the tree was quite young. However, fire alone is not able to form a xylem scar while the trunk is covered by a bark layer; something must first damage the bark near the base of the trunk. The obvious culprit was a porcupine.

Porcupines are common at Nylsvley (about one pair plus their immature offspring per km^2). Most *Burkea africana* saplings bear porcupine feeding scars 2–5 cm in diameter just above the root collar. The porcupines preferentially eat the basal bark of young specimens of *D. rotundifolia*, *B. africana* and, to a lesser extent, *Terminalia sericea* in the late dry season.

Yeaton (1988) proposes that the following sequence of events leads inexorably to the death of the tree. A small piece of basal bark and the underlying cambium of a young tree is removed by a porcupine, forming a scar of dead wood. The exposed dry wood is ignited by a fire, and burns a short distance into the tree. Successive fires, over a period of 20 or more years, enlarge the scar, causing lop-sided development of the tree. The chimney effect of the scar accelerates the hollowing-out process. Eventually a windstorm, often some time after the fire, causes the stem to snap. This does not inevitably kill the tree, but in many cases it does. In any case, it greatly reduces the tree biomass. In the fire which Yeaton (1988) observed, about 2% of the mature individuals of *Burkea africana* and *Terminalia sericea* (greater than 16 cm trunk diameter) were felled; however, they constituted a much greater proportion of the tree basal area.

Three interacting factors, operating a long time apart, are therefore responsible for tree mortality: porcupines, fires and windstorms. Tree mortality following the fires studied at Nylsvley was low because they were not followed by windstorms. If fire is excluded, is there an upper limit to the woody biomass which can be supported at Nylsvley? A patch of *Burkea africana* woodland on the adjacent farm Mosdene had been protected from fire for 35 years when Rutherford & Kelly (1978) measured the basal area and stem radial increment in 1974. The total basal area was 10.64 m^2 ha^{-1}, which is 57% higher than the mean basal area in similar vegetation at Nylsvley, measured in 1974 (Smith 1974). The radial increment over the 1974–5 season was 4.7%, compared with 7.5–12.5% per season at Nylsvley (calculated using data from Lubke & Thatcher 1983), suggesting that tree growth will ultimately be limited by intertree competition.

Effects on the herbaceous layer

O'Connor (1985) reviewed the findings with respect to grass layer composition and yield of the many fire experiments conducted in southern Africa over the past half-century. In general, fire frequency had a weak negative long-term effect on total grass basal cover. In relatively high-rainfall areas on infertile soils, basal area declined slightly in unburned swards, perhaps partly because of the increase in woody biomass. On fertile soils in dry areas, grass basal area decreased on both annually burned and fire-protected plots. Changes in composition occurred in most of the experiments, but the variability between years and replicates was high, making it difficult to distinguish between changes caused by fire and those caused by interannual variability in rainfall.

The regrowth of the herbaceous layer following a back burn (one where the flame front moves towards the wind) at Nylsvley was monitored by Grossman, Grunow & Theron (1981), and also reported in Grossman (1980) and Gandar (1982c). The biomass accumulation rate of the herbaceous layer was substantially lower on the burned sites (0.29 g m^{-2} d^{-1}) than on the unburned sites (0.73 g m^{-2} d^{-1}) for the first 2 months after the burn, but measured over the entire growing season the aboveground productivity on the burned sites was nearly double that on the unburned sites (0.315 vs 0.152 g m^{-2} d^{-1}). By the end of March there was no difference in green biomass between the sites, but the necromass was twice as high on the unburned site as on the burned site, since it included accumulations from previous years. The initial retardation of growth on the burned plots appeared to be due to the lower leaf area at the beginning of the season, although drought stress in November could have contributed to the delay.

The contribution of forbs to the herbaceous layer production was higher on the burned than unburned treatment. Total grazeable herbaceous layer production appears to have been increased 88% by the burning treatment: from a minimum estimate of 49.2 g m^{-2} on the unburned plots to a minimum estimate of 92.3 g m^{-2} on the burned plots. The grass leaf area duration, integrated over the entire 1978–9 season, was 42% higher on the unburned than on the burned plots, since the former started off with 14.0 g m^{-2} and grew rapidly in the early part of the season. Total forage on offer was therefore higher in the unburned than the burned plots, but its acceptability was much lower (see the comments on herbivore behaviour below).

Unburned plots which were clipped to stubble level showed a 58% increase in aboveground productivity relative to unburned unclipped plots, suggesting that two-thirds of the increase in productivity following fire could be explained by the removal of necromass, which shades the new growth. The remaining third is probably attributable to the nutrient flush released from the ash.

Forbs contributed about one third of the post-burn herbaceous layer production. The burning treatment increased the productivity of the preferred forage species more than it increased that of the non-forage species, resulting in a beneficial change in the sward composition after the burn. The basal area of the unpalatable fibrous grass *Eragrostis pallens* decreased significantly ($p < 0.05$) from 1.52% before burning, to 0.69% eighteen months later. No decline occurred on the unburned plots. Tuft mortality of *E. pallens* due to the fire was 51%. The basal area of the palatable species

Digitaria eriantha did not decrease following the fire, and the total basal area of both burned and unburned plots was relatively constant over the entire period, at around 5%.

Yeaton *et al.* (1986) compared the grass layer structure, composition and dynamics in three stands at Nylsvley which differed only in fire history. Species richness was highest on the stand which was burned annually (a firebreak), lower on a stand last burned 5 years previously, and lowest on a stand last burned 12 years previously. Total basal area increased with time since the last fire, mostly due to increasing dominance by *E. pallens* and *D. eriantha*. In the case of *D. eriantha*, the increase was due to a trebling of plant density. In *E. pallens* the plant density increased from the annually burned stand to the 5-year stand, and decreased again to the 12-year stand. The overall increase was sustained by an increase in mean tuft diameter. The density decrease in the stands was attributed to self-thinning, supported by data from a regression analysis of nearest-neighbour pairs which indicated increasing intensity of competition with time since the last burn. *Eragrostis pallens* produced more tillers per unit basal area in the annually burned stand, but a smaller percentage was grazed, possibly because more desirable fodder was on offer.

Effects on nutrient cycling and the soil

The effects of the fire regime on the soil properties of southern African ecosystems, including savannas, have been reviewed by Cass *et al.* (1984). Since the inorganic nutrients are mostly returned to the soil in the ash, there is seldom a significant change in the total amount of available basic cations, but sometimes an upward shift in their distribution in the profile. Carbon, and elements bound to it, such as nitrogen, phosphorus and sulphur, are partially or completely released to the atmosphere in wildfires, and therefore have the potential to show significant changes in the soil following alterations in the fire regime. The nitrogen, phosphorus and sulphur returned to the soil in the ash is in an available, inorganic form. Grass growing immediately after a fire typically has an above-average nitrogen content, and primary production in the first season after burning is typically enhanced (see above).

Repeated burning has been suggested as the cause of the reputedly lower than average organic matter content of savanna soils. The reverse effect was shown in pine 'savannas' of the coastal plain of southeastern North America (Daubenmire 1968). In Nigerian savannas, light early dry season burns increased the soil organic matter, but hotter late dry season burns reduced it (Moore 1960). Trapnell *et al.* (1976) showed a 13–16% increase in soil carbon in the top 15 cm of a *miombo* woodland in Zambia when annually burned, and a 0–8% decrease in nitrogen. Jones *et al.* (1990) showed an increase in organic matter and nitrogen following 30 years of fire protection in savannas on clayey soils of the eastern Transvaal. The situation is obviously fairly complex, and depends on fire intensity, soil type, the degree of stimulation of production by fires, and changes in the tree-to-grass balance in the savanna in response to the altered fire regime.

There is no evidence of loss of humified organic matter from the soil as a short-term consequence of burning at Nylsvley. In Chapter 11 it is calculated that only a small proportion of the primary production is consumed by fire. The low organic matter content of Nylsvley soils is ascribed to the relatively low primary productivity, the high temperatures and the sandy nature of the soil. Organic matter is stabilised in the soil largely in association with clays, which constitute only 8% of the Nylsvley soil.

Lumps of charcoal are frequently encountered in the soil profile, especially in the top 30 cm. The charcoal has a C:N ratio of more than 120, making it very resistant to decomposition. The charcoal fragments in the top 50 cm of soil in the infertile savanna amount to 42.5 g m^{-2}, or 19% of the unhumified carbon in the soil, and about 1% of the total soil carbon. In the fertile *Acacia* patches, which are thought to be former village sites, there is 43.0 g m^{-2} of charcoal, suggesting that its main origin is trees burned during wildfires, rather than domestic middens. Rutherford (1981) reported that 24% of the standing dead wood was felled by the September 1978 fire, compared with 19% which had fallen of its own accord in the control block. Of the dead trees toppled in the fire, 84% were completely burned.

Meredith (1987) investigated the potential for nutrient leaching in the Nylsvley soil following a fire, and found no increase relative to unburned controls. There was also no significant difference in microbial biomass, soil respiration or microbial propagule counts between the burned and unburned sites after the September 1978 fire. Respiration from the rhizosphere increased slightly in the burned sites between days 5 and 30 after the burn. This could result either from the decomposition of dead roots, or from the increased metabolic activity as root reserves were mobilised for translocation to the resprouting leaves (Gandar 1982c).

There is circumstantial evidence that nutrient availability increases for a short period after a fire. The herbaceous layer productivity increases, and many studies have shown post-fire increases in the forage nutrient content (Mes 1958; Tainton, Groves & Nash 1977). This is partly due to the fact that all the post-fire leaves are young, and therefore high in protein. Other causes are the mineralisation of organically bound elements by the fire, stimulation of mineralisation in the soil due to higher temperatures and water content, and the fact that the carbon assimilation organs of the plants receive a setback during a fire, but the nutrient assimilation organs are less affected. The forage nutrient content has not been measured after fire at Nylsvley, but herbivore behaviour suggests it is much more acceptable after burning.

Effects on insects and small mammals

Frequently burned savannas have lower biomass and numbers of both insects and small mammals than savannas protected from fire (Gillon 1983). The cause is most likely the decrease in food availability and cover and the unfavourable microclimate after the fire, rather than direct mortality during the fire. The population size and composition of both groups of organisms alters substantially immediately after a burn, but it is unclear what proportion of this change is due to mortality, and what proportion is due to dispersal. Mortality need not be due to thermal injury alone. Smoke asphyxiation and predation during and after the fire are also important causes of death. Dean (1987) noted that insectivorous birds (and seed-eaters, to a lesser extent) were attracted to recently burned areas at Nylsvley.

Larvae of the beetles in the families Scarabaeidae and Curculeonidae live in the soil. Emergence of the adults was monitored for 6 months after burning the broad-leafed savanna (Gandar 1982c). There was no significant difference between the numbers trapped on the burned and unburned plots until 2 months after the fire, when emergence peaked on the unburned plots, but remained constant on the burned plots. It appears that the fire did not directly kill the larvae, but reduced root production or microclimatic changes on the burned plot made it a less favourable habitat for them. No dead termites were found in colonies excavated immediately after the fire, but the attack rate on toilet-roll baits decreased in the burned areas for the month after burning (Ferrar 1982c).

The number of arboreal insects in *Ochna pulchra* and *Dombeya rotundifolia* trees decreased substantially after the fire, although the intertree variability was high (Gandar 1982c). Most of the species collected before the fire are capable of flight, so the potential for escape and subsequent return is high, particularly since the burn was conducted at night. Insect-eating birds flock to a daytime fire, and could have a substantial impact on the fleeing insects (Dean 1987).

At the time of the burn (September), most of the grasshoppers were also in the winged adult form. Figure 8.3 shows the change in grasshopper biomass within the mosaic of burned and unburned patches. The overall biomass declined 30% immediately after the fire, relative to a completely unburned savanna, and continued to decline. The mortality of nymphs was extremely high in the months after the burn; predation by birds was a likely cause.

The number of small mammals trapped increased after the burn, but this does not necessarily indicate an increase in population densities (Gandar 1982c). The rodents could have been hungrier, and therefore more attracted to the baited traps. There was a significant decrease in the number of black-eared climbing mice (*Dendromus melanotis*) trapped after the burn, and an increase in bushveld gerbils (*Tatera leucogaster*) and pouched mice (*Saccostomus campestris*). The climbing mouse feeds on grass seeds, which are in short supply after a fire, while the other two feed on tree seeds, which survive the burn. *Mus munitoides* increased after the fire despite being a grass seed-eater, perhaps because it also eats insects.

Interactions between fire and grazing

The number of impalas observed on the burned areas increased steadily from day 10 after the fire until day 30, peaking at 0.51 impala ha^{-1}, after which they declined slowly over a period of 3 months to the pre-burn densities of 0.06 ha^{-1} (Gandar 1982c). Impala at Nylsvley typically show a four- to fourteen-fold preference for the fine-leafed over the broad-leafed savanna patches, but this preference was reversed when the broad-leafed

Figure 8.3. Biomass of grasshoppers in a small-scale mosaic consisting of 89% burned patches and 21% unburned patches, following a fire in September 1978 (Gandar 1982c).

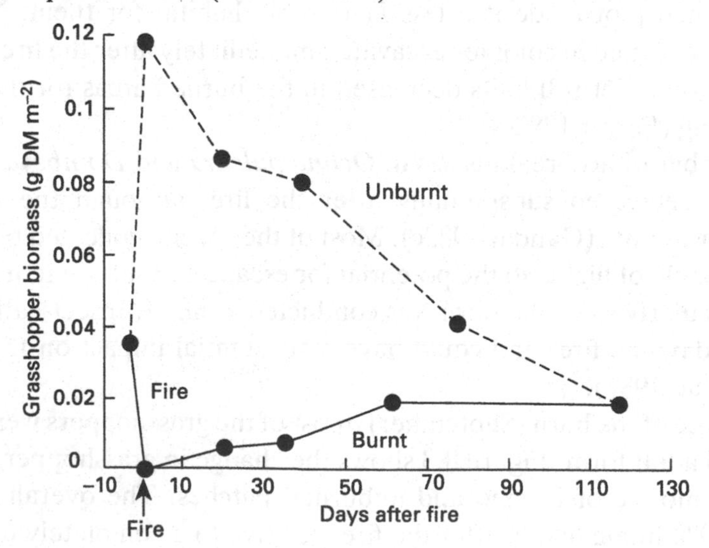

savanna was burned. The impala density in the fine-leafed savanna was 0.39 ha^{-1} before the burn, dropping to 0.02 ha^{-1} during the peak of impala concentration in the burned broad-leaf patches.

When burned patches were clipped every 8 weeks to simulate grazing, the biomass accumulation rate was reduced relative to that on burned but unclipped plots (0.251 vs 0.315 g m^{-2} day^{-1}), but was nevertheless still higher than on unburned plots with or without grazing simulation (0.240 and 0.152 g m^{-2} d^{-1}: Grossman *et al.* 1981).

Summary

In this savanna, and probably savannas in general, fire is the main agent which directly controls the balance between trees and grass. However, fire acts in conjunction with other less obvious factors, such as porcupines and windstorms, and the fire frequency and intensity are a function of rainfall, soil type and the existing tree biomass.

The woody biomass and density at Nylsvley are gradually increasing, probably as a result of changes in the burning regime during the last half-century. Evidence from fire-protected plots suggests that the woody biomass can reach twice its present level before stabilising due to intertree competition. In the long term, occasional burning does not have a major detrimental effect on nutrient cycling or on the survival of savanna organisms, but it does increase both the palatability and productivity of the herbaceous layer for a short period after the burn. Fires result in a temporary redistribution of faunal communities; initially away from the burned patch, but later strongly towards recently burned areas.

9

Herbivory

African savannas are famous for the diversity and abundance of large mammalian herbivores that they support. However, over large parts of Africa the wild herbivores have been replaced by cattle, sheep and goats. Domestic livestock and wildlife form the basis of the present-day economic use of savannas, through meat production and tourism respectively. There has been considerable debate over the relative merits of 'indigenous' ungulates versus 'introduced' cattle (Dasmann 1964; Johnstone 1975; Walker 1979; Goodman 1985) in terms of economic return, protein production efficiency, ecological impact and disease tolerance. The Nylsvley programme included studies of both domestic and wild herbivores, often in comparative experiments.

The Nylsvley study also permitted comparisons to be made between mammalian and invertebrate herbivores, and within the invertebrates, between caterpillars and grasshoppers. Among the mammalian herbivores, the differences between predominantly grass-eaters (grazers) and predominantly tree-leaf eaters (browsers) are of scientific and commercial interest.

The work on herbivory performed at Nylsvley focused on two broad areas of interest. The first relates to secondary production in an infertile savanna system, and includes concepts such as forage consumption, carrying capacity, secondary production and conversion efficiency. These topics are addressed in this chapter. The second focus relates to plant–herbivore interactions, and is dealt with in Chapter 15. In brief, this chapter describes who eats what and how much, while Chapter 15 asks why.

Many people contributed to the research reported in this chapter. Major studies were undertaken by Clark Scholtz (Lepidoptera), Mark Gandar (grasshoppers), Ibo Zimmerman and Sybille Grundlehner (cattle), Rob Munro, Susan Cooper and Norman Owen-Smith (impala, kudu and goats).

To distinguish between plant biomass, which is usually expressed on a dry mass basis, and animal biomass, usually on a fresh mass basis, the units g DM and g FM will be used.

The historical trend of ungulate stocking rates at Nylsvley

Before the arrival of white settlers in the early nineteenth century, the Nylsvley area probably supported a wildlife population similar in number and diversity to equivalent present-day conservation areas in Africa. East (1984) used wildlife census data from parks throughout the savanna regions of Africa to obtain predictive equations for large mammal biomass, as a function of soil fertility and annual rainfall. The Nylsvley area has soils of greatly varying fertility (see Chapter 4): in East's terms, about 30% would be fertile, 30% moderately fertile, and 40% infertile. This, and the presence of permanent water, allows Nylsvley to support a herbivore population greater than might be expected from the infertile soils of the Savanna Biome study site. The predicted large herbivore biomass for the Nylsvley area is 3768 kg km^{-2}, whereas for the study site alone it is 2412 kg km^{-2} (East 1984). During the period for which Nylsvley was a productive cattle ranch (1945–73) it sustained a cattle biomass of about 5645 kg km^{-2} (peaking at 7900 kg km^{-2}), with no obvious long-term deleterious effects (George Whitehouse, personal communication).

Wildlife, which had virtually disappeared since the turn of the century, began to return during the 1930s. The impala herd on Nylsvley reached about 500 in 1974 (Carr 1976). Following the proclamation of the reserve in that year, species thought to have been indigenous to the area in the pre-Colonial period were reintroduced, with the exception of the very large herbivores (elephant, rhinoceros, buffalo and hippopotamus) and carnivores (Figure 9.1). Breeding of the locally rare antelope species, roan and tsessebe, is among the reserve management objectives, so the populations of other herbivores are controlled by periodic culling or live capture. The levelling-off of the herbivore populations after 1980 is therefore not indicative of a resource-imposed 'carrying capacity' limit. The large mammalian herbivore numbers in July 1990 are shown in Table 9.1 (J. Coetzee, Warden, Nylsvley Nature Reserve).

Rates of herbivory at Nylsvley

The main pathways of disappearance of plant material in the broad-leafed savanna are depicted in Figure 9.2. The interannual variation in the magnitude of the pathways is large, and there are significant uncertainties in their estimation. Therefore extrapolation from the Nylsvley broad-leafed

Table 9.1. *The large mammalian herbivore populations in the Nylsvley Nature Reserve (3127 ha) in 1990. The numbers are controlled by culling and capture at a level well below that which they would reach in an unmanaged situation*

Species	Number Nylsvley Reserve	Mean mass[a] (kg)	Biomass ($kg\ km^{-2}$)	Energy requirement[b] ($MJ\ d^{-1}$)	Feeding class[c]
Impala	250	40	320	14.0	80% grazer, 20% browser
Tsessebe	85	105	285	24.6	Grazer
Wildebeest	44	145	204	37.3	Grazer
Reedbuck	88	40	113	14.0	Grazer
Warthog	100	60	192	21.1	Omnivore
Bushpig	20	60	38	21.1	Omnivore
Zebra	6	270	52	61.0	Grazer
Roan	45	220	317	42.1	Grazer
Waterbuck	122	130	507	34.6	Grazer
Kudu	84	125	336	27.9	Browser
Giraffe	18	850	489	136.0	Browser
Duiker	30	10	10	3.2	Browser
Steenbok	20	10	6	3.2	Browser
Total			2869		
Cattle (pre 1974)	600	350	6715	67.3	Grazer

[a]Smithers (1983), based on mean female mass.
[b]Meissner (1982), young female with calf.
[c]Grazer = predominantly grass-eating; browser = mainly tree leaf-eating; omnivore = eating bulbs, rhizomes, fruits and grass.

savanna pattern to savannas in general must be done with caution. However, some general conclusions can be drawn: a relatively small proportion of the primary production in an infertile savanna is consumed by herbivores; and invertebrate herbivory is very important, especially with respect to woody plants.

Large mammalian herbivores

The consumption of plant material by large mammalian herbivores can be estimated from the stocking density and energy requirements. The details of the calculation are given in Box 9.1. For grazers, the estimate is 18.2 g DM $m^{-2}\ y^{-1}$, and for browsers 6.1 g DM $m^{-2}\ y^{-1}$. There are two problems in relating these values to other African savannas: first, the large mammal biomass at Nylsvley is kept artificially low; and secondly, the distribution of herbivory within the landscape is uneven. For the areas of the study site on infertile soils, these two sources of error tend to cancel each other and the estimates are probably both reasonably accurate and representative. For the fertile patches there are insufficient data to make a reliable assessment.

The consumption estimates represent 22% of the aboveground grass production for the broad-leafed savanna reported in Chapter 10, and 3.4% of the woody plant leaf production. For favoured sites and species, the situation is completely different. No quantitative estimates are available of the proportion of grass production consumed by grazers in the fine-leafed savanna patches, but observations suggest that they are closer to the 50–80% range reported for fertile savannas by McNaughton (1979) and Drent & Prins (1987). Both cattle and impala graze preferentially in the fine-leafed savanna, especially in winter (Carr 1976; Monro 1979; O'Connor 1977;

Figure 9.1. Historical trend of large mammalian herbivore biomass in the Nylsvley Nature Reserve. The data were obtained from the warden, and are based on regular aerial and ground counts. Two major culling events are shown. Live capture and culling occurred throughout the period.

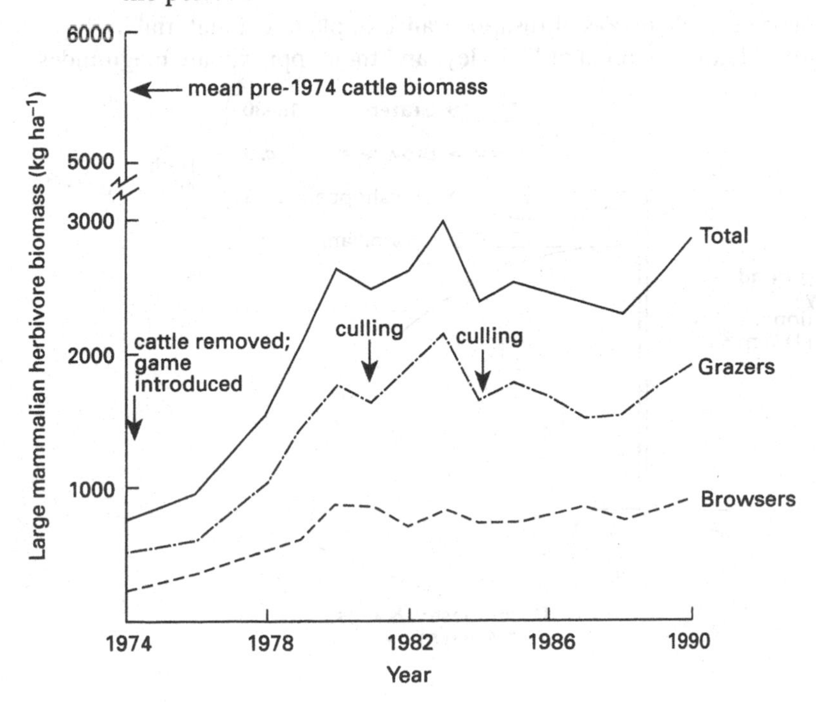

Skinner, Monro & Zimmerman 1984), and the grass is always cropped conspicuously short there, despite the higher grass production (Chapter 12). Browsing is concentrated into the zone below 2.5 m in height, and on the more favoured species. The impact of browsing on individual plants can therefore be greater than the overall estimate suggests.

Grass-eating insects

The dominant insect herbivores of the herbaceous layer are grasshoppers (Acridoidea). There is substantial variation in grasshopper biomass within and between years, and between the broad-leafed and fine-leafed savannas. In the broad-leafed savanna, the calculated consumption of herbaceous layer plant material in one year of intensive study was 9.4 g DM m^{-2} y^{-1} (Gandar 1982b; see Box 9.1), which is about 6% of aboveground grass production. A further 3.6 g DM m^{-2} y^{-1} passed from the green plant tissues to the litter layer through inefficient feeding. The combined consumption and wastage attributable to grasshoppers in the fine-leafed savanna was 40.6 g DM m^{-2} y^{-1}. Sinclair (1975) estimated that grasshoppers in the Serengeti removed 19–48 g DM m^{-2} y^{-1} from the herbaceous layer, a similar range to that estimated for the fine-leafed savanna at Nylsvley.

Figure 9.2. Pathways of disappearance of plant leaf material in the broad-leafed savanna at Nylsvley, and their approximate magnitudes.

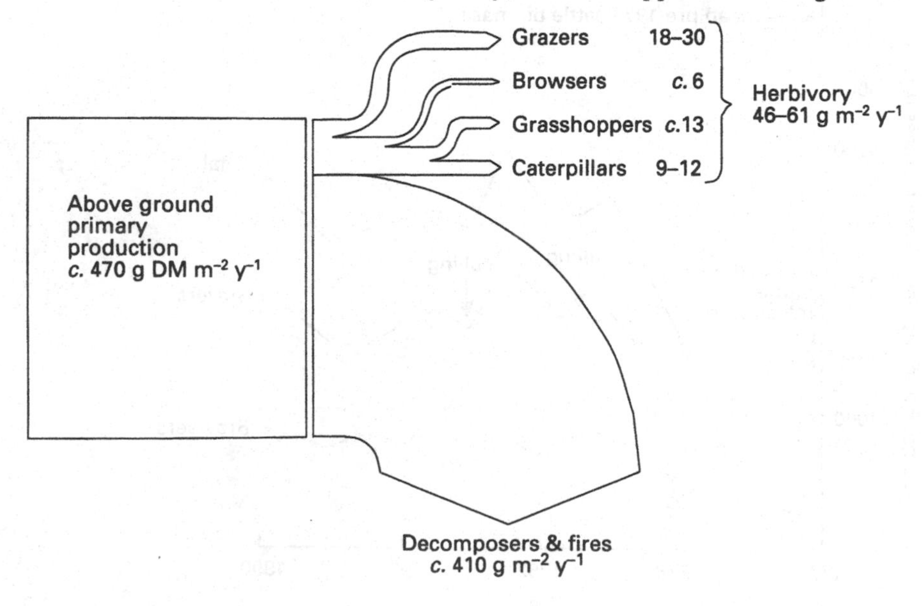

Tree-leaf eating insects

Most of the dominant tree species in the broad-leafed savanna (*Burkea africana*, *Ochna pulchra*, *Terminalia sericea* and *Dombeya rotundifolia*) are subject to occasional defoliation by lepidopteran larvae. The main defoliating species are an emperor moth, *Cirina forda*, which feeds principally on *B. africana*, and a noctuid moth *Sphingimorpha chlorea*, which eats several species (Scholtz 1976). When outbreaks occur, which is irregularly every 2–4 years, the target species may be almost completely defoliated over large areas of the study site. *Cirina forda* has two generations per year. The first outbreak is large, and occurs early in the wet season. There may be a second, smaller defoliation in the late summer. Leaf consumption by lepidopteran larvae in non-outbreak years is about 2.2 g DM m^{-2} y^{-1} (1.7% of leaf production: see Box 9.1). This value is probably an underestimate, since the biomass was measured on branches below 2.5 m, while the Lepidoptera tend to be concentrated on the upper parts of the canopy. The estimates for outbreak years are very rough, but are probably in the region of 17–43 g DM m^{-2} y^{-1} (33% of leaf production). The uncertainty lies in the variety of defoliating caterpillars. Each has its own host species, frequency and degree of outbreak. The outbreak pattern has not been studied in sufficient detail, nor for a sufficient period, to estimate the overall loss accurately. While *C. forda* has a mild outbreak (20% defoliation of *B. africana*) approximately every second year, *S. chlorea* has had only one observed outbreak (1979), during which it defoliated 80% of *B. africana* over a vast region. Assuming that moderate outbreaks occur about once every 2–4 years, the consumption by lepidopteran larvae, averaged over outbreak and non-outbreak years, is 9–12 g DM m^{-2} y^{-1}, or 6–8% of leaf production. Litterfall and shoot production are only slightly depressed during outbreak years (Rutherford 1984), indicating that compensatory growth must follow defoliation.

Beetles and their larvae (Coleoptera)

The next most numerous insect herbivores in the broad-leafed savanna are beetle larvae (Holm, Kirsten & Scholtz 1976). They consume an insignificant portion of the leaf material, compared with the lepidopteran larvae. The main impact of coleopteran larvae on the vegetation is probably through their activities as seed predators (particularly on the leguminous trees) and as root herbivores (Levey 1977). Sap-sucking insects (Hemiptera) are also common in both the broad-leafed and fine-leafed savannas at Nylsvley, but there is no information on the amount of material that they consume.

BOX 9.1. Estimates of herbivory at Nylsvley. These values are only approximate, because the intensity of feeding varies from place to place and year to year. They are intended to give an order-of-magnitude indication of the contribution by various classes of herbivores.

Large mammalian herbivores

Forage consumption = herbivore energy requirements/forage energy content × (100/% digestability)

Grazers

Total energy content of grass = 19.57 kJ g^{-1} Gandar (1982b)
Grass digestibility by ungulates 55% (Zimmerman 1980b)
Mean grazing biomass, 1978–1990 = 1.96 g FM m^{-2} (0.522 g DM m^{-2})
Sum of grazer energy requirements = 196 kJ m^{-2} y^{-1}

Annual rate of grass consumption = 18.2 g DM m^{-2} y^{-1}.

Browsers

Total energy content of browse = 19.57 kJ g^{-1} (assume as for grass)
Browse digestibility = 55% (Owen-Smith & Cooper 1989)
Mean browser biomass = 0.91 g FM m^{-2} (0.237 g DM m^{-2})
Browser energy requirements = 66 kJ m^{-2} y^{-1}

Annual rate of browse consumption = 6.1 g DM m^{-2} y^{-1}

Grasshoppers (Acridoidea) (Gandar 1982a, b)

In energy terms

Consumption = Grasshopper production × (100/ecological efficiency)

Broad-leafed savanna, 1977

Mean annual grasshopper biomass = 0.073 ± 0.014 g DM m^{-2}
Annual grasshopper production = 0.439 g m^{-2} (10.3 kJ m^{-2})
Ecological efficiency = 5.7%
Grass energy content = 19.57 kJ g^{-1}
Grasshopper energy content = 22.89 kJ g^{-1}

Grass consumption = 9.4 g DM m^{-2} y^{-1}
Grass removed from plant but not ingested = 3.6 g DM m^{-2} y^{-1}

BOX 9.1. *(Cont.)*

Fine-leafed savanna, 1977

Mean annual grasshopper biomass = 0.228 ± 0.058 g DM m^{-2}
Assuming the same rate of consumption per unit biomass, and the same percentage wastage:
Grass consumption = 29.4 g DM m^{-2} y^{-1}
Grass wastage = 11.2 g DM m^{-2} y^{-1}

Caterpillars (Lepidoptera) (Scholtz 1976)

In non-outbreak years (Scholtz 1976, 1982):

Mean caterpillar biomass = 0.04 gDM m^{-2}
Leaf consumption = 2.2 g DM m^{-2} y^{-1}

In outbreak years:

Consumption = peak leaf biomass × proportion of leaf biomass in affected species × percentage defoliation of affected species

Peak leaf biomass = 110 g DM m^{-2}
Proportion of leaf biomass susceptible to outbreak = 79% (Rutherford 1979)
Percentage defoliation of affected trees = 20–50%
Consumption = 17–43 g DM m^{-2} y^{-1}

Scenario 1:
mild outbreaks every two years
Consumption = (2.2 + 17)/2 = 9.8 g DM m^{-2} y^{-1}

Scenario 2:
severe outbreaks every four years
Consumption = (2.2 + 2.2 + 2.2 + 43)/4 = 12.4 g DM m^{-2} y^{-1}

Belowground herbivory

Virtually nothing is known about belowground herbivory at Nylsvley. At the end of the growing season (May) the biomass of root-feeding coleopteran larvae in the broad-leafed savanna was 8.15 g FM m^{-2} (about 2.4 g DM m^{-2}; R. J. Scholes, unpublished data), and in the fine-leafed savanna it was 7.15 g FM m^{-2} (2.15 g DM m^{-2}). Therefore the belowground herbivorous insect biomass is 10–60 times that found aboveground. If their appetite matches their abundance, then they constitute the major herbivores in both of these ecosystems. This is possible, given that more than half of the plant production occurs below the ground. Nematodes frequently dominate the belowground herbivory in grassland ecosystems (Scott 1979); nothing is known about their biomass or impact at Nylsvley.

Competition between classes of herbivore

If the invertebrate herbivores are responsible for a large proportion of the total herbivory, do they reduce the amount of food available to the large mammalian herbivores? In the infertile savannas there is probably little competition, first, since the overall amount of herbivory is quite small, and secondly, because the target resource is slightly different for insects and mammals. The grass species least favoured by cattle, *Eragrostis pallens*, is a major food source for grasshoppers (Gandar 1982b). The trees defoliated by caterpillars are of low acceptability to mammalian browsers, while the main browse species are not subject to insect outbreaks. Some of the underlying reasons for these differences are dealt with in the next chapter. However, there are also similarities in the feeding patterns of insects and mammals. Both are responsive to forage quality: for instance, both grasshoppers and antelope are attracted to recently burned patches (Gandar 1982c). Therefore, in the fertile savannas, where the rates of insect and mammal herbivory are much higher, the possibility exists for competition between different types of herbivore.

It is very apparent when a grass sward has been grazed by large mammals. If insect grazers are so important, why is it not equally obvious? First, insects feed on individual leaves, not on whole tufts. Therefore the effect of their grazing is hidden in the sward. Secondly, the impact of mammals tends to be concentrated at certain times of the year (particulary in rotational grazing management systems), while insect grazing is more evenly distributed. Finally, there may be an interaction between mammal and insect grazing. Gandar (1982b) noted that the decline in grasshopper biomass coincided with the removal of cattle from the study area. In the dry times of the year, the grasshopper biomass inside a mammal-proof exclosure was lower than in areas where cattle grazed (Gandar 1980, 1982b). In summer the situation was reversed. It is suggested that the new growth resulting from defoliation by cattle is attractive to grasshoppers at the time of the year when forage quality is limiting.

There is some evidence from the same experiment (Gandar 1980), that feeding by grasshoppers has a stimulatory effect on grass production. The mean green biomass per unit basal area of favoured grasses was higher in plots with grasshoppers than in plots where grasshoppers had been killed by spraying with insecticide. Plots with both cattle and grasshoppers showed a decrease in green biomass.

Ecological and production efficiency

The energy contained in material removed from the plant by a herbivore can be absorbed by the herbivore or excreted. Part of the absorbed energy is lost through respiration, and the remainder is incorporated in the growth of animal tissues (Figure 9.3). Ecological efficiency is defined as the ratio of the energy incorporated into growth to the amount of energy ingested. It therefore reflects both the digestive efficiency and the metabolic efficiency. The relative contributions of metabolic versus digestive processes to the overall efficiency can be illustrated by defining a further index, production efficiency, as the ratio of growth to digested energy.

Insect herbivores show higher ecological and production efficiencies than mammalian herbivores do, since the latter expend energy in maintaining a constant body temperature. The ecological and production efficiencies measured in insects at Nylsvley are lower than those reported for locations in temperate grasslands (5.2–13.7 and 25.6–44.9 respectively), for no obvious reason (Gandar 1982b). The values for grasshoppers in the savanna site at Lamto, Ivory Coast are 8.5 and 42.5 (Gillon 1973).

The difference in ecological efficiency between the large-bodied Africander cattle and the small-bodied impala is too small to use as ammunition in the debate regarding the merits of wild versus domestic stock. The ecological efficiencies for large mammals measured at Nylsvley are high relative to the value of 1.6% for African ungulates in general (Phillipson 1973).

Secondary production

Few of the studies undertaken at Nylsvley had the determination of secondary production as an objective, and for the large mammalian herbivores the exercise is biased by the fact that they are maintained at artificially low numbers. However, it is possible to make some broad estimates, which are of interest given that the agricultural economy in savannas is largely based on animal production (Box 9.2). The total aboveground secondary production is in the region of 1.4 g DM m^{-2} y^{-1}, or about 0.34% of the aboveground primary production on an energy basis. The belowground secondary production is unknown.

Figure 9.3. Partitioning of energy consumed by four types of herbivore into intake (I), excretion (E), production (P) and respiration (R). The fraction not excreted is assimilated (A) by the digestive system. In the case of grasshoppers, a significant fraction of the material removed from the plant is not ingested, but falls to the ground during the feeding process. P/I, ecological efficiency; A/I, assimilation efficiency; P/A, production efficiency.

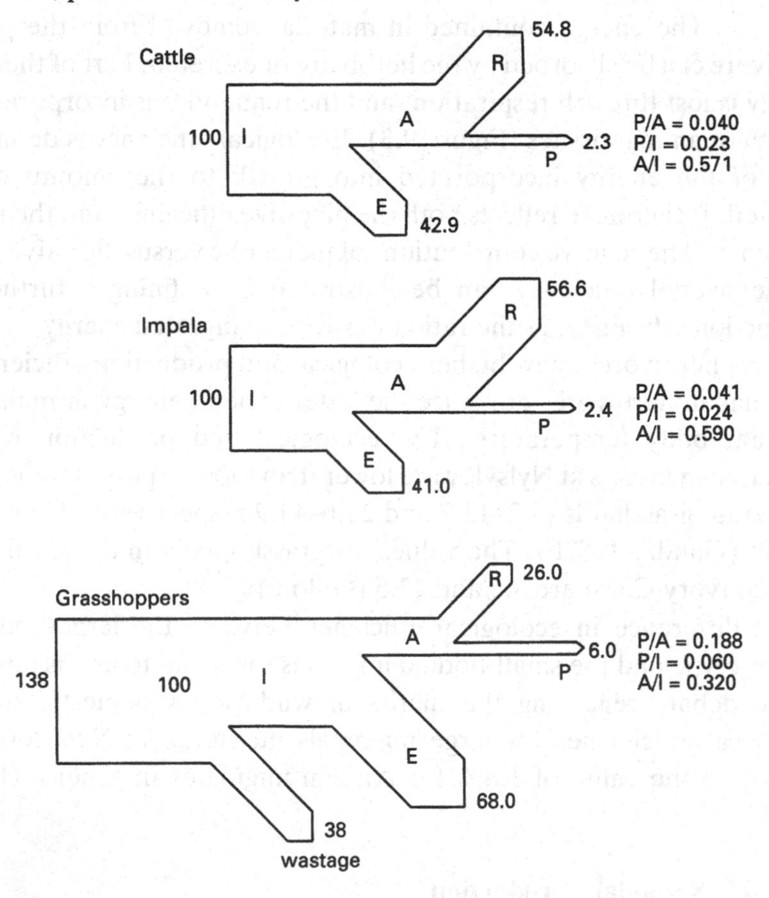

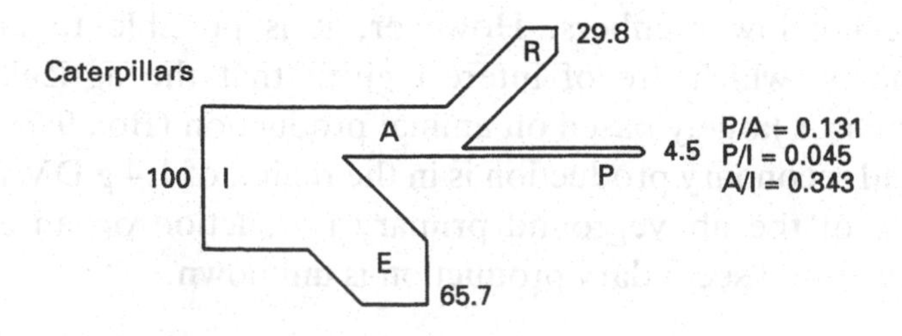

BOX 9.2. Estimation of the secondary productivity of the broad-leafed savanna at Nylsvley.

In energy terms:

Secondary production = Plant consumption × (ecological efficiency/100)

Large mammalian grazers

Consumption = 18.2 g DM m^{-2} y^{-1} (see above)
Energy content of grass = 19.57 kJ g^{-1} (Gandar 1982a)
Energy content of wild ungulate tissue = 6.28 kJ g FM^{-1} (Skinner, Monro and Zimmerman 1984)
Water content of herbivore tissues = 71%
Ecological efficiency = 2.3%
Secondary production = 1.30 g FM m^{-2} y^{-1} (0.38 g DM m^{-2} y^{-1})

Large mammalian browsers

Consumption = 6.1 g DM $m^{-2}y^{-1}$
Assume the same energy content and efficiency as for grazers
Secondary production = 0.44 g FM m^{-2} y^{-1} (0.13 g DM m^{-2} y^{-1})

Grasshoppers

Secondary production = 0.44 g DM m^{-2} (Gandar 1982a)

Lepidopteran larvae

Consumption = 9–12 g DM m^{-2} y^{-1}
Ecological efficiency (mass basis) = 4.2%
Secondary production = 0.42–0.56 g DM m^{-2} y^{-1}

Constraints on secondary production

Does primary production limit the biomass and production of wild herbivores in savannas? The strongly positive relationships between rainfall, primary production and herbivore biomass in African savannas suggests that it does (Coe *et al.* 1976): closer examination shows, however, that the linkage is much weaker in savannas on infertile soils than in fertile savannas (Bell 1982; East 1984). Soil fertility influences primary production, but also has a profound impact on the forage value of that production. The herbivore biomass on fertile savannas can be several times higher than on infertile savannas receiving the same rainfall. The primary production difference is much less: therefore the proportion of primary production converted to herbivore biomass must be much higher in the fertile than in the infertile savannas.

As a rule of thumb, the productivity of large mammalian herbivores in fertile savannas is limited by food quantity, while that in infertile savannas is limited by quality. The interesting feature of the Nylsvley situation is that both the infertile and fertile savanna patterns occur within the same landscape. Large mammalian herbivores make use of both to satisfy their resource needs.

At Nylsvley, approximately 10% of the broad-leafed savanna aboveground primary production is currently consumed by herbivores (Figure 9.2). If the large mammal biomass was at its pre-Colonial level, herbivory would rise to about 15% of primary production. This range is neither unusually low nor high for terrestrial ecosystems in general, but is toward the lower end for African savannas (Sinclair 1975). It is much lower than the values reported for the fertile, grass-dominated savannas of East Africa, where up to 80% of the aboveground grass production may be consumed by grazers (McNaughton 1979; Drent & Prins 1987). In the infertile *miombo* savannas on the other hand, consumption drops to less than 5% of aboveground production (Malaisse *et al.* 1975). The key issue with respect to constraints on secondary production in infertile savannas is what controls forage acceptability, not the relationship between primary and secondary production.

Owen-Smith (1982) proposed another broad rule: that grazing large mammals in savannas face a period of protein shortage during the dry season, while browsers must overcome a period of energy shortage at that time. This difference between 'quantity' constraints on browsers and 'quality' constraints on grazers appears to be quite general in savannas (Ellis & Swift 1988). Owen-Smith & Cooper (1989) used a simple animal energy and nutrient balance model, in conjunction with observed feeding rates and forage nutrient contents, to show that kudu at Nylsvley are able to satisfy their protein requirements at all times of the year, but that they are unable to meet their energy requirements for a 2-month period in late winter (Figure 9.4). This is because the forage consumed by browsers has a high protein content relative to grass, even in the dry season, but owing to the high proportion of deciduous plants in the savanna, forage is seasonally scarce. The phosphorus requirements of the kudu may also not have been satisfied for part of the year, but the model for phosphorus requirement by kudu is much less certain than for nitrogen and energy, and the animals may be able to remobilise phosphorus stored during times of excess.

Applying a similar model to cattle, which are grazers, using data from the Nylsvley study by Zimmerman (1978), shows that they have an adequate supply of energy all year round (Figure 9.4). Their protein intake, however, drops below maintenance levels during the late dry season. Standing dead grass is both accessible and retains a high energy content, although the digestibility of the energy declines as the fibre content increases and the protein level decreases. The forage protein content drops due to the translocation of nitrogen out of the old leaves before they senesce. At higher

Figure 9.4. Seasonal course of protein and energy balance in a browser (kudu) and grazer (cattle). The graph for kudu is simplified from Owen-Smith & Cooper (1989), while that for cattle is based on data presented in Zimmermann (1980b). The solid line is the observed intake, while the dashed line is the approximate threshold for maintenance.

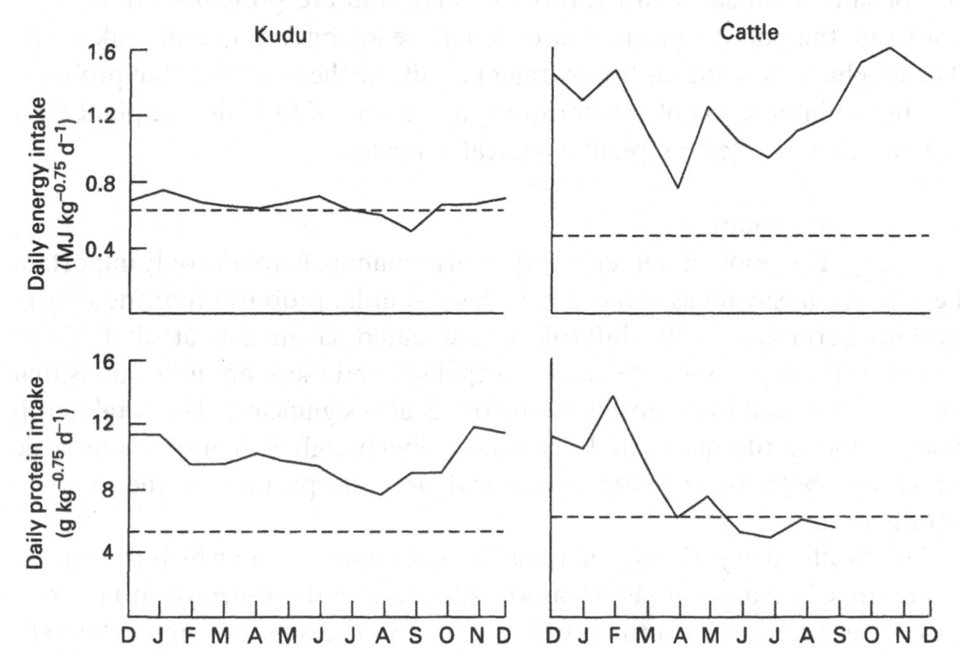

stocking rates than were imposed at Nylsvley, grazers could be under both energy and protein limitation.

Combining the two generalised rules presented above leads to a prediction for the seasonal pattern of large herbivore distribution in the Nylsvley landscape. Grazers should have an increased preference for the 'high-quality' fine-leafed savanna in the dry season, while obligate browsers should have an increased preference for the denser broad-leafed savanna at that time. Observations bear this out. This balancing of nutritional demands by varying use of different landscape facets is a key to the maintenance of those patches (Chapter 12), and to the success of large mammal herbivores at Nylsvley. The pattern of feeding by browsers is not proportional to the available leaf mass, and the impact on particular species varies with the time of the year (Owen-Smith & Cooper 1989). Therefore certain key browse resources, such as palatable evergreen leaves below 2.5 m in the late dry season, can be in short supply despite the overall low consumption of leaf material.

The constraints on the growth of invertebrate herbivore populations at Nylsvley are unknown. Various factors have been suggested to control lepidopteran outbreaks in temperate forest systems, including the weather, predation, disease, plant defence responses and density-dependent feed-backs on the population growth rate (Myers 1988). The densities of insecti-vorous birds and wasps parasitic on lepidopteran larvae are relatively high in the broad-leafed savanna (Tarboton 1980), and are probably sufficient to constrain the populations in non-outbreak years, but not in outbreak years. The mechanism of the outbreak trigger awaits further research, but probably lies in the interaction of temperature, the timing of rainfall, the plant C:N balance and its effect on plant chemical defences.

Summary

The popular perception that large mammals are the only important herbivores in savannas is incorrect. A substantial proportion of the above-ground herbivory in the infertile broad-leafed savanna is attributable to insects such as grasshoppers and caterpillars, and there are indications that the belowground invertebrate herbivory is also significant. The total herbi-vory in the fertile fine-leafed savannas is much higher than in the infertile savannas, both in absolute terms and as a proportion of the primary production.

The productivity of large mammalian herbivores, upon which the human economy of savanna areas is based, is limited by the shortage of key food resources during critical times of the year. In the case of grazing mammals, the limitation is mostly due to the low protein content of the grass in the dry season, while browsers struggle to satisfy their energy requirements at this time, since most of the tree leaves have fallen. Both grazers and browsers move between the energy-rich but protein-poor broad-leafed savannas and the protein-rich but energy-poor fine-leafed savanna patches in order to balance their nutritional requirements. Secondary productivity at Nylsvley is therefore sustained by the landscape-scale patchiness in nutrient distribution.

Part III

The carbon cycle

10

Primary production

One of the first unifying ideas in ecosystem theory was that an ecosystem could be characterised by describing the path and pattern of energy flow through it (Lindeman 1942). By reducing all the interactions within an ecosystem to the common currency of energy exchange it became possible to make comparisons within and between ecosystems. Energy transfers are difficult to quantify directly. Carbon flows came to be a surrogate measurement for energy, since most of the useful energy in an ecosystem is transferred as chemical energy stored in carbohydrates. Carbon assimilation by plants through the process of photosynthesis is an especially important ecosystem parameter, since it places a limit on the amount of energy available to sustain the entire ecosystem. It is conventionally expressed in terms of dry matter produced by plants per unit area over a given period of time (g DM m^{-2} y^{-1}), and referred to as primary production. The measurement of primary production was a key element of the International Biological Programme, which strongly influenced the initial work at Nylsvley.

The accurate and complete measurement of primary production is an extremely onerous task. When the labours of several scientists for several years are required to produce a single number, with disappointingly large confidence limits and doubtful generality, disillusionment inevitably sets in. The emphasis on energetics was a major criticism of the IBP: could the scientific effort have been directed more productively? It is now accepted that energy is only one of several currencies which can be used to describe ecosystems, and that many key ecosystem processes (such as plant establishment and death, and aspects of animal behaviour) are not well described by either energy or material transfers.

It is only in the decades after the IBP studies ended that the value of this effort has come to be appreciated. Sufficient primary production data are now available to begin to make comparisons between different ecosystems and to detect some general patterns. Furthermore, the growing concern over CO_2 accumulation in the atmosphere has made detailed studies of the carbon cycle invaluable in their own right. It is now apparent that carbon plays an central role in the cycling of other elements (notably nitrogen, phosphorus and sulphur) and in the maintenance of soil fertility and structure.

This chapter will examine primary production at three levels of resolution. At the crudest level, the savanna will be regarded as an absorbing and reflecting surface for radiant energy, with little internal structure. The data are derived from micrometeorological studies of the radiation budget. The next section will detail the primary production by the major vegetation components of a savanna, at the scale of a vegetation patch measured over the period of a year. The data are derived from repeated biomass measurements in the field. The third section will focus on the short-term and small-scale processes which control photosynthesis, and is based on laboratory and field gas exchange studies.

Because ecosystem energetics, primary production and the carbon cycle are so closely linked, units relating to all three will be used in this chapter, and can be interconverted. The carbon and energy content of different tissues varies slightly. As a rough conversion guide, 1 g DM is equivalent to 0.4 g C and 20 kJ of energy. Table 10.1 contains typical carbon contents of plant material from Nylsvley.

This chapter draws on the work of many researchers, but the foundations were laid by three in particular: Trevor Harrison for the work on energetics, Mike Rutherford on primary production, and Pam Ferrar on gas exchange.

The energy balance of broad-leafed savanna

Savannas, like most ecosystems, ultimately derive their energy from the sun. The subtropical regions of the world (which characteristically support savannas) receive very high total solar radiation: higher than any other terrestrial biome with the exception of the circumtropical deserts. The low cloud cover in subtropical regions more than compensates for the reduced radiation resulting from being a small distance away from the equator. Summation of the radiant energy that would be received over the period of a year by a horizontal surface at the latitude of Nylsvley but above the atmosphere (S_e) gives a value of 11 915 MJ m^{-2}. Of this, 39% is absorbed or reflected by clouds, dust and atmospheric gases, leaving 7316 MJ m^{-2} as incoming solar radiation at the top of the canopy (Chapter 3; Harrison 1984). About 12.5% is reflected by the canopy and the soil (this value, the shortwave reflection coefficient, is also known as the albedo), leaving 6401 MJ m^{-2} to be dissipated by the vegetation.

Table 10.1. *The carbon content of various plant components at Nylsvley. The values were determined by wet oxidation using a Walkley–Black procedure modified for complete oxidation (Nelson & Sommers 1975)*

	Carbon content (%)	
Plant part	Minimum	Maximum
Woody plants		
Stem wood	45.01	46.96
Current twigs	42.73	44.73
Coarse roots	42.69	42.69
Fine roots	40.00	40.08
Flowers	40.00	40.00
Fruits and seeds	40.00	45.26
Live leaves	45.78	45.90
Leaf litter	42.00	44.49
Grasses		
Live leaf	43.40	45.15
Standing dead leaf	40.23	40.80
Fine roots	27.20	44.83
Crown	40.21	41.75

On average, 580 mm of water are evaporated from broad-leafed savanna per year (Chapter 6), which requires 1427 MJ m^{-2} (assuming that the latent energy of evaporation at 19 °C is 2.46 MJ kg^{-1}). A small amount of energy (about 11.2 MJ m^{-2}, or 0.17%) is converted to chemical form by the process of photosynthesis. The rest (4964 MJ m^{-2}) leads to warming of the vegetation canopy and the soil. This heat is re-radiated as longwave radiation, or is carried away by convection in the atmosphere. The difference between the radiant energy received by the vegetated surface and the amount reflected and re-radiated is known as the net radiation (R_n). On an annual basis it averages 4132 MJ m^{-2} in the broad-leafed savanna. This means that about 2270 MJ m^{-2} is lost by re-radiation, and about 2694 MJ m^{-2} by convection.

Several generalities emerge from this overall energy budget. First, only a very small amount of the available solar radiation finds its way into trophic structure of the ecosystem. This is a first indication that primary production in this system is unlikely to be limited by insufficient radiant energy. Next, the ratio of the energy lost by convection to the energy dissipated through the evaporation of water (the Bowen ratio) is 1.90. This value is high relative to other ecosystems (such as forests, temperate grasslands or croplands), indicating that water limitation is likely to be a major constraint on primary production. Finally, the large convection component in the energy budget means that shading by tree canopies will not have a major impact on evaporative demand in the subcanopy area, since sufficient energy can be supplied by the hot air moving beneath the canopy to sustain a high evaporation rate.

Energy budgets for the broad-leafed savanna have been worked out at a shorter timescale (every hour over a 4-day period) by de Jager & Harrison (1982), and at the spatial scale of individual subhabitats by Harrison (1984), with similar conclusions.

The main variables in the above scheme which are influenced by the biology of the system are the absorptivity and reflectivity of the canopy. The three-dimensional structure of the canopy controls the mix of leaf types and angles exposed to the radiation; species composition and phenology influence the reflective, transmissive and absorptive properties of those leaves. The soil at Nylsvley is light in colour, and has a higher reflectivity than is typical for green leaves. Given the relatively low leaf area index (about 1.2 at maximum) of this savanna in relation to forests and temperate grasslands, it would therefore be expected that the albedo at Nylsvley would be on the high side of the 0.10–0.21 range typical of those formations. It is likely that the measured albedo at Nylsvley (0.12) is an underestimate, for unknown reasons. The albedo shows a seasonal pattern related to the green leaf area. The lowest values occur at the time of peak biomass (April, 0.111), and the highest at the end of winter (August, 0.145). There is a diurnal variation of 0.05, with the minimum at noon (Harrison 1984).

The vegetation canopy in savannas is organised at several scales. First, there is the layering of tree canopies, shrubs and herbaceous plants. Secondly, the tree and shrub layers are discontinuous, and the proportion and size of gaps varies considerably. Since tree canopies generally have a lower reflectivity than grass canopies, and in the case of this savanna, the leaf area index within the canopy spread is about 2.25, while between canopies it has a maximum of around 0.6, the albedo would be expected to decrease as the tree cover increases. Thirdly, the leaf orientation, size and clustering within the canopy is a species characteristic which is under strong environmental influence.

The crown characteristics of savanna trees, including some from Nylsvley, were studied by van der Meulen & Werger (1984a, b). The mean foliage gap size, leaf size and light throughfall of the dominant trees in the broad-leafed savanna fell within the ranges typical of 'tropical' localities, such as *miombo* woodland, rather than those typical of hotter, more arid climates. The throughfall ranged from 19.1 to 40.1%, with a system mean probably near the 26.7% recorded for *Burkea africana*. The crown architecture tended to be mono-layered (pagoda-like) with some irregular, multi-layered species.

Eller *et al.* (1984) examined the optical properties of a variety of tree leaves from a savanna locality close to Nylsvley. For a wide range of leaves, they found that absorbance (integrated across the spectrum from 300 to 3000 nm) was close to 58%, but that the contributions of transmission and reflection to the non-absorbed 42% varied between species. On average, 20% was transmitted, and 22% reflected. Succulence increased the absorbance, while hairs on the leaf surface decreased absorbance and increased reflectance. Eller (1984) showed that the optical properties of *Ochna pulchra* did not change substantially with leaf age, despite the change from bright red foliage in spring to dark green foliage in summer. These results also point to a higher albedo than was reported for Nylsvley by Harrison (1984).

Primary production in broad-leafed savanna

The conventional method of estimating the primary production of a terrestrial ecosystem is to measure the increase in dry mass of each important vegetation component over a period of time. Since vegetation is spatially variable, this method requires a large number of samples at each sampling time. In order to detect the change over and above the natural variation in biomass, the time interval needs to be in the order of weeks. To account for the variation in primary production with the season of the year, the measurements must be taken throughout the year, and preferably for several years.

Savannas have many plant components which need to be individually measured in order to obtain an overall production estimate.

Production by woody plants

The main production components in woody plants are wood, terminal shoots, leaves and fine roots. Although the terminal shoots are destined to become part of the wood mass, it is customary to measure them separately, since they are partially deciduous and form a significant part of the diet of mammalian browsers. It is convenient to measure the above-ground wood separately from the belowground wood (coarse roots). It is not possible to sort tree fine roots from grass fine roots reliably. The calculated contribution by each of these sources is based on the ratio of carbon isotopes in the mixed root sample; more details are provided in the section on fine root production.

In the broad-leafed savanna, three woody plant species (*Burkea africana*, *Terminalia sericea* and *Ochna pulchra*) account for 78% of the woody plant leaf area, and by inference, a similar proportion of the primary production (Rutherford 1979). The annual increase in wood mass was estimated by measuring the change in girth over the period of a year, and relating this radial increment to the equivalent wood mass via the allometric relations derived for the major species by Rutherford (1979). The potential errors in this approach include the prediction error of the allometric equations, and the changes in girth due to changes in the water status of the plant (Rutherford 1984). The first problem is reduced by taking a large and representative sample, and the second by measuring the stems at a time of similar water status every year (in this case, midwinter), or by measuring the increment over several years. When the latter approach is used, tree mortality during the measurement period leads to a potential underestimate. In particular, there are pronounced decreases in the woody plant basal area following fires.

Several sources of this type of data are available for the broad-leafed savanna at Nylsvley (Table 10.2). The percentage increases calculated from Lubke & Thatcher's (1983) data cover a period of slightly higher rainfall than those from the litterfall plot. The mean tree biomass is very similar in the two studies. The savannas cited for comparison are either drier than Nylsvley, or have a considerably higher woody biomass, both of which would lead to lower radial increments. A long-term average wood biomass increment for Nylsvley is probably around 6% per annum, which translates to wood production of 89 g DM m^{-2} y^{-1} using the Rutherford (1979) biomass data, which is in turn based on Lubke's (1976) woody plant census data. Since the woody plant biomass has grown substantially since 1975, the current value will be significantly higher.

Rainfall and competition with other trees have been proposed as the main factors controlling tree growth in savannas (Scholes 1990a). Rutherford (1984) suggests, on the basis of 3 years' data, that wood production in the broad-leafed savannas at Nylsvley is unaffected by rainfall in the current or previous year, except during a drought year.

The coarse root increment estimate is based on 2 years of monthly measurement on four roots of *Burkea africana* using spring-loaded bands (Rutherford 1984). The mean annual radial increment was 2.5%. There is no obvious reason why coarse root increment should be so much less than

Table 10.2. *Annual wood production expressed as a percentage of the standing wood biomass, at Nylsvley and comparable savannas. In most cases it is estimated by measuring the annual radial increment and converting to biomass*

Authority	Location and part	Annual increment (%)
Nylsvley		
Rutherford (1984)	*Burkea africana* & *Ochna pulchra* stems 1977–9	4.4
Rutherford (1984)	*Burkea africana* & *Ochna pulchra* coarse roots 1977–9	2.5
Lubke & Thatcher (1983)	Stems of all species in 4 transects 50 × 320 m	8.71 SD = 1.04
P. G. H. Frost (unpublished data)	Litterfall plot; all species in 220 × 50 m area 1978–83	9.14
Other African savannas		
Rutherford & Kelly (1978)	Mosdene, 10 km north of Nylsvley. Stems of all species, composition similar to Nylsvley 1974	4.70
van Vegten (1984)	Eastern Botswana (500 mm MAP). Stems of all species, largely fine-leafed 1950–63	4.05
van Vegten (1984)	As above, 1963–75	3.75
Malaisse *et al.* (1975)	Zaïre, *miombo* woodland, stems of all species	2.90
Endean (1967)	Zambia, *miombo* woodland, stems of all species	2.20

aboveground wood increment, estimated as 4.4% per year over a period of 4 years including the 2 years in which roots were studied. Rutherford (1983) indicates that the root:shoot ratio declines in *Ochna pulchra* as the plant increases in size.

The coarse root biomass value used to calculate coarse root production for this chapter was 460 ± 72 (SE) g m^{-2} (R. J. Scholes, unpublished data: 600 cores 40 mm in diameter to a depth of 0.5 m, extrapolated to a depth of 1 m using root depth-distribution data). This value is in close agreement with the 496 g m^{-2} measured by Rutherford (1983) under *Ochna pulchra* clones, and the 491 g m^{-2} measured by van Wyk (1977) in three hundred 75 mm cores to 1 m in areas not immediately beneath tree canopies. It is substantially less than the 850 g m^{-2} obtained for the average of three 1 × 1 × 1 m pits excavated by van Wyk, or the 1231 g m^{-2} for forty-three 75 mm cores obtained beneath tree canopies. Coarse root production was calculated to be 11.5 g DM m^{-2} y^{-1} assuming a coarse root biomass of 460 g m^{-2} and an annual increment of 2.5%. Using a less conservative biomass estimate of 584 g DM m^{-2} (area-weighted mean of van Wyk's data), and an annual percentage increment equal to the aboveground wood increment (6%), coarse root production could be as high as 35 g DM m^{-2} y^{-1}.

There are several possible approaches to calculating leaf and current season's twig production. If the falling litter of woody plant origin is caught in baskets and weighed, its long-term average should equal woody plant aboveground production less herbivory. This is an inefficient way to measure wood production, but is a reliable method for production of leaves and reproductive structures. The Nylsvley litterfall study used one hundred 0.5 × 0.5 m baskets, randomly distributed in an area 220 × 60 m and sampled monthly. The mean leaf litterfall over the period 1976 to 1981 was 138.3 g DM m^{-2} y^{-1}, SD = 13.7, while flowers and fruits averaged 10.0 g DM m^{-2} y^{-1}, SD = 5.3. Herbivory of woody plant leaves is mostly due to outbreaks of lepidopteran larvae, with a small amount taken by mammalian browsers (see Chapter 9). During non-outbreak years, the losses to herbivory are about 3.4 g DM m^{-2} y^{-1}, but the long-term average, including outbreaks, is about 18 g DM m^{-2} y^{-1}.

Twigs and bark are also present in the litterfall sample, but the sample variability is high (27.5 g DM m^{-2} y^{-1}, SD = 6.0) . Another method for estimating twig production is to tag individual twigs and monitor their growth (Rutherford & Panagos 1982). Leaf production can be measured in the same way. Twig production over a 4-year period was estimated to be 27.1 g DM m^{-2} y^{-1} by this method, and woody plant leaf production was estimated to be 126.4 g DM m^{-2} y^{-1}.

Since the woody plants in the broad-leafed savanna are 97% deciduous (on a leaf mass basis), the peak leaf standing crop provides a minimum estimate of leaf production. Estimates of leaf and twig production in the broad-leafed savanna at Nylsvley, derived from various sources, are given in Table 10.3. The litterfall data are considered most reliable for leaf and flower production, and the marked twig data for current shoot production.

Fine root production is generally estimated by sequential harvest techniques, although there may be substantial problems associated with this approach (Singh *et al.* 1984; Vogt *et al.* 1986). The essence of the problem lies

Table 10.3. *Annual woody plant leaf and current season's twig production in broad-leafed savanna at Nylsvley, as estimated by three methods*

	Production ($g\ m^{-2}$)	
Method	Leaves	Current shoot
Litterfall traps	138.3	26.7
Marked twigs	126.4	27.1
Peak standing crop	110.0	23.6

in distinguishing the root growth signal from the random variability of the sample. In the case of the Nylsvley study, the seasonal signal was sufficiently clear that the problem did not arise. Fine root production was estimated as the difference between the peak and trough standing crops (Figure 10.1), thereby avoiding the problematic sorting of fine roots into live and dead categories. The standing crop estimates were determined monthly for 15 months, from forty 38 mm diameter cores taken to a depth of 50 cm. (R. J. Scholes, unpublished data). The fine root masses were corrected for ash content (14%) and the proportion of roots below 50 cm depth (16.5%). The total fine root production to 1 m depth was calculated to be 498 $g\ m^{-2}\ y^{-1}$, SE = 47. This is an underestimate, since some fine roots are inevitably lost during the washing and sorting process, and because it ignores root decay during the 5-month period between the trough and peak root standing crops.

Savanna trees and forbs have a C_3 photosynthetic pathway, which results in a characteristic $\delta^{13}C‰$ in their tissues of 27.4, while savanna grasses have a $\delta^{13}C‰$ of 13.6. This isotopic discrimination can be used to indicate what proportion of the fine root production is attributable to trees and forbs, and what to grasses. From the seasonal trend of $\delta^{13}C$ it is estimated that 35% (175 $g\ m^{-2}\ y^{-1}$) of the fine root production is due to C_3 plants, predominantly trees, but with a small contribution by shrubs and forbs. This probably represents a slight overestimation for the savanna as a whole, since these root samples were taken in areas of 50% above-average tree biomass.

Production of fine roots is sometimes estimated by assuming that the ratio of below- to aboveground production is equal to the ratio of below- to aboveground biomass. Rutherford (1984) applied this approach at Nylsvley using a root:shoot ratio for woody plants of 0.78. He then subtracted the estimated coarse root production to obtain a fine root production estimate of

148.8 g m^{-2}. This major underestimate illustrates the fallacy of the initial assumption, compounded by a root:shoot ratio which did not include fine root mass.

A third approach to estimation of belowground production is the direct observation of the roots through the glass wall of a 'rhizotron' (Rutherford & Curran 1981; Rutherford 1983). This approach showed that the fine roots had a mean longevity greater than one year, which supports the conclusions of the sequential coring study.

The relative timing of growth in each woody-plant production component (Figure 10.2) is described for the broad-leafed savanna by Rutherford (1984), based on direct observation. Shoot growth (current season's twig plus leaf) is 80% complete within the first 8 weeks after growth commences. Wood growth continues fairly steadily throughout the wet season. Fine root growth commences about 10 weeks after shoot growth, and continues to the end of the season.

Production by grasses and forbs

Grasses contribute about 80% of the herbaceous layer production. Three grass species (*Eragrostis pallens*, *Digitaria eriantha* and *Panicum maximum*) contribute about three-quarters of the total grass basal area and dominate grass production. For grass plants, leaf and fine root turnover and crown growth are the major production components. Crown growth has not

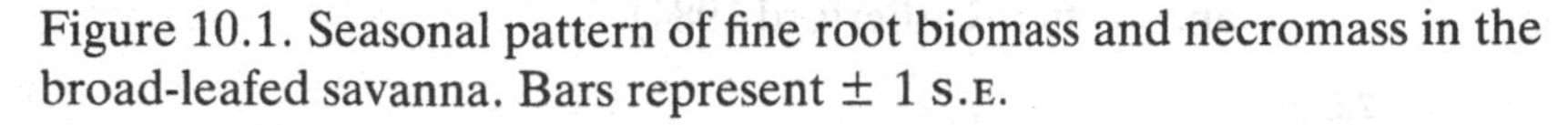

Figure 10.1. Seasonal pattern of fine root biomass and necromass in the broad-leafed savanna. Bars represent ± 1 s.e.

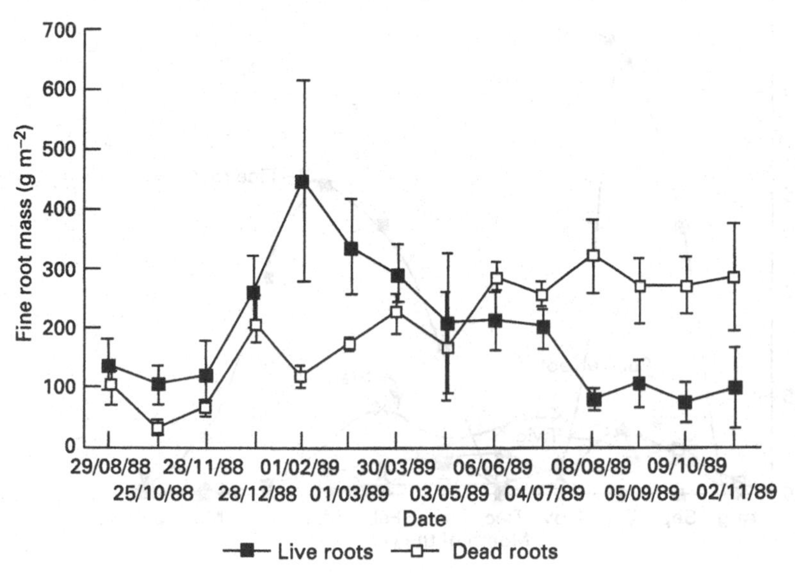

been adequately measured at Nylsvley, since all of the studies were harvested at 5 cm above the ground level. The production estimates will therefore be referred to as the grazeable production.

The grazeable production estimates for the broad-leafed savanna were obtained by sequential harvest techniques. There are several ways in which to calculate production from a time series of biomass and necromass data (Long *et al.* 1989). The method used in this review is the summation of biomass and necromass changes, where both are greater than zero. In practice, since the biomass and necromass curves at Nylsvley are fairly smooth in most years, the production is given by the difference between the peak standing crop and the standing crop at the beginning of the growing season. This approach can be applied to all the data sets, and yields a conservative estimate, since it does not consider the loss of necromass to the litter layer. Over the 4-month period between growth initiation and peak standing crop, and in the absence of trampling by herbivores, these losses are relatively small (about 14 g DM m^{-2}: R. J. Scholes, unpublished data).

Grunow *et al.* (1980) measured the grazeable production over three growing seasons (July 1974 to June 1977). In the 1974–5 season, thirty 0.5 m^2 quadrats were clipped to ground level every 3 weeks in summer and 6 weeks

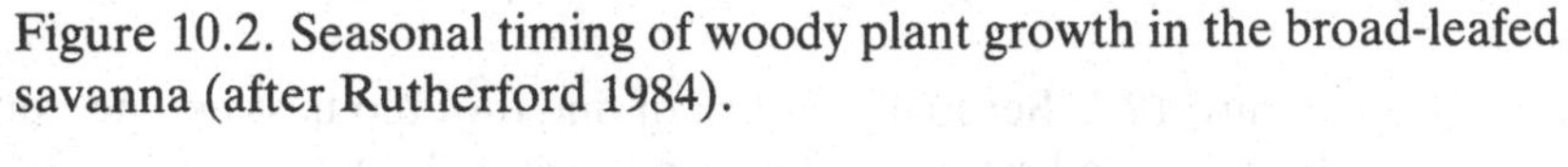

Figure 10.2. Seasonal timing of woody plant growth in the broad-leafed savanna (after Rutherford 1984).

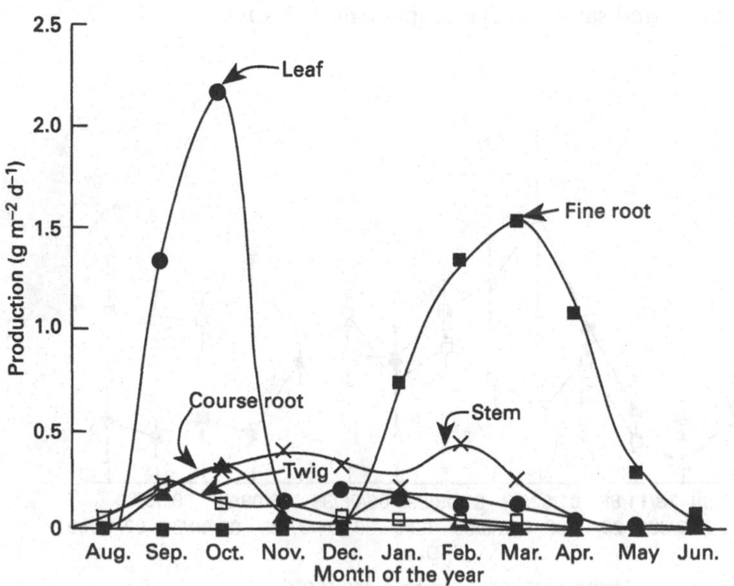

in winter. From 1975 to 1977 seventy quadrats were clipped, of which 30 were within movable (mammalian) herbivore exclusion cages. The material was sorted into forb and grass (crown, live leaf and dead leaf) components, dried and weighed. A smooth function, which included a growth and mortality term, was fitted to the estimates of biomass and necromass over time. The integral of the growth term was taken to be the herbaceous layer grazeable productivity. Except for the smoothing, this approach is conceptually similar to the one outlined above. They calculated the grazeable production to be 76 g DM m^{-2} y^{-1}, including a 10 g m^{-2} y^{-1} allowance for insect herbivory, based on an average biomass curve for the three seasons. However, averaging biomass over the seasons before calculating production makes no biological sense, since the time course of biomass accumulation is different in every season, and leads to a substantial underestimate of productivity. If their method is applied to each season independently, and the three production estimates are then averaged, the mean herbaceous layer grazeable production is 127 g m^{-2} y^{-1}, SD = 28, including insect herbivory.

Grass harvest data were also collected by Randall & Cresswell (1983) during the 1981/2 season. The biomass increment over the season was 75 g DM m^{-2}, and the necromass increment (excluding the first sample, which is obviously an error) was 45 g DM m^{-2}, giving a grazeable herbaceous layer productivity of 120 g DM m^{-2} y^{-1}, without the insect herbivory correction.

Both the above studies were located in areas where the tree cover was relatively low (28%; the average is about 40%). Grunow & Bosch (1978) indicate that the grazeable biomass is on average 44% higher in the open areas between trees than under tree canopies. Grossman *et al.* (1980) showed that the availability of prefered forage species was very similar in both habitats.

Grazeable production data were also collected by M. C. Scholes & R. J. Scholes (unpublished data) in a portion of the broad-leafed savanna where the tree biomass is about twice the average for Nylsvley. Grazeable production, calculated from the peak standing crop, was 135 g DM m^{-2}, SE = 9.4, excluding insect herbivory.

Herbaceous layer production in many savannas is related to rainfall. The production values calculated in each of the above studies, all corrected for 10 g DM m^{-2} y^{-1} insect herbivory, are shown in relation to the growing season rainfall up until the date of peak standing crop in each study (Figure 10.3). There is a weakly significant relationship between rainfall and herbaceous layer production at Nylsvley, based on these data ($n = 6, r^2 = 0.48, p < 0.1$). The slope is 0.28 g m^{-2} mm^{-1}, which is low relative to semi-arid grasslands (Rutherford 1980). This is a reflection of the nutrient limitations at Nylsvley.

Assuming an average growing season rainfall of 480 mm, the long-term average grazeable production is probably about 157 g DM m^{-2} y^{-1}.

This is still an underestimate of aboveground herbaceous layer production, since it does not include growth of the tiller bases (the crown), and because the calculation method is inherently conservative. There is a vague seasonal trend in the average stubble (i.e. crown) standing crop values given by Grunow *et al.* (1980). The lowest value (132 g m^{-2}) occurs in September, before growth begins. The highest values occur over the months in which peak biomass is achieved (December to March). The average value for these months is 302 g m^{-2}. Therefore crown growth could add 170 g m^{-2} y^{-1} to the aboveground herbaceous production.

Belowground production by grasses has been calculated using the same data and techniques applied to fine root production by woody plants (R. J. Scholes, unpublished data). Since 65% of the total fine root production is due to grasses, grass root production is estimated to be 325 g m^{-2} y^{-1}. This value

Figure 10.3. Herbaceous layer aboveground grazeable production in the broad-leafed savanna at Nylsvley in relation to rainfall. The values reported by Walker and Knoop (1987) are peak biomass clippings only, and are excluded from the regression, as was the Grossman (1980) burnt plot. The other data points are based on the sum of standing crop increments throughout the growing season. $y = 0.28x + 15$.

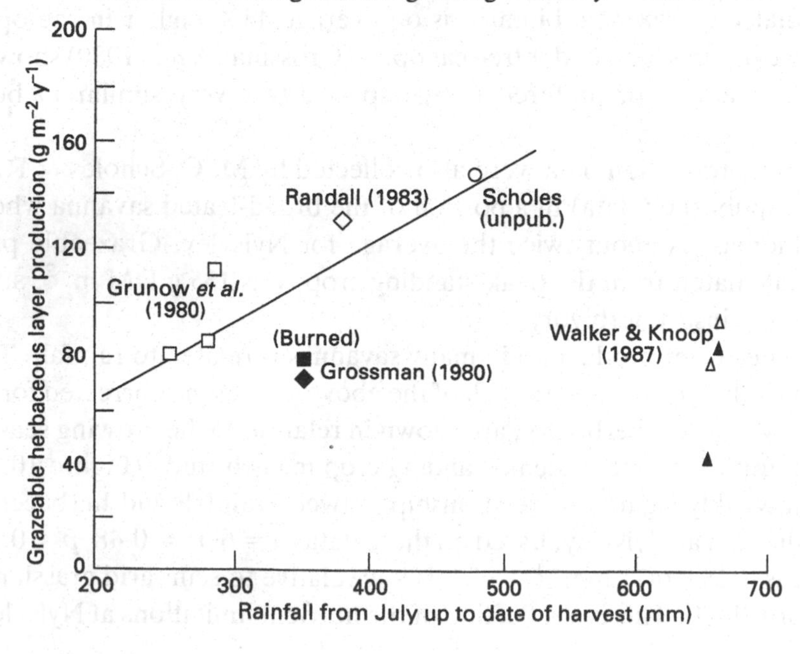

Table 10.4. *Best and upper-limit long-term estimates of primary production in the broad-leafed savanna at Nylsvley. The interannual variation is an approximate guide only, based on twice the interannual standard deviation, expressed as a percentage of the mean. Approximately 95% of all years should fall within the range defined by the mean ± the interannual variation*

Component	Primary production ($g\ DM\ m^{-2}$) Best estimate	Maximum estimate	Interannual variation (%)
Woody plants			
Wood growth	89.0	117.9	40
Current twigs	27.1	27.1	40
Leaves	156.3	156.3	20
Flowers and fruit	10.0	10.0	100
Coarse roots	11.5	35.0	40
Fine roots	175.0	175.0	40
Subtotal	468.9	521.3	35
Herbaceous plants including grasses			
Leaf and culm	157.0	171.0	50
Crown	0.0	170.0	30
Roots	325.0	325.0	50
Subtotal	482.0	666.0	50
Total	950.9	1187.3	43

is based on data from a single year, in which the rainfall amount and distribution were close to the long-term average.

Summary: production estimates

The production values presented above are summarised for the whole system in Table 10.4. The best available estimate of long-term production for each component was used. The standard errors are within 10% of the mean in most cases where it is possible to calculate the sample variability. The greatest uncertainty is associated with coarse root and grass crown production. In both cases the most conservative estimate has been used. Woody plant leaf production has been corrected for 18 g m^{-2} y^{-1} total herbivory, and herbaceous aboveground production for 10 g m^{-2} y^{-1} insect herbivory.

Carbon assimilation measured by gas exchange

The factors that control the rate of photosynthesis per unit leaf area include radiant flux density ('light intensity'), leaf temperature, atmospheric CO_2 concentration, the type, concentration and activity of photosynthetic enzyme systems in the leaf, leaf anatomy, and stomatal resistance (Farquhar & von Caemmerer 1984). Several of these factors vary on a daily or shorter-term basis. Exploration of these controls therefore requires techniques that are much more sensitive and responsive than biomass sampling. Gas exchange measurements satisfy this need. However, gas exchange studies are practical only for short periods of time and on a small scale; often only on a portion of a single plant leaf.

Net photosynthesis measurements

Several projects have measured the rate of net photosynthesis of species found in the broad-leafed savanna at Nylsvley. The methods and equipment which they used varied greatly, and so did the environmental conditions under which the measurements were made. The earliest work was carried out in the laboratory, either on young plants grown in pots or liquid culture in growth chambers, or on mature shoots removed from the field several hours previously. The radiant flux density at which these measurements were taken was substantially lower than full sunlight at Nylsvley (up to 2500 μmol PPFD m^{-2} s^{-1}). A caravan was later equipped with gas analysis instruments for use as a field laboratory. More recently, portable gas analysis equipment became available for field use. The net photosynthetic rates reported in these studies, converted to common units, are presented in Table 10.5.

The variation in the reported values is striking. In particular, the laboratory measurements are substantially lower than the field measurements. This is due to the lower irradiance at which the plants were grown and measured, or the stress induced by cutting and transporting stems to the laboratory. The most recent field measurements are comparable to the rates reported elsewhere, but still relatively low: for instance Ludlow & Wilson (1971) reported rates of 44.1 μmol CO_2 m^{-2} s^{-1} for *Panicum maximum*, while Wilson (1975) reported 12.6–23.3 μmol CO_2 m^{-2} s^{-1}.

Influence of radiant flux density

The influence of radiant flux density on photosynthetic rate was reported by Ferrar (1980) and Cresswell *et al.* (1982) for two woody plants and two grasses, grown in growth chambers. *Terminalia sericea* approached its maximum photosynthetic rate at about 850 μmol PPFD m^{-2} s^{-1}, which is equivalent to about one third of full sunlight at Nylsvley. The other woody plant, *Grewia bicolor*, only approached saturation at 1400 μmol PPFD m^{-2} s^{-1}. The two grass species, *Digitaria eriantha* and *Eragrostis pallens*, had not reached light saturation at 1400 μmol PPFD m^{-2} s^{-1}, the maximum radiant flux density achievable in the laboratory. Blackmore (1992) measured the light response curve in the field using neutral filters of different densities, and showed that *Cenchrus ciliaris*, a C_4 grass species growing in the more fertile sites at Nylsvley, approached saturation at 2000 μmol PPFD m^{-2} s^{-1}. The initial slope of the light response curve, known as the quantum efficiency of photosynthesis, was 0.0175, 0.0179, 0.0157 and 0.0192 μmol CO_2 per μmol PPFD for *Terminalia sericea*, *Grewia flavescens*, *Digitaria eriantha* and *Eragrostis pallens* respectively. The last two values are approximate, since they were calculated *post facto* from figure 14 of Cresswell *et al.* (1982). The C_3 plants would be expected to have slightly lower quantum efficiencies than the C_4 plants. The quantum efficiency for *Cenchrus ciliaris* in the field was 0.0727 μmol CO_2 per μmol PPFD.

Influence of temperature

Ferrar (1980) showed that both the quantum efficiency and the maximum photosynthetic rate of broad-leafed savanna trees were influenced by the temperature at which the plant was grown. In *Terminalia sericea* and *Grewia bicolor* both were increased when the plants were grown at 31 °C during the day and 26 °C at night relative to a 25 and 21 °C regime. The relationship between photosynthetic rate and leaf temperature is presented for these two woody species and the two dominant grasses (*Digitaria eriantha* and *Eragrostis pallens*) in Cresswell *et al.* (1982). The temperature optima for photosynthesis were 26 °C for the trees, and 30 °C for the grasses at 1400 μmol PPFD m^{-2} s^{-1}, and decreased with decreasing radiant flux density.

Influence of water availability

The field photosynthetic rates collected by A. K. Baines (unpublished data: see Table 10.5) when the soil was moist and when it was dry show the strong control exerted by water supply. Water stress influences the photosynthetic rate through stomatal closure. Ferrar (1980) showed that the

Table 10.5. *Rates of net photosynthesis measured on species occurring in the broad-leafed savanna at Nylsvley. The maximum rate is given in those cases where it is not possible to calculate the mean*

Species	Net photosynthesis ($\mu mol\ CO_2\ m^{-2}\ s^{-1}$)			Author
	Maximum	Mean	SD	
Woody plants				
Terminalia sericea	6.77			1
	5.82			2
	13.90			5
	7.72			3
		21.00	5.10	4
Grewia flavescens	7.07			2
	7.90			5
	9.08			3
	63.18			6
	12.50			7
		23.0	7.1	4
Burkea africana	10.90			3
	28.63			6
	8.40			7
		19.6	3.5	4
Ochna pulchra	4.14			8
	13.20			3
	55.20			6
	13.6			7
		11.80	4.7	4
Grasses				
Digitaria eriantha	0.69			9
	2.01			10
	7.9	2.11		11
	14.52			12
	27.20	19.47	5.71	6
	25.19	15.20	14.10	7
		26.1	4.20	4
Panicum maximum	2.77			10
	18.61			14
	5.90			15
	6.63	2.37		11
		27.90	1.09	4

Table 10.5. *(Cont.)*

Species	Net photosynthesis (μmol CO_2 m^{-2} s^{-1})			Author
	Maximum	Mean	SD	
Eragrostis pallens	2.20	1.31		11
	16.57			13
	33.40	22.47	7.47	6
	18.20	13.20	5.00	7
		25.78	5.30	4

Authors and notes

1. Ferrar (1980). Pot grown, Leaf temperature = 31 °C, 1400 μmol m^{-2} s^{-1}.
2. As above, leaf temperature = 25 °C.
3. Cresswell *et al.* (1982). Field measurement, no water stress.
4. Blackmore (1992). Field measurement using LiCor 6000 IRGA, PPFD = 2190 μmol m^{-2} s^{-1}, $n = 5$.
5. P. Ferrar (undated report). Single shoot.
6. Baines (1989). Field measurement, wet soil. LiCor 6000. $n = 6$.
7. As above, dry soil, $n = 4$.
8. Ludlow (1987). Cut shoot, 150 μmol m^{-2} s^{-1}.
9. Wolfson & Cresswell (1984). Pot-grown on 200 mg l^{-1} NO_3-N.
10. Rey (1982). Pot grown, 1100 μmol m^{-2} s^{-1}.
11. Randall (1983). Field measurement.
12. Unknown source.
13. Cresswell *et al.* (1982). Pot grown, 31 °C, 1400 μmol m^{-2} s^{-1}.
14. Ariovich & Cresswell (1983). 350 μmol m^{-2} s^{-1}, liquid culture with 200 mg l^{-1} NO_3-N.
15. As above, with 20 mg l^{-1} NO_3-N.

photosynthetic rate declined sharply in *Terminalia sericea* and *Grewia flavescens* when the leaf water potential dropped below −1 MPa, and was negligible at −2 MPa. When the plants had been 'hardened' by prior exposure to water stress, the photosynthetic rate declined below −1.5 MPa, was about half of the unstressed rate at −2 MPa, and one quarter at −3 MPa. Stomatal resistance showed the inverse response. A further discussion of the relationship between photosynthetic and transpiration rates can be found in Chapter 6.

The woody plants, with a C_3 photosynthetic pathway, have average field-measured photosynthetic rates about 18% lower than the grasses. The latter have a variety of C_4 pathways, which are theoretically better adapted to high temperatures, high irradiance and water stress. Within each category, the range of average values, given the same conditions of measurement, is quite small. The large difference between the maximum value and the average reflects diurnal and seasonal patterns. The daily pattern reflects the curve of radiant flux density. Photosynthesis is highest in the mid-morning, when the radiant flux densities are already high, the leaf temperature is optimal and the plant is fully hydrated. There is frequently a decline in photosynthesis around midday due to transient water stress, stomatal closure and excessive leaf temperatures, followed by recovery later in the afternoon (Randall 1983; K. A. Baines, unpublished data).

Influence of leaf nitrogen content

The photosynthetic rate of grasses, in particular, shows a strong decline as the season progresses (Randall 1983). This is caused by a combination of physiological changes in each leaf as it ages and changes in the nitrogen nutrition of the plant. The photosynthetic enzyme RUBP-carboxylase makes up a large portion of the organic nitrogen content of leaf tissue. Under light-saturated conditions, the rate of photosynthesis is constrained by the RUBP-carboxylase concentration (Farquhar & von Caemmerer 1984). Thus there is a causal link between the leaf nitrogen content and the photosynthetic rate (Monson 1989). Near the beginning of the season, once mineralisation of organic matter has begun to release inorganic nitrogen into the soil, the leaf nitrogen content is high (20–25 mg N per g DW). As the season progresses, the carbon assimilation rate rises, but the nitrogen mineralisation rate remains fairly constant. Despite nitrogen translocation from ageing to new leaves within the plant, the mean leaf nitrogen content falls to 3–7 mg N per g DW. The photosynthetic rate falls simultaneously. This trend can be observed in Randall's (1983) data, although she was unable to establish a statistically significant link between leaf nitrogen content and photosynthetic rate.

Wolfson & Cresswell (1984) and Wolfson (1988) showed that the leaf nitrogen content and photosynthetic rate of *Digitaria eriantha* were influenced by nitrogen supply to the plant, but declined over a 4-month period, even in the plants with a high nitrogen supply. The rate of photosynthesis per unit leaf mass or leaf area was lowest in the plants receiving the most nitrogen, but the leaf area was highest. Therefore, on a whole-plant basis, the photosynthesis of the high nitrogen plants was highest. The form in which the nitrogen was supplied (NH_4 or NO_3) made little difference to the observed photosynthetic rates, but the plants supplied with NO_3 grew faster and bigger than those supplied with an equal amount of nitrogen in the form of NH_4.

Influence of pathogens

Rey (1982) investigated the effect that infection by viral and fungal pathogens have on the photosynthetic rates of *Burkea africana* and *Digitaria eriantha*. Although a significant decrease in the leaf photosynthetic rate occurred in both species following infection, she concluded that the impact on primary production was negligible. This was because the total leaf area affected was small and of relatively short duration.

Gross photosynthesis and dark respiration

Gas analysis techniques cannot detect the respiration which occurs simultaneously with photosynthesis. The measured carbon assimilation rate is therefore known as net photosynthesis. The respiration which occurs while the leaf is illuminated (photorespiration) can be estimated in two ways: as the difference between the gross and net respiration; or by assuming that photorespiration proceeds at essentially the same rate as the respiration measured when the leaf is completely darkened. Gross respiration has been measured on several broad-leafed savanna plants at Nylsvley using ^{14}C labelling techniques (Table 10.6). The variability of the results, as well as the variability of the net photosynthesis data, precludes the estimation of photorespiration by this method. The fact that the highest estimates of gross photosynthesis are lower than the mean net photosynthesis illustrates a deficiency in the technique.

The dark respiration for *Digitaria eriantha* and *Eragrostis pallens* grown in pots was in the range 0.7–1.4 μmol CO_2 m^{-2} s^{-1}, at 30 °C, or 5–10% of the maximum net photosynthetic rate (Cresswell *et al.* 1982). Rey (1982) reported dark respiration in *Digitaria eriantha* measured in the field to be 1.2–1.5 μmol CO_2 m^{-2} s^{-1}. In *Terminalia sericea* and *Grewia bicolor* it was about 1.1 μmol CO_2 m^{-2} s^{-1}, or 15–18% of maximum net photosynthesis at 30 °C, but was only 9–12% of net photosynthesis at the photosynthetic optimum of 25 °C. The dark respiration rate increased exponentially with increasing temperature in all cases.

Table 10.6. *Gross photosynthesis estimates for plants growing in the broad-leafed savanna at Nylsvley, as determined by ^{14}C pulse labelling*

Species	Gross photosynthesis ($\mu mol\ m^{-2}\ s^{-1}$)	Source
Burkea africana	2.2	1
	14.2	2
Terminalia sericea	7.3	2
Ochna pulchra	4.5	2
Grewia flavescens	9.5	2
Panicum maximum	1.4	1
	0.5	3
Digitaria eriantha	2.5	1
	0.5	3
Digitaria eriantha	0.3	3

Sources: 1. Rey (1982). 20 second pulse with ^{14}C.
2. Cresswell *et al.* (1982).
3. Randall (1983). Mean whole season.

Soil respiration

The rate of CO_2 generation by the soil ('soil respiration') is an integrated measure of respiration by roots, soil fauna and soil microbes. If it is measured above the litter layer, then in the long-term, for a steady-state system, it should be equal to the net primary production by the system plus the carbon losses from the root not reflected in growth. If the litter layer is cleared away before the measurement is taken, then it should equal the total belowground carbon allocation, including root growth, respiration and rhizodeposition (exudates, sloughed and lysed cells, which can account for 10–30% of the belowground allocation). As such it can be a useful check on production estimates (Raich & Nadelhoffer 1989).

Anderson (1976) performed a pilot study on soil respiration in the broad-leafed savanna. The CO_2 evolution in five 0.01 m^{-2} cuvettes was monitored with an Infra-Red Gas Analyzer (IRGA). Towards the end of the growing season (18–20 April), with the soil temperature at 20 °C and the soil water content 8.2%, the soil respiration rate (without litter) was 5.8 g CO_2 m^{-2} d^{-1}. After rain one month later, the rate was 9.7 g CO_2 m^{-2} h^{-1} at the same soil temperature. In late winter (8–10 August), when the soil temperature was 15 °C and the water content was 1.1%, the rate declined to 1.8 g CO_2 m^{-2} d^{-1}. When the roots were sieved out of the soil to a depth of 35 cm, the soil respiration rate declined 35% relative to an undisturbed control. The top 10 cm contributed 37% of the total soil respiration. Carbon dioxide evolution from litter was 0.39 mg CO_2 g^{-1} h^{-1}. The mean litter mass at the time was 424 g m^{-2} (SD = 60), giving a litter respiration rate of 4.0 g CO_2 m^{-2} d^{-1}.

Soil respiration was measured monthly from April 1976 to April 1977 by Bezuidenhout (1978), using the method described by Coleman (1973). The CO_2 generated over a 4-hour period in 15 cm diameter cuvettes, pushed 10 cm into the ground, was trapped in 25 ml of 0.1 N KOH in a 3.5 cm diameter bottle. Ten replicates were used. Note that both the deep insertion and the small trapping bottle diameter (< 6% of the cuvette area) have been shown to lead to substantial underestimates of soil respiration (Raich & Nadelhoffer 1989). The time course of soil respiration measured beneath the canopy of *Burkea africana* and between tree canopies is shown in Figure 10.4. Note the eight-fold increase in soil respiration between the dry, cool winter and the warm, wet summer. Over the period of a year, the soil plus litter respiration beneath the canopy was 957 g CO_2 m^{-2}, while between the tree canopies it was 681 g CO_2 m^{-2} (70%). For respiration by root-free soil alone, the between canopy value was 66% of the beneath canopy value of 447 g CO_2 m^{-2}. The difference in soil respiration between the habitats results from the

Figure 10.4. Seasonal pattern of soil respiration in three subhabitats of broad-leafed savanna at Nylsvley (data from Bezuidenhout 1978).

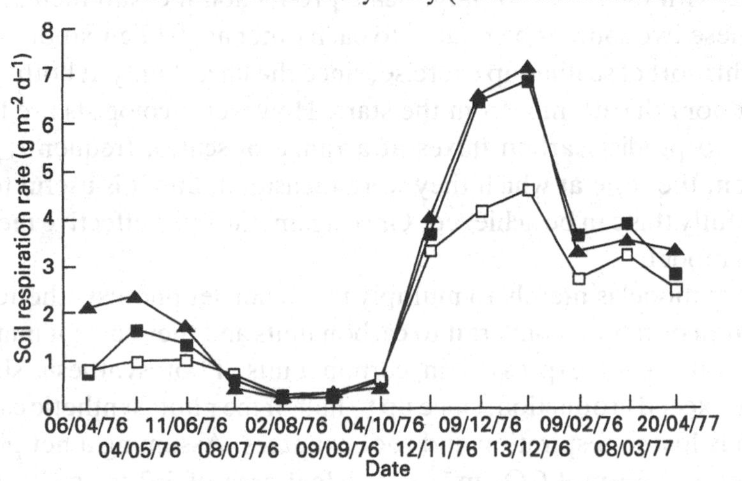

difference in the quantity of organic inputs which they receive, which is also reflected in different soil organic matter levels and litter masses.

Samples taken at the same time as the *in situ* soil respiration measurements, but at different depths and incubated in the laboratory showed a decline from 5.40 $\mu g\ CO_2\ g^{-1}\ d^{-1}$ at 0–5 cm, to 2.65 at 5–15 cm and 0.95 at 15–30 cm for the beneath-canopy habitat. Samples from between canopies showed the same trend at 14% lower absolute rates. These rates, measured at 25 °C and field water content, are equivalent to 4.28 g $CO_2\ m^{-2}\ d^{-1}$, in agreement with Anderson's (1976) measurements, but 250% higher than the rates measured in the field by Bezuidenhout (1978). The higher values in the laboratory could partly be attributable to soil disturbance and higher incubation temperatures, but mainly indicate underestimation by his *in situ* method.

Bezuidenhout (1978) concluded that soil temperature and water content accounted for more than 80% of the variation of *in situ* measurements of soil respiration, and that of these controls, water content was dominant at Nylsvley. He also measured respiration with and without roots, and with and without prior treatment with an insecticide, to eliminate soil fauna. Based on these data, he concluded that 29% of the total respiration was caused by organisms in the litter layer, 21% due to roots and rhizosphere organisms, 7% due to soil invertebrates, and 43% due to soil microorganisms. These percentages were relatively consistent across subhabitats.

Integrating gas exchange and production measurements

Do the results collected at the small scale by gas analysis techniques agree with the results of large scale production measurements? Can data from these two sources be related to each other at all? Few studies have attempted this sort of scaling-up exercise, since the uncertainty at both scales appears to doom the attempt from the start. However, ecologists are being called upon to predict carbon fluxes at a range of scales, frequently very different from the scale at which they were measured, and it is useful to see how successfully this can be achieved. Once again, the most effective method is through a model.

The crudest model is merely to multiply the mean net photosynthetic rate by the leaf area duration, convert it to carbon units and then see if it matches the production values expressed in carbon units. Photosynthesis should comfortably exceed production, since up to half of the photosynthetic carbon assimilation is lost to respiration and root exudates. Assuming a net photosynthetic rate of 25 $\mu mol\ CO_2\ m^{-2}\ s^{-1}$, a leaf area of 1.2 $m^2\ m^{-2}$, a leaf duration of 150 days, and a 45% tissue carbon content, the annual carbon assimilation in the broad-leafed savanna could not exceed 4666 g C m^{-2}. The carbon content of annual production (Table 10.4) is between 428 and 557 g C m^{-2}. Thus the rate of carbon assimilation measured by gas analysis measurements is sufficient to account for the observed primary production, allowing for deviations from the maximum photosynthetic rate due to the diurnal pattern of irradiance and temperature, leaf shading, water stress and losses through respiration.

Similar calculations can be applied to soil respiration and production. Soil plus litter respiration should exceed production by the amount of root respiration and rhizodeposition. Using the annual time course of respiration provided by Bezuidenhout (1978), but increasing the values by a factor of 2.62 to bring the peak summertime values into line with Anderson's (1976) peak estimates, and assuming 40% canopy cover by trees, the annual soil plus litter respiration is 2073 g CO_2 m^{-2}, or 565 g C m^{-2}. Considering that the broad-leafed savanna is not an equilibrium system (the woody biomass is accumulating), and the uncertainties in the soil respiration correction, the degree of agreement between respiration and production values is reassuring.

Summary

The total net annual primary production in the broad-leafed savanna at Nylsvley is about 1 kg DM m^{-2} y^{-1} (10 tons per ha), of which just over half occurs belowground. The primary production is almost equally contributed by trees and grasses, and represents about one third of 1% of the radiant energy incident on the savanna. The low efficiency of conversion relative to other terrestrial ecosystems is due mainly to the large part of the year during which there is insufficient water for carbon assimilation to occur. The low stomatal conductances, thought to be an adaptation to a dry environment, lead to an inherently low maximum photosynthetic rate, even when water is present. Low temperatures during the early and late part of the growing season constrain primary production to a small degree, as may low leaf nitrogen contents late in the growing season.

11

Decomposition

Building nutrients into organic molecules is one half of the cycle of life; disassembling the molecules is the equally important other half. The rate of material cycling within an ecosystem can be controlled by either process. Primary production in many ecosystems is limited by the rate at which organically bound elements such as nitrogen, phosphorus and sulphur are released for uptake by plants, through the processes of decomposition of plant litter and the mineralisation of soil organic matter. Most of the energy fixed by primary production is not passed to herbivores and so on up the trophic ladder, but is liberated by decomposing organisms. The process of decomposition, and the organisms which accomplish it, are therefore key components in any ecosystem.

The disappearance of organic litter is a result of two processes: *comminution*, in which the particles are broken up into smaller and smaller pieces and mixed with the soil through the action of the weather and animals; and *oxidation*, in which the long-chain organic molecules are converted to carbon dioxide and inorganic residues (Swift, Heal & Anderson 1979). Oxidation releases the nutrient elements bound in the organic molecules, and is therefore also referred to as *mineralisation*. Since comminution can have important effects on the site and rate of oxidation the two processes must be considered together as part of the overall decomposition process, but failure to distinguish them conceptually can lead to misinterpretation of field and laboratory data.

Litter in savannas can follow three pathways to decomposition: via microbial oxidation; via the gut of a detritivore (where the digestion is partially performed by microbes); and via combustion in a fire. All three pathways result in incomplete oxidation, and the remaining carbon compounds are more resistant to decomposition than the original substances. In the case of microbial oxidation, the residual compounds are high molecular weight organic polymers of unknown structure, collectively known as the humic fraction. The faeces of detritivores are frequently physically protected from decay, either because they form microaggregates in the soil (in the case of earthworms), or because they become part of the nest chamber wall (in soil-feeding termites). The charcoal remaining from fires is very resistant to microbial decay owing to its low nitrogen content. These partial oxidation products are in turn subject to the same three pathways to further oxidation.

The decomposition data reported in this chapter are mostly due to the work of Jurg Bezuidenhout. The termite information is from a study by Paul Ferrar, the mesofauna work was by E. A. Ueckerman, and the dung beetle study by S. Endrödy-Younga. The decomposition studies at Nylsvley have already been synthesised both in the form of a model and in several papers (Furniss *et al.* 1982; Morris, Bezuidenhout & Furniss 1982; Morris 1983).

The role of soil fauna

Soil-dwelling animals are functionally grouped according to body size. Those animals larger than the soil pores (longer than 2 mm is the arbitrary limit) must burrow in order to live in the soil, and are called the soil macrofauna. Animals between 0.2 and 2 mm in length live within the soil pores, and are called the mesofauna. Creatures smaller than this are very susceptible to dehydration, and are confined to the water films around soil particles. They are known as the microfauna.

Only 5–8% of the carbon efflux from Nylsvley soils is due to respiration by soil invertebrates. This estimate was obtained by Bezuidenhout (1978) by measuring soil respiration (including the litter layer) before and after insecticide treatment. Since some portion of the soil fauna is herbivorous, the proportion of litter and soil organic matter oxidised through digestion by soil fauna must be less than 8%. The indirect impact of soil fauna on the decomposition process can, however, be quite large, through their actions in reducing the size of litter particles and transporting them into the soil.

Arthropod macrofauna are the main agents in fragmenting litter and redistributing it spatially at Nylsvley. The small fragments have a high surface area to volume ratio, and therefore would be expected to be more available to microbial attack than whole leaves, particularly if the latter are sclerophyllous and protected by a thick cuticle. Bezuidenhout (1980) tested this hypothesis in laboratory incubations at optimal temperature and litter water content, and found that for leaf fragments from Nylsvley down to 1 mm diameter, size had no influence on the decomposition rate. This suggests that access to the tissue is not a major constraint to microbial oxidation.

The small litter fragments work their way down through the litter layer, which can be many leaf layers thick in the broad-leafed savanna. Peter Frost (unpublished data) estimated the litter layer to contain four to five identifiable annual strata. The fragments are mixed into the mineral soil by the activities of litter-feeding arthropods. This relocation alters the microclimate of the fragments, by increasing the humidity and stabilising the temperature, and probably increases their decay rate as a consequence. This untested hypothesis is included in the litter decay models constructed for Nylsvley (Furniss *et al.* 1982). Sensitivity analysis of those models shows that the steady-state litter respiration is unaffected by this assumption, although the equilibrium amounts of litter standing crop are altered.

Termites

A feature which differentiates decomposition in tropical from temperate terrestrial ecosystems is the importance of termites in the tropics, and particularly in Africa. Primitive wood-feeding termites occur in temperate and even Palaearctic ecosystems, but the 'higher termites' (Termitidae) are confined to warm climates. A specialist subfamily of the Termitidae, the fungus-growers (Macrotermitinae) is restricted to the Ethiopian and Oriental biogeographic realms, suggesting that they evolved in Africa during the Tertiary Period. Another specialist subgroup of termites, the soil-feeders, are most diverse in the Ethiopian realm. They also occur in the Oriental and Neotropical realms, but not in Australia. By contrast, earthworms dominate the macrofauna in temperate regions. They can be very prominent in moist savannas, but are largely absent from savannas receiving less than about 1000 mm of rainfall per year (Lavelle 1983). The main macrofaunal groups involved in litter decomposition in dry savannas are termites, millipedes, dung beetles (Scarabaeinae), coleopteran larvae, ants, and cockroaches.

There are at least 21 species of termites at Nylsvley, in 15 genera and two families (Ferrar 1982a, e). This termite diversity is comparable with other savannas, although one third lower than the wet West African savannas (Ferrar 1982a; Josens 1983). They form two main functional groups (Table 11.1): the litter feeders, which eat mostly dead surface litter, wood and dry dung; and the soil feeders, which ingest soil and presumably consume a fraction of the soil litter and the humus it contains. Some litter feeders use the litter to grow fungus 'gardens', on which the termites feed. Soil feeders contribute 68% of the termite biomass in the broad-leafed savanna at Nylsvley, excluding gut contents. Litter feeders make up the remaining 32%. This is consistent with the general pattern in African nutrient-poor savannas (Josens 1983). The fine-leafed savanna at Nylsvley is conspicuously low in

Table 11.1. *Termite species, functional groups, populations and biomass at Nylsvley (Ferrar 1982a, b, d)*

Species	Food	Habitat	Density (m^{-2})	Biomass[a] (g DM m^{-2})
Family Kalotermitidae				
Bifiditermes sibayiensis	wood	various	1366	0.26[b]
Family Termitidae				
Subfamily Apicotermitinae				
Aganotermes oryctes	soil	various		
Astralotermes brevior				
Subfamily Macrotermitinae				
Allodontermes rhodesiensis	litter	*vlei* edge		
Macrotermes natalensis	litter	various		
Microtermes cf. *albopartitus*	litter	mostly FLS[d]	64	0.01
Odontotermes badius	litter	various		
Subfamily Nasutitermitinae				
Fulleritermes coatoni	litter?	various	2	0.0005
Trinervitermes dispar	litter?	mostly BLS[e]	387	0.16
Trinervitermes rhodesiensis	litter	various		
Subfamily Termitinae				
Amitermes hastatus	soil?	*vlei* edge		
Cubitermes pretorianus	soil	mostly BLS[e]	1238	0.44[c]
Cubitermes muneris	soil	rocky hill		
Cubitermes testaceus	soil	rocky hill		
Lepidotermes lounsburyi	soil	BLS[e]		
Microcerotermes brachygnathus	soil	rocky hill	245	
Microcerotermes sp. B	soil	rocky hill		
Microcerotermes sp. C	soil	rocky hill		
Promirotermes spp.	soil	BLS[e]		

[a]Dry mass is 30% of fresh mass. Excludes gut contents in the case of soil feeders, where the gut contents make up 72% of the total dry mass.
[b]Includes all soldierless termites.
[c]Includes all large soil feeders.
[d]FLS = Fine-leafed savanna.
[e]BLS = Broad-leafed savanna.

termites in general (0.16 vs 3.17 g DW m^{-2} in the broad-leafed savanna, annual mean value excluding gut contents: Ferrar 1982d) and soil-feeding termites in particular (0.06 vs 2.16 g DW m^{-2}). A relatively low litter-feeding termite biomass has been suggested to be a characteristic of nutrient-rich

African savannas, for reasons which are unknown (Scholes 1990b). The prominence and diversity of humivorous termites at Nylsvley contradicts Bodot's (1967) suggestion, repeated by Josens (1983), that they are poorly represented on sandy, low organic matter soils.

The harvester termite, *Hodotermes mossambicensis*, which is unusual in that it can consume large quantities of live grass, is not found on the sandy soils of the Nylsvley study site. It is common on nearby fertile, clayey soils. This is consistent with its distribution elsewhere in Africa, and suggests that sandy soils are unsuitable for the diffuse system of galleries which this species constructs.

Wood (1978) has collated data on the foraging rates of litter-feeding termites. Estimates range from 7 to 30 mg DM of litter per g FM of termite per day. Applying his laboratory-derived mean of 10 mg g^{-1} d^{-1} to the Nylsvley termite biomass estimates indicates an annual litter consumption of 3.9 g m^{-2} y^{-1}. The consumption rate by soil-feeding termites is much more difficult to estimate. Furniss *et al.* (1982) used the relation between body mass and respiration rate developed by Wiegert & Coleman (1970) to calculate a carbohydrate assimilation rate of 4.8 g m^{-2} y^{-1}; quite close to the 3.35 g m^{-2} y^{-1} quoted by Peakin & Josens (1978) for another soil feeder, *Cubitermes exiguus*. The assimilation efficiency by termites is relatively high (54–93% according to Wood 1978), so organic matter intake is probably 5–10 g m^{-2} y^{-1}. Given that the organic matter content of the Nylsvley soil is low (about 2% in the 0–20 cm horizon), and about half of this is material which is probably resistant to termite digestion, at least 500–1000 g m^{-2} y^{-1} of soil must be ingested by termites per year to satisfy their energy requirements. This is equivalent to a soil layer 0.33–0.66 mm thick, or complete turnover of the 0–20 cm horizon in 300 to 600 years.

Furniss *et al.* (1982) used their model to investigate the consequences of allowing soil-feeding termites to digest organic polymers of varying degrees of resistance, such as cellulose, lignin and humus. To reproduce an acceptable termite respiration rate while allowing a 5% assimilation rate for lignin and humus, the soil intake would have to be 2000 g m^{-2} y^{-1}, equivalent to a soil layer 1.5 mm thick. The discrepancy between this calculation and the one given above results from the assumption by Furniss *et al.* that all the organic matter in the soil was humified: in fact, in the surface horizons, a large proportion of the organic matter is in the form of small, unhumified litter fragments.

The combined organic matter consumption by soil- and litter-feeding termites at Nylsvley is therefore 9–14 g m^{-2} y^{-1}. This range is consistent with the value of about 12 g m^{-2} y^{-1} given for a dry Sahelian savanna and 27 g m^{-2}

y^{-1} for a dry East African savanna (Josens 1983, using data from Lepage and Buxton). The above- and belowground primary production in the broad-leafed savanna at Nylsvley is about 950 g m^{-2} y^{-1}, of which about 50 g m^{-2} y^{-1} is consumed aboveground by herbivores. Clearly only a very small proportion of the litter (less than 1.5%) is consumed by termites; the rest must be oxidised by microbes outside termite guts, or by fire. The importance of termites as an agent of decomposition in dry savannas has perhaps been overstated. They may nevertheless have important ecological effects on the spatial redistribution of litter and nutrient elements.

Ants

The average ant biomass over the period of a year in the broad-leafed savanna at Nylsvley was 0.167 g DM m^{-2} (Kirsten 1978). Twenty-six species were collected from the canopies of the dominant tree species alone (*Burkea africana*, *Terminalia sericea* and *Ochna pulchra*: Grant 1984). The ants include predacious, seed-, leaf- and litter-eating species. Annual forage consumption for all ant types was calculated to be 8.5 g m^{-2} y^{-1}, of which 62% was due to ants of the genus *Camponotus*. About two-thirds of the forage was dry matter, including dead insects, leaf litter, seeds and faeces. The rest was honeydew collected from aphids. This suggests that ants are at least as important as litter-feeding termites as agents of comminution and redistribution of litter in this savanna. However, given the six-fold difference in biomass between ants and litter-feeding termites, this seems unlikely. The ant foraging rate given by Kirsten (1978) is probably an order of magnitude too high.

The larger soil-dwelling ants, were estimated to have a biomass of 760 mg FM m^{-2}, and a density of 800 m^{-2} (R. J. Scholes, unpublished data). Small soil-dwelling ants (Hymenoptera) are technically mesofauna, but will be included here for convenience. Ueckermann collected them while sampling for acarid mites. Their average biomass was 11 mg FM m^{-2} and they maintained densities around 300 m^{-2} throughout the year.

The effects of soil macrofauna on ecosystem function

Ferrar (1982c) indexed the activity of litter-feeding termites in various subhabitats by recording the rate of mass loss from toilet-roll bait. Both the rate at which the rolls were located, and the rate at which they were consumed, were consistently higher in the fine-leafed savanna than the broad-leafed savanna, despite the former having a litter-feeding termite biomass only one third that of the latter. The dominant species in the fine-leafed savanna, *Microtermes* sp. (probably *albopartitus*, a fungus-grower), was a much more active forager than the *Microcerotermes brachygnathus* which dominated in the broad-leafed savanna. Wood & Sands (1978) report that fungus-growers forage at about twice the rate per unit termite biomass than non-fungus-growing litter feeders. The patchy attack pattern on the baits suggested that the minimum foraging radius around a nest was about 5 m. The seasonal pattern of foraging differed between species: *Microcerotermes* was active all year, while *Microtermes* foraged only in summer. Foraging in a recently burned patch of broad-leafed savanna was sharply reduced for 2 months after the fire, until a grass cover had redeveloped. The reduction was not apparently attributable to loss of litter (the fire was cool) or to termite mortality. It was attributed to the increased risk of predation on the exposed ground surface. Frost (1987) showed that both Pallid and Black Flycatchers concentrated their feeding activity on burnt patches for some weeks after the fire. Their feeding behaviour suggested that termites were the main prey.

Termite foraging concentrates material, collected over a wide area, into the termite nesting galleries. Vegetation and soil fertility changes associated with termite nests have been widely reported from African savannas (Wood 1976; Malaisse 1978; Griffioen & O'Connor 1990), and are exploited by subsistence farmers. The locally enhanced crop growth on the sites of old termite nests is easily visible from the air in fields which have been in cultivation for many years in the Nylsvley region. In the large mound-building termites (e.g. *Microtermes*) part of this enrichment is due to the fine mineral particles brought up from the subsoil for nest construction, and cemented with saliva. Large mounds at Nylsvley are restricted to the fringe of the *vlei* (low-lying hydromorphic grassland on vertic clay) where clay is present in the subsoil. Nests of litter-feeding termites contain organic matter in the form of forage reserves, fungus gardens and faeces. Two small *Trinervitermes rhodesiensis* mounds excavated by Ferrar (1982b) contained 3 and 16 g of litter respectively, mostly of grass and tree-leaf origin. The most commonly reported fertility change associated with termite mounds is an increase in the concentration of basic cations. This is probably the result of the wicking effect of water evaporated in the mound ventilation system, rather than biological accumulation.

In the sandy upland regions of the Nylsvley site, only the smaller mounds of *Cubitermes* are found. Ferrar (1982b) found that seven other termite species and two ant species shared the nests with *Cubitermes*. The density of *Cubitermes* nests is 380 ha^{-1}, with a mean mass of 5.3 kg each (SE = 0.95, $n = 11$; Ferrar 1982b). *Cubitermes* is a soil-feeding termite, and the gallery walls of their nests are made up of their faeces, which is a mixture of soil and resistant organic matter. The nests make up about 0.07% of the topsoil (0–30 cm) mass, and have a slightly higher organic carbon content than the bulk soil. Since the termites must excrete less carbon than they ingest in order to live, the higher concentration of carbon in the wall than in the soil is further evidence that most of the energy requirement of soil-feeding termites comes from the digestion of small fragments of litter in the soil rather than humified soil organic matter.

Methane production by termites

Termites and ungulates both rely on anaerobic gut symbionts to achieve the digestion of cellulose in their diets. Consequently, both groups of organisms produce methane as a byproduct of digestion. Methane has recently become prominent as an important 'greenhouse gas': that is, one responsible for the maintenance of the global surface temperature. The atmospheric concentration of methane is increasing at 0.8% per annum, leading to speculation about its major sources and sinks. Since termites and ungulates are both predominantly found in savannas, there is a suspicion that the vast savannas of the world could be responsible for part of the increase (Zimmerman *et al.* 1982). A study at Nylsvley indicated, contrary to expectations, that the broad-leafed savanna was a net sink for methane rather than a source (Seiler, Conrad & Scharfe 1984), due to the methane-consuming properties of the soil. Termites are not currently believed to be major global methane sources (Rasmussen & Khalil 1983; Collins & Wood 1984; Zimmerman, Greenberg & Darlington 1984).

In the Seiler *et al.* (1984) study, the flux of methane from termite nests was highly correlated with the carbon dioxide flux, and ranged from 0.0067 to 0.87% (g C per g C) of the CO_2 flux, depending on the termite species. There was a diurnal pattern of methane and carbon dioxide efflux, with a minimum in the early morning and a maximum in the afternoon, but little variation between days. By assuming an assimilation efficiency of 35% for termite populations, and a global organic matter consumption by termites of 7.3×10^{15} g DM y^{-1}, Seiler *et al.* (1984) calculated a global methane flux due to termites of $2\text{–}5 \times 10^{12}$ g CH_4 y^{-1}. This is nearly two orders of magnitude lower than the Zimmerman *et al.* (1982) estimate.

When Seiler *et al.* (1984) measured the atmospheric methane exchange in a cuvette placed over the soil away from a termite mound, however, they found a net decomposition of methane by the soil. The average rate was 52 μg CH_4 m^{-2} h^{-1}, or 455 mg CH_4 m^{-2} y^{-1}, comparable to values reported for other ecosystems (Mosier *et al.* 1991). If this rate is representative of the world's savannas, then the methane decomposition by savanna soils exceeds methane production attributable to termites by a factor of 5–10 times. Applying the methane : carbon dioxide ratio measured by Seiler *et al.* (1984) to the termite foraging rate measured at Nylsvley (about 14 g DM m^{-2} y^{-1}) indicates a methane efflux from termite nests of 4 to 10 mg CH_4 m^{-2} y^{-1}, which is two orders of magnitude less than the rate of methane decomposition in the soil.

The second potential methane source in savannas is from the guts of ungulates. Calculations from data presented by Boomker & van Hoven (1983) for *in vitro* digestion, using kudu rumen fluid from Nylsvley, indicate that 9 mg CH_4 is liberated per gram of forage consumed, assuming 65% digestibility. Therefore, given an ungulate forage intake of 22.3 g DM m^{-2} y^{-1}, a rough estimate of methanogenesis from this source is 200 mg CH_4 m^{-2} y^{-1}. Thus, ungulates are likely to be a significant methane source, but the savanna as a whole remains a net sink.

There are several groups of methane-consuming bacteria: in aerobic soils the main methanotrophs appear to be the same bacteria that mediate nitrification. Grassland and forest soils fertilised with nitrogen change from being sinks to being sources of methane (Steudler *et al.* 1989; Mosier *et al.* 1991). The dominance of the mineralisation and nitrification processes in the nitrogen cycle of the broad-leafed savanna (Chapter 7) could explain the net methane decomposition measured there.

The dung-associated fauna

Several termite species include dry dung in their diet, but consumption of fresh dung is the speciality of the dung beetles (subfamilies Scarabaeinae and Aphodiinae of the Scarabaeidae: Coleoptera). Dung varies greatly in composition, consistency and the size of the pellets. The diversity of African ungulates is paralleled by a diversity of dung-associated insects. For instance, the genera *Onthophagus* and *Aphodius* are among the largest in the animal world. There are 75 species of dung-associated beetles at Nylsvley (Endrödy-Younga 1982). They are representative of the dung fauna of dry southern and East Africa, rather than that of wet tropical Africa. Dung beetles are active only in summer, and this is reflected in the seasonal course of the rate of dung removal (Figure 11.1). The rate of dung removal from naturally dropped faecal pellets is higher than indicated in this experiment: in summer, virtually 100% of pellets are consumed (Endrödy-Younga 1979). Both temperature and rainfall influence the activity patterns and population dynamics of dung beetles. At Nylsvley, dung-beetle species producing two generations per year are more than twice as common as those with only a single generation, and there is one species which shows three peaks of adult abundance within one year. The generation time of dung-beetles at Nylsvley varies from 2 to 12 months.

How can the coexistence of so many species with an apparently similar niche be explained? It appears that dung beetles are rather eclectic eaters. The species differ in the type and age of dung which they prefer, their mode of feeding, and seasonal activity peak. Forty-six taxa showed a marked preference for cattle dung baits over human faeces, rotten banana and decaying meat. There was considerable overlap in the species attracted to the human faeces, decaying meat and, to a lesser extent, rotten fruit. The Nylsvley genera are almost equally distributed between the three basic patterns of dung feeding. 'Diggers' burrow under the dung-pat, and carry portions of the dung down into larval chambers constructed in the soil below it. Dung-rollers form a portion of the pellet into a ball and roll it to a tunnel excavated some distance away. The remaining species feed at the soil–dung interface. Dung-rollers are responsible for about 15% of the dung removal, and typically reject pads older than 2 days. Those species which colonise the interior of the pad will persist as long as the interior remains moist.

There are no available dung fauna biomass data for Nylsvley, nor any estimate of the amount of dung that they redistribute. A rough calculation based on the mammalian herbivore density indicates an annual flux of about 10 g DM m^{-2} y^{-1} of dung, of which about 80% is redistributed by dung beetles (although a smaller proportion is actually consumed by them). The rate of dung deposition in the fine-leafed savanna is several times higher than

Figure 11.1. Seasonal pattern of dung removal by dung beetles. Several one-litre fresh cattle dung samples were exposed at each time, and the removal after three days was expressed as a percentage (data from Endrödy-Younga 1982).

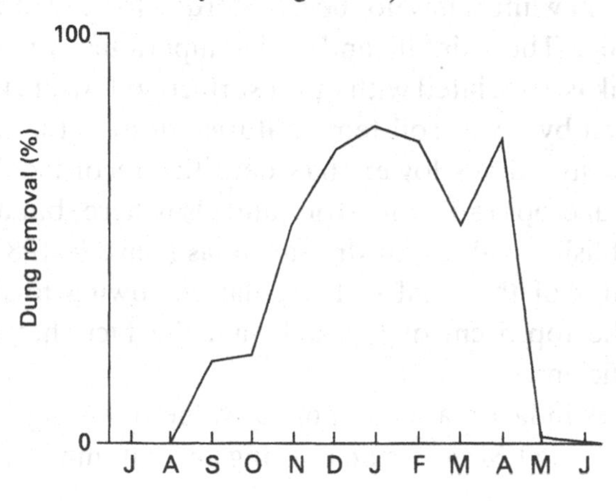

this value, due to the preference of ungulates for the fine-leafed patches. This is reflected in a greater biomass of dung-beetle larvae in the soils of the fine-leafed savannas.

The soil mesofauna

The soil mesofauna at Nylsvley is dominated by mites (Acari) and Collembola. The research conducted on soil mites at Nylsvley was primarily taxonomic. It resulted in the description of five new species and a new genus (Olivier 1976; Olivier & Theron 1988, 1989a, b). Comparatively little is known about the ecology of these animals in savannas. Feeding studies and inferences from the structure of the mouthparts indicate that they include fungivorous, algivorous, saprophytic, nematophagic, omnivorous, predatory and parasitic species (Theron 1974; Loots 1974; Walter 1988). Ueckermann (1978) sampled the Acari from the top 5 cm of soil and litter in the broad-leafed savanna at Nylsvley over the period of a year. Forty 6.5 cm diameter undisturbed cores were taken each month, and the Acari driven out, through a 1.5 mm mesh, using a heat gradient. The number and biomass of mites, small insects (predominantly Collembola and Hymenoptera), spiders and Pseudoscorpionidae were recorded.

The Acari in the litter and soil of the broad-leafed savanna at Nylsvley are dominated by the Trombidiformes (also called the Prostigmata), which are widespread in arid areas, and thought to be a primitive group. Thirty-two families were recorded. The Nanorchestidae were the most prominent, with *Nanorchestes* dominant in summer, and *Speleorchestes* in winter. The Oribatae (Cryptostigmata) were the next most prominent group (17 families), followed by the Mesostigmata (5 families) and the Astigmata, which were rare.

The seasonal pattern of mite density is shown in Figure 11.2. The double peak, one in summer and one in winter, has not been recorded for any other soil faunal group at Nylsvley. The rainfall and soil temperature record suggests that the summer peak is associated with moist surface soil, while the winter peak may be permitted by lower soil temperatures, despite the dry soil. The densities are about five times lower than densities recorded for grassland, forest and perennial crops in South Africa and elsewhere, but are in broad agreement with published values for dry savannas (Lavelle 1983). The values are underestimates of the total soil populations owing to the restriction of sampling to the top 5 cm of the soil, and the fact that no extraction method is fully efficient.

Biomass estimates for mites indicate a mean annual value of 48 mg FM m^{-2}, with a summer peak of 91 mg FM m^{-2}, and a spring and autumn low of around 20 mg FM m^{-2}. Even assuming that these estimates are half the true biomass, there is a maximum of 0.06 g DM m^{-2} of mites, or 2% of the termite biomass. It is therefore difficult to imagine a major role for Acari in the decomposition of litter at Nylsvley, despite the higher metabolic rate found in smaller organisms. Walter (1988) suggests that the Nanorchestidae are fluid feeders, and therefore probably suck out the contents of fungal hyphae and soil algae. Using the inferred feeding classes reported in Ueckermann (1978), about 56% of the acarid biomass at Nylsvley consists of phytophages (including fungiphages), 19% are predatory or parasitic, and the remaining 25% are omnivorous or have unknown diets.

Apart from the mites, the other small arachnids in the soil are the spiders and pseudoscorpions. Both groups are predatory, and contribute an annual mean of 12 and 24 mg FM m^{-2}, respectively, to the soil fauna biomass. The larger soil-dwelling spiders, not sampled in this study, are important predators in this ecosystem (Heidger 1988). The Collembola are arthropods living on decaying organic matter. They occurred in much smaller numbers than the Acari (the peak was 1765 m^{-2} in early summer), but due to their larger body size, the mean annual biomass of 107 mg FM m^{-2} was double that of the mites.

The soil microfauna includes protozoa and small nematodes, and can be enumerated using plating and extraction techniques. A nematode study was initiated at Nylsvley, but not completed.

Figure 11.2. Seasonal pattern of mite density in the broad-leafed savanna at Nylsvley. The sampling method changed slightly after the March sample (data from Ueckermann 1978). ———, total for all taxa; –·–, Nanorchestidae; – – – Oribatae; · · · · ·, Mesostigmatae.

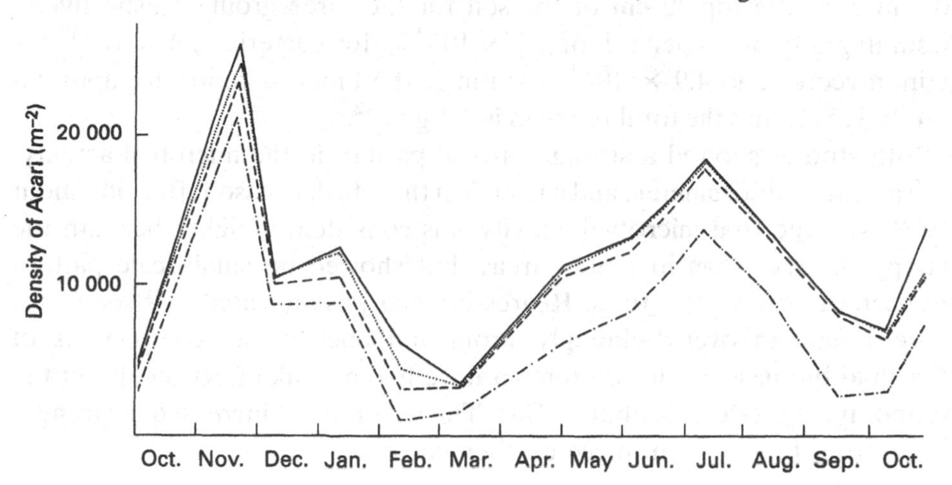

The soil microbes

The soil microbial biomass is not a reliable indication of the activity of soil microorganisms, and furthermore varies greatly according to the method used in its estimation. This is illustrated by comparing the values reported by Bezuidenhout (1978) with those obtained by Mary Scholes (unpublished data). Bezuidenhout (1978) applied both direct methods (plate counts) and indirect methods (soil respiration and soil ATP assays) to the estimation of the soil microbial biomass. He calculated it to be 1–2 g DM m^{-2}, with reasonably good agreement between the indirect methods, but poor correlation between indirect methods and direct counts. Mary Scholes used a single-fumigation technique (Anderson & Ingram 1989) to estimate the microbial biomass in the subcanopy habitat of the broad-leafed savanna. During the dry season it was steady at 10 g C m^{-2} (about 20 g DM m^{-2}), and the peak during the wet season reached 200 g C m^{-2} (about 400 g DM m^{-2}).

The Bezuidenhout (1978) estimates are very low by comparison with other ecosystems, while the later estimates agree with the rule-of-thumb that the microbial biomass contributes about 5% of the soil organic carbon in most soils. Despite the low biomass reported in the Bezuidenhout (1978) study, the short generation time of microbes can result in a high microbial productivity. He calculated the productivity to be between 40 and 200 g DM m^{-2} y^{-1}. Given the likelihood that his biomass estimates were too low, the microbial productivity is probably nearer to the larger of these values. This is possible since about 80% of the primary production of 950 g m^{-2} y^{-1} is processed by the soil microbial biomass.

The direct counts done by Bezuidenhout indicated that the soil microbial propagules were dominated by bacteria, followed by actinomycetes and fungi. Counts in the between-tree subhabitat averaged 145.2, 61.3 and 0.6×10^{10} m^{-2} in the top 30 cm of the soil for the three groups, respectively. Assuming a biomass per cell of 2.5×10^{-13} g for bacteria, 2.4×10^{-12} for actinomycetes and 4.9×10^{-11} for fungi, the biomass ratios are approximately 1:5:1, and the total biomass is 2.1 g m^{-2}.

Both studies showed a strong seasonal pattern in the microbial activity, with a peak in midsummer, and a trough in the late dry season. Bezuidenhout (1978) showed that microbial activity was consistently higher beneath the canopy of trees than in grassy areas, but showed no small-scale pattern between or below grass tufts. Regression analysis indicated that soil water content had an overwhelmingly strong influence on all components of microbial biomass, while soil temperature had a weak effect on all but the Actinomycetes (Bezuidenhout 1978). The water effect increased in strength in the order fungi, Actinomycetes, then bacteria.

Controls on the decomposition rate of leaf litter

There are two widely used methods of estimating the decay rate of leaf litter (Woods & Raison 1982). The first assumes that litterfall is relatively constant between years, and relates the mean amount of litter on the soil surface to the average annual litterfall to obtain a mean litter turnover rate. The second places known amounts of newly fallen litter of a given type into coarse-mesh bags, and monitors the loss of mass of the bag and its contents over time.

Accumulations of tree-leaf litter on the soil surface are very obvious in the broad-leafed savanna, despite the occasional fires. The litter is patchily distributed, with the average value below tree canopies about three times higher than that between canopies. The leaf litter standing crop has been variously quoted as 1242 (SE = 57, n = 4: R. J. Scholes & M. C. Scholes, unpublished data), 304 (Grunow *et al.* 1980), 970–1340 (Bezuidenhout & Morris 1978) and 1011 gm^{-2} (Huntley & Osborne 1977, quoted in Morris *et al.* 1982). The variation is due to differences in the location and date of collection, and the definition of the lower size threshold for litter. The material in the litter layer is 90–95% of woody plant origin.

It is clear that the litter standing crop represents the accumulation of several years of litterfall, since the average annual leaf, flower and fruit litterfall is 150 gm^{-2}. Given this rate of litterfall and a standing crop of 1200 gm^{-2}, and assuming a first-order exponential decay function ($M_t/M_0 = e^{-kt}$) and a steady-state system, the value for the bulk decomposition constant, k, is 0.13 y^{-1}. This is equivalent to a litter half-life of 5.2 years. The litter inputs and the litter standing crop in the broad-leafed savanna are not strictly at equilibrium, since the woody biomass is gradually increasing and the periodic fires remove part of the accumulated litter mass.

The rate of decay is three to five times higher for grass litter than for tree litter in the broad-leafed savanna. Using litter bags with a mesh size of 2 mm, Bezuidenhout (1980) found that the litter of the two dominant grasses lost about 60% of their mass in the first year, while the tree litter lost only 11–26%, depending on species, over the same period (Table 11.2). The various chemical constituents of the litter (sugars, celluloses, hemicelluloses, lignin, protein and potassium) decayed at different rates, resulting in a relative increase in the proportion of lignins with time. Potassium was lost much faster from grass litter than from tree litter. R. J. Scholes & M. C. Scholes (unpublished data) used litter bags with a 5 mm mesh and obtained similar results to those given by Bezuidenhout (1980) for grass (*Digitaria eriantha*) and tree (*Burkea africana*) litter. The decay rate of all the species measured is low in relation to rates recorded in terrestrial ecosystems in general (Singh &

Table 11.2. *The chemical composition and decomposition rates of seven leaf litter types from the broad-leafed savanna at Nylsvley (Bezuidenhout 1980, Cooper 1985)*

Species	Decay constant (K) year^{-1}	Percentage of dry mass						
		Nitrogen	Sugar	Hemi-cellulose	Cellulose	Lignin	Total phenolics	Condensed tannin
Trees and shrubs								
Burkea africana	0.132	1.78	7.5	6.6	64.4	19.6	11.8	5.8
Ochna pulchra	0.123	1.36	7.8	8.6	62.7	18.9	10.9	11.0
Combretum molle	0.349	1.37	7.5	2.1	67.1	20.6	18.4	1.0
Terminalia sericea	0.231	1.30	8.0	5.3	75.8	10.3	22.4	12.3
Grewia flavescens	0.323	1.93	3.7	6.6	73.2	14.3	4.7	3.0
Grasses								
Eragrostis pallens	1.044	0.66	4.7	25.4	51.6	17.1		
Digitaria eriantha	1.178	0.66	3.0	28.9	57.6	9.4		

Gupta 1977; Morris 1983). Decomposition at Nylsvley is about eight times slower than that measured in infertile savannas in Zimbabwe receiving 930 mm rainfall per annum (Peter Frost, personal communication) and 20 times slower than reported for *miombo* woodland in Zaïre (Mallaise *et al.* 1975), which receives 1200 mm per annum. The slow decay rate is attributed to the combination of a dry climate, which allows decomposition to progress only sporadically, and a decomposition-retarding leaf litter chemistry.

Bezuidenhout (1980) found a marked seasonal pattern in microbial biomass (as measured by ATP content) and respiration from litter organisms, but little correlation between these activity measures and the number of microbial propagules. In one-year-old litter, the ratio of actinomycete to bacterial propagules was 1:1000, while fungi:bacteria propagules was 1:8. This represents far fewer actinomycetes, and many more fungi, than is typical for topsoil. Eighty percent of the variation in litter respiration rate measured in the field could be explained by changes in the litter water content.

A simulation model was constructed to account for the differences in decay rate between litter types (first version, Morris *et al.* 1978; final version Furniss *et al.* 1982). It was based on the principle that the various chemical constituents of the litter (sugars, hemicellulose, cellulose and lignin) decay at different rates. The differences between litter types were therefore attributable to their different initial chemical composition. The decay rates are controlled by the moisture content and temperature of the litter, which are inferred from climate data, and are influenced by the size of the litter particles. The model traces seven chemical components in four litter types through three age classes and three particle size classes. The action of soil fauna is included through simulation of the effects of litter-feeding and soil-feeding termites.

This model was never validated against field data, but provided several insights into the decomposition process. Adjusting the various estimates of comminution and decay rates had little long-term effect on the soil organic matter (SOM) pool, since this was controlled by the proportion of decomposition-resistant (lignin and lignin-like polymers) material in the litter. Allowing the soil-feeding termites partially to digest this resistant material increased the SOM C:N ratio and decreased the amount of SOM. The observed SOM pool of about 5000 g C m^{-2} is incompatible with the 800-year mean turnover time for SOM assumed in the model, which was based on data from temperate grasslands. Furniss *et al.* (1982) suggest that a mean SOM turnover time of around 300 years would be more appropriate for this savanna.

Several indices have been proposed for predicting the decomposition rate of leaf litter from measurements of the leaf chemistry: for instance, the C:N ratio or the lignin:nitrogen ratio. The litter decomposition data from Nylsvley do not conform to these simple indices. The slowly decaying tree-leaf litter has a relatively high nitrogen content, since it is of legume origin. It also has a high lignin content, but then so does *Eragrostis pallens*, a fast-decomposing grass. It appears necessary to include decomposition-retarding substances in the predictor equation as well as decomposition-enhancing factors. Combining the data on carbon fractions from Bezuidenhout (1980) with the tannin, protein and phenolic data from Cooper (1985) allows the construction of an index in which all the readily decomposable factors are in the numerator, and the resistant or retardant factors are in the denominator:

$$\text{Decay Rate Index} = \frac{(\text{sugar} + \text{hemicellulose} + \text{protein})}{(\text{lignin} + \text{condensed tannin})}$$

All the values in this equation are expressed in g g^{-1}. The litter decomposition rates are illustrated as a function of the index in Figure 11.3.

Figure 11.3. The decomposition rate constant, *k*, of several litter types from the broad-leafed savanna as a function of a decomposition index based on leaf chemistry (details in the text). The factors in the numerator of the index accelerate the decay rate, while those in the denominator slow it down. All are expressed as percentage dry weight of the litter (data from Bezuidenhout 1980).

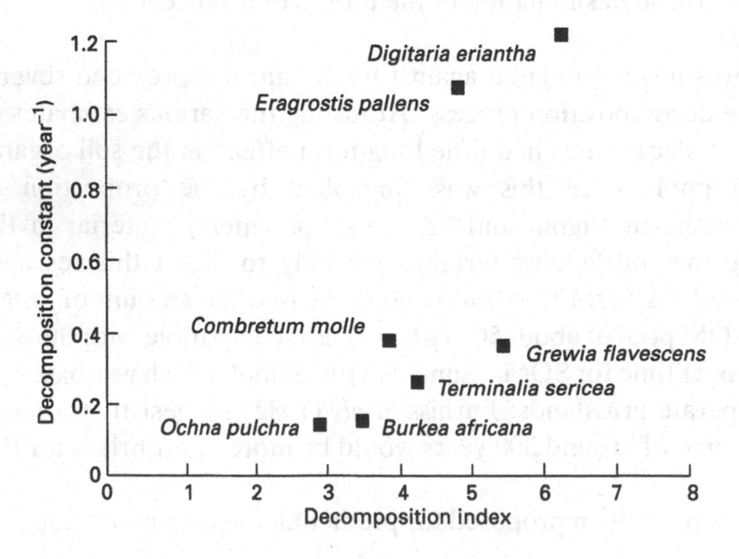

Decomposability and digestibility

Since decomposition and digestion in the ungulate rumen are both microbially mediated processes, it is reasonable to expect that leaf types which are slow to decompose are probably relatively indigestible as well. Comparing the seasonal acceptance values given by Cooper (1985) for the leaves of woody plants with the *in situ* decomposition rate data of Bezuidenhout (1980) indicates that the species with the most recalcitrant litter (*Burkea africana* and *Ochna pulchra*) are not favoured by kudu, impala or goats, while the most rapidly decaying leaves (*Combretum molle* and *Grewia flavescens*) are highly favoured by browsers. *Terminalia sericea* is intermediate in both cases.

This raises an interesting question: did the leaf constituents which retard digestion evolve for that purpose, or did they evolve to protect the leaf against microbial attack, either before or after leaf fall? It has been suggested (Zucker 1983) that slowly decomposing leaves are an advantage in nutrient-poor environments, since they reduce the possibility of nutrient loss due to rapid decomposition early in the rainy season, before the plants are able to take them up. There are three main weaknesses to this argument. First, altering the decay rate only reduces the nutrient release rate temporarily. Once a sufficiently large litter layer has accumulated, the overall release rate is once again high, although it is likely to be better spread over time. Secondly, the accumulated litter is susceptible to oxidation by burning, which also represents a significant avenue of loss. Thirdly, nutrient retention is a system-level property, unlikely to be selected for at the level of individual organisms. It is more likely that the benefits of secondary chemicals are gained while the leaf is still on the plant.

Condensed tannins have been shown to limit the intake of browse by ungulates at Nylsvley, but it is ultimately the deciduousness of the vegetation which constrains the numbers of browsing ungulates, and therefore their impact. Condensed tannins, being bound to the cell wall, are not considered to affect insect herbivores, which cannot digest the cell wall material. They should reduce cellulose digestion by termite gut symbionts, which may explain the relatively small proportion of the litterfall which is consumed by termites in the broad-leafed savanna.

A variety of microbial pathogens has been observed on live leaves in the broad-leafed savanna (Rey 1982), and have been shown to reduce the photosynthetic potential of the leaf. The overall impact on primary production is negligible, since the proportion of leaf area affected is relatively small. Could protection of the living leaf against microbial pathogens be the main role of condensed tannins in savannas?

The pathways of decomposition

A crude model of annual litter accumulation and decomposition was constructed, using the data presented in this chapter, Chapter 8 and Chapter 10. If assumptions are made regarding the frequency of fires and the proportion of the litter layer which is consumed by a single fire, it is possible to estimate the fraction of the litterfall which is oxidised by microbes, soil fauna and fire. When litter nitrogen contents are included in the model, and further assumptions are made regarding the proportion volatilised in a fire, the nitrogen losses through burning can be estimated.

The results of this exercise show that for a 5-yearly fire regime and 50% litter layer combustion (both reasonable estimates for Nylsvley), only 13% of the litter is oxidised by fire, and 2% by soil fauna. The remaining 85% passes through the microbial decomposition pathway. These estimates are not substantially changed by altering the assumptions within a realistic range: for instance, a 3-year fire regime with 100% combustion results in 22% oxidation by fire and 75% by microbes.

The direct loss of nitrogen as a result of burning is proportionately lower, since only part of the nitrogen is lost from the burned portion of the litter. The magnitude of the loss depends on the fire intensity. One of the reasons for the relatively low pyrogenic losses of carbon and nitrogen is that a large proportion of the 'litterfall' occurs belowground, where it is protected from fire. This has implications both for the nitrogen cycle and for the accumulation of soil carbon. Burning at a typical frequency for dry savannas (triennially or greater) is unlikely to deplete the system nitrogen pool, or significantly reduce the soil carbon content. This finding is supported by long-term burning trials in savannas (Moore 1960; Trapnell *et al.* 1976; Jones *et al.* 1990).

Summary

The decomposition rate of plant litter at Nylsvley is very slow: it has a mean turnover time of about 5 years. This is partly because of the dryness of the climate, and partly because the litter contains secondary chemicals which make it resistent to decay.

About 80% of the primary production at Nylsvley passes to the decomposer organisms, rather than to herbivores. Of this, less than 2% is processed by termites, which are the most active component of the soil macrofauna. The soil microbes perform almost all the decomposition, with comminution apparently playing a small role. About 10% of the primary production is oxidised in fires.

We have already argued that primary production in the broad-leafed savanna at Nylsvley is constrained by the rate of nutrient assimilation (particularly nitrogen), rather than carbon assimilation. The process which controls the rate at which nitrogen becomes available to the plant is decomposition. Therefore it is the rate at which organic compounds can be disassembled which limits the primary production at Nylsvley, not the rate at which they can be assembled.

Part IV

Community and landscape pattern and change

12

Rich savanna, poor savanna

Ecologists have traditionally subdivided the African savannas into functionally different types along a moisture gradient: wet savannas, mesic savannas and arid savannas, for example. The species composition, vegetation structure and key ecological processes in an arid savanna are clearly different from those in a wet savanna, but there is a gradual transition between these extremes. The distinct changes in composition, structure and function which are apparent in the field are usually related to soil changes; which can lead to water availability differences, or fertility differences, or both. On a continental scale, the moisture gradient and the soil fertility gradient are correlated to a large degree: areas of high rainfall have leached, infertile soils; while arid regions tend to have relatively fertile soils. For this reason, the first level of subdivision of African savannas proposed by Huntley (1982) is between the arid/eutrophic types on the one hand, and the moist/dystrophic types on the other. While this is a valid scheme in a general sense, it tends to obscure the fact that both arid dystrophic and moist eutrophic savannas do occur, although they may not be large in extent.

Scholes (1990b) argued that the changes caused by soil fertility are both more discrete and more fundamental than the changes caused by water availability. It is therefore more useful to distinguish between fertile and infertile, rather than between wet and dry, as a first level in the ecological classification of savannas. Much of the motivation for that argument was provided by a comparative study of fertile and infertile savannas at Nylsvley.

Thus far, this book has concentrated on the infertile broad-leafed savanna at Nylsvley, dominated by *Burkea africana*. This chapter contrasts the ecology of the broad-leafed savanna with that of adjacent fertile, fine-leafed savanna patches dominated by *Acacia tortilis*. Since both occur under the same climate and on soils derived from the same parent material, the comparison is not confounded (as it is in most African examples) by textural or water availability differences. The contrast between the nutrient-poor broad-leafed savanna at Nylsvley, and the nutrient-rich fine-leafed savanna derived from it, formed the basis of an experiment in the Tropical Soil Biology and Fertility Programme (TSBF: Ingram & Swift 1989).

The origin of the nutrient-rich patches at Nylsvley

The fine-leafed savanna patches at Nylsvley are discrete and easily recognised on the ground or from aerial photographs. They occupy about 17% of the land surface of the study site, and are arranged in a semicircle around the low hill, Maroelakop (Blackmore, Mentis & Scholes 1990; Figure 12.1). The dominant trees of the fine-leafed savanna, *Acacia tortilis* and *A. nilotica*, and the characteristic grass, *Eragrostis lehmanniana*, are frequently associated with disturbance. Therefore the fine-leafed savannas were mapped as 'disturbed sites' in the initial phytosociological survey of the study site (Coetzee *et al.* 1976). It was soon noted that traces of previous inhabitation were everywhere in the fine-leafed patches, but rare in the broad-leafed savannas. This one-to-one relationship of archaeological remnants with fine-leafed savannas was confirmed in an extensive survey undertaken by Fordyce (1980).

Figure 12.1. Location of fine-leafed savanna patches within the broad-leafed savanna at Nylsvley (redrawn after Blackmore *et al.* 1990 and Walker *et al.* 1986).

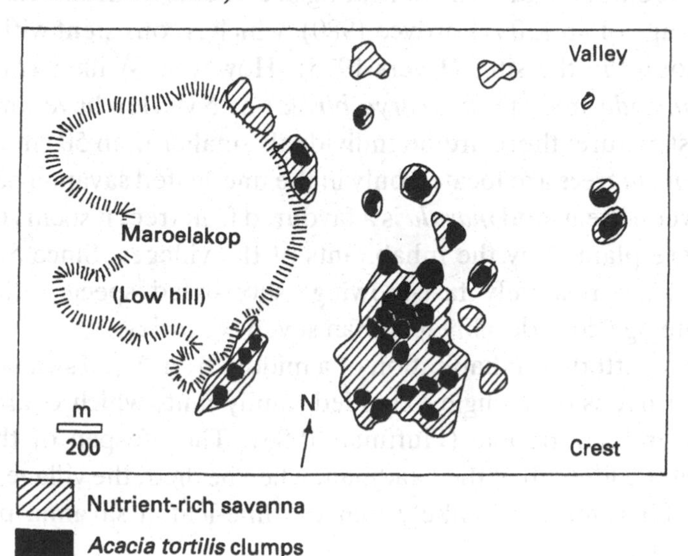

Two theories have been advanced to account for the origin of the fine-leafed savanna patches at Nylsvley. The first, attributed to Tinley (1981), proposes that the downslope movement of silt and clay particles, derived from mudstone inclusions in the Waterberg sandstone of Maroelakop, created fans of marginally richer soil around the base of the hill. Tswana tribespeople recognised the superiority of these areas for crop production, and settled on them, leading to further enrichment and changes in the vegetation. The second theory was that the Tswanas settled around the base of the hill for other reasons, such as the protection and vantage point it afforded. Their activities, such as firewood collection and the practice of herding the cattle into a central pen (*kraal*) at night for protection against predators, but allowing them to graze widely during the day, resulted in nutrient accumulation within the village site.

Blackmore *et al.* (1990) discounted the 'geomorphological' theory on the basis that some fine-leafed patches occur upslope from Maroelakop, or remote from any possible source of silt and clay enrichment. Furthermore, the increase in silt and clay content in the soils of the fine-leafed savannas is relatively slight, and could result from the mud which the villagers would have imported for construction of hut walls. They calculated that cattle penning and firewood collection were adequate mechanisms to account for the observed nutrient accumulations, and therefore the second theory could not be rejected on this basis.

The village sites have not been inhabited in living memory. There are no surface traces of hut walls or middens, as would be expected if they were occupied this century. Radiocarbon dating of charcoal from an excavated site indicated an age of AD 1300 (Fordyce 1980), which is consistent with the style of pottery found in the sites (Evers 1975). However, Walker *et al.* (1986) noted that *marula* trees (*Sclerocarya birrea*) at Nylsvley have an unstable population structure: there are no individuals smaller than 50 cm diameter. Since the *marula* trees are located only in the fine-leafed savanna patches, or in lines between them, and *marula* is a favoured fruit tree, it seems likely that the trees were planted by the inhabitants of the villages. Since *S. birrea* is thought to be a relatively fast-growing, short-lived species, this would indicate a site age considerably less than seven centuries.

The typical pattern of inhabitation of a middle Iron Age Tswana village is based on the needs of a single extended family unit, which could number between 35 and 115 people (Huffman 1986). The lifespan of the village depended on the lifespan of the headman; when he died, the village moved to a new site. Therefore, it is likely that the fine-leafed savanna patches at

Nylsvley resulted from many successive small villages over a long period of time, rather than a large settlement at a particular moment.

Differences in vegetation composition and structure

Typical views of the broad-leafed and fine-leafed savanna at Nylsvley are shown in Figure 12.2*a* and *b*. Figure 12.3 represents a transect from broad-leafed savanna, through a fine-leafed patch, and back into broad-leafed savanna.

The transition from the broad- to fine-leafed savanna occurs over a horizontal distance of only a few metres, particularly if gauged from the changes in the herbaceous layer (Figure 12.3*b*). The *Acacia* trees tend to form a clump in the middle of the patch, leaving a treeless band around the edge. The *Acacia* clumps appear to coincide with the location of the former cattle pens (which are centrally located in the village), since they are associated with the highest nutrient levels in the soil, traces of 'fossil' dung, and a compacted 'village horizon' in the soil profile. The herbaceous layer changes are well correlated with changes in the soil chemistry: the enrichment with phosphorus, calcium, magnesium and potassium is particularly clear (Figure 12.3*c*).

The woody plants in the fine-leafed savanna are lower, sparser and less diverse than those in the broad-leafed savanna (Table 12.1), and have a lower root:shoot ratio. The high root biomass in the broad-leafed savanna is mostly due to the contribution of the coarse roots (those greater than 2 mm in diameter), since the mean fine root biomass is similar in both types. The herbaceous layer is more diverse in the fine-leafed than broad-leafed savanna, since it includes a large number of forbs in the former. The herbaceous standing crop is usually lower in the fine-leafed savanna, despite 50% higher productivity, owing to the heavy grazing it supports.

The morphology of individual species typical of the two savanna types show consistent differences. The dominant trees in the broad-leafed savannas have leaves (or leaflets, in the case of compound leaves) with dimensions in excess of 2 cm. Few are armed with thorns or prickles. In the fine-leafed savanna, the dominant trees have leaflets about 2 mm long, and are conspicuously thorny.

Differences in productivity

The Tropical Soil Biology and Fertility programme 'Minimum Experiment' conducted at Nylsvley involved four pairs of 20 × 30 m study plots, with one member of each pair in broad-leafed savanna, and the other in

Figure 12.2. A comparison of the gross structure of (*a*) broad-leafed and (*b*) fine-leafed savanna at Nylsvley.

Figure 12.3. A transect through a fine-leafed savanna patch within broad-leafed savanna. (*a*) Woody plants: solid symbols are fine-leafed trees; (*b*) herbaceous layer indicator species; (*c*) soil chemistry (redrawn from Blackmore *et al.* 1990).

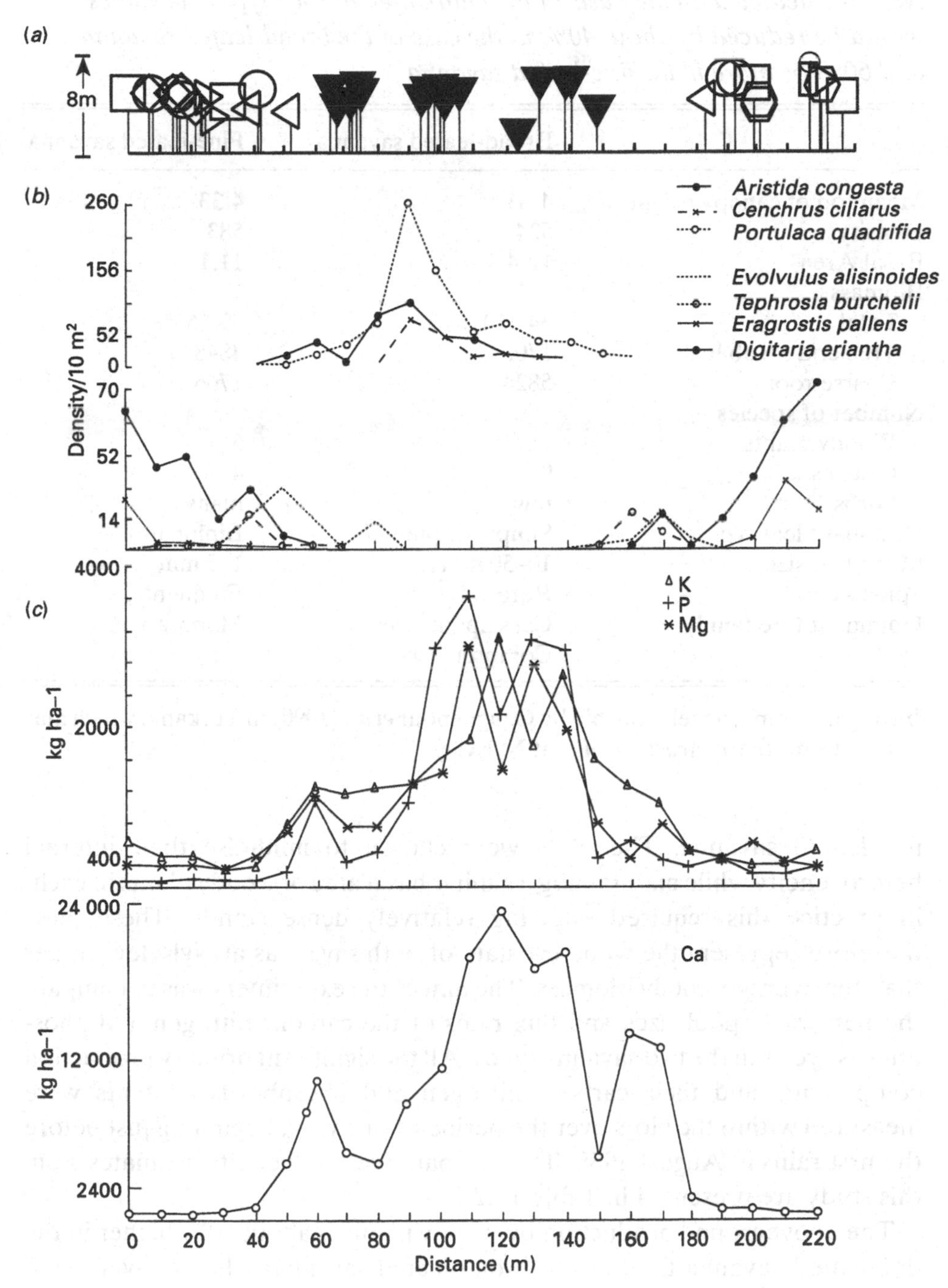

Table 12.1. *A comparison of the woody plant characteristics in a broad-leafed and fine-leafed savanna at Nylsvley, based on four 20 × 30 m quadrats in each type. The quadrats were located in the densest parts of each type; if calculated on the basis of the entire area of each type, the values would be reduced by about 40% in the case of the broad-leafed savanna, and 60% or more in the fine-leafed savanna*

	Broad-leafed savanna	Fine-leafed savanna
Mean top of canopy height	4.83	4.33
Density	524	583
Basal Area	12.4	11.1
Biomass		
Total	34 334	59 653[a]
Tree and shrub leaf	1706	4848
Coarse root	5824	1706
Number of species		
Woody plants	11	5
Grasses	9	4
Forbs	few	many
Dominant leaf type	Simple or pinnate	Bipinnate
Mean leaf size	10–50 mm	1–5 mm
Spinescence	Rare	Frequent
Dominant tree family	Caesalpiniaceae Combretaceae	Mimosaceae

[a]From an allometric relationship by Coughenour *et al.* (1990) in Turkana, which may not be accurate for *Acacia tortilis* at Nylsvley.

fine-leafed savanna. The plots were chosen to minimise their internal heterogeneity while maintaining a similar basal area of woody plants in each: in practice this required selecting relatively dense stands. These plots therefore represent the woodiest state of both savannas at Nylsvley, rather than the average woody biomass. The aim of the experiment was to compare the pathways, pool sizes and flux rates of the carbon, nitrogen and phosphorus cycles in the two savanna types. All the significant primary production components and their carbon, nitrogen and phosphorus contents were measured within the plots over the period of one year, beginning just before the first rains in August 1988. The comparative productivity estimates from this study are presented in Table 12.2.

The aboveground production by woody plants is about 50% higher in the fine-leafed savanna than in the broad-leafed savanna, while aboveground

Table 12.2. *Components of primary production over the period 1 October 1988 to 30 September 1989, in four pairs of fine-leafed and broad-leafed savanna plots. These plots represent the subcanopy habitat in both savanna types, and therefore have higher woody plant production, and possibly lower grass production, than the average for the savanna types*

	Broad-leafed savanna		Fine-leafed savanna	
	Mean	SD	Mean	SD
Aboveground production (g m^{-2} y^{-1})				
Tree leaf	265	38	399	89
Tree twig fall	57	85	81	34
Grasses & forbs	120	8	199	12
Belowground production (g m^{-2} y^{-1})				
Fine roots	497	59	493	66

herbaceous layer production is about 70% higher. Fine root production is not significantly different in the two savannas.

Biogeochemical differences

Both the broad-leafed and fine-leafed savannas are dominated by leguminous trees, but there is no evidence that they support nitrogen-fixing symbioses in either case. Nevertheless, the flux of nitrogen and phosphorus through the fine-leafed savanna system is substantially higher than through the broad-leafed savanna, as measured by the total nitrogen in all the litterfall fluxes (Table 12.3). This is partly because of the greater mass of litterfall in fine-leafed savannas, but is largely attributable to the higher mean nitrogen and phosphorus content in the litterfall in the fine-leafed savanna. This is particularly true for grass litter (Figure 12.4). The sustained high nutrient content of the grasses in the fine-leafed savanna, even when dead, goes a long way to explaining the concentration of herbivory on those patches.

The nutrient-enriched sites offer further circumstantial evidence that the savannas at Nylsvley are primarily nitrogen, rather than phosphorus, limited. The phosphorus pools in the fine-leafed savannas are many times higher than

Table 12.3. *The flux rates and pool sizes of nitrogen and phosphorus within the subcanopy habitats of fine-leafed and broad-leafed savannas*

	Nitrogen		Phosphorus	
	Broad-leafed savanna	Fine-leafed savanna	Broad-leafed savanna	Fine-leafed savanna
Pool sizes ($g\ m^{-2}$)				
Soil	306.2	340.6	88.2	120.3
Herbaceous plants	1.1	1.5	0.2	0.2
Woody plants	18.7	59.2	1.3	6.5
Litter	5.6	8.1	0.2	0.6
Total	331.6	409.4	89.9	127.6
Fluxes ($g\ m^{-2}\ y^{-1}$)				
Tree leaf litterfall	1.99	7.26	0.16	0.55
Herbaceous leaf turnover	0.29	1.02	0.06	0.18
Fine root turnover	5.72	6.26	0.21	0.33

Figure 12.4. Nitrogen content of live and recently dead grass leaves in the fine- and broad-leafed savannas, as a function of the time of year. A leaf nitrogen content of greater than 1% is regarded as the minimum for the maintenance of ungulate rumen fauna.

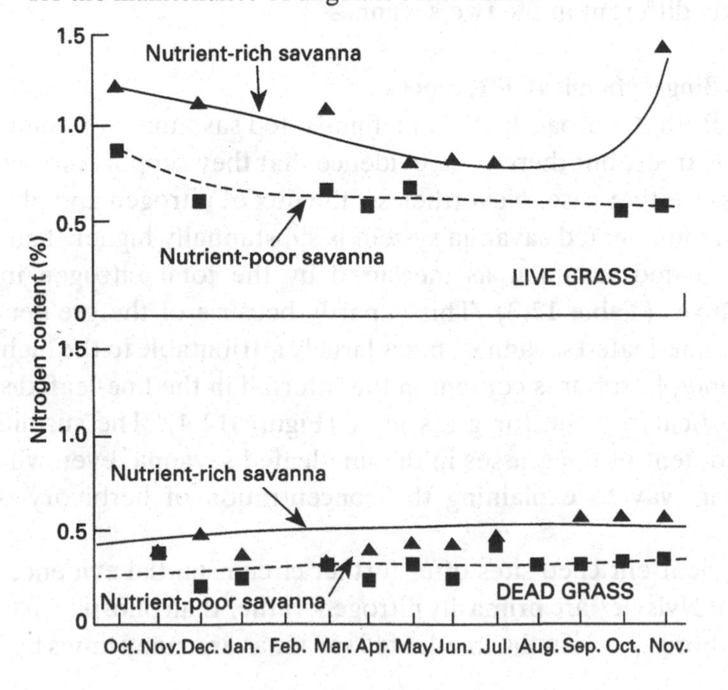

those in the broad-leafed savannas, but the fluxes through the vegetation are not enhanced to the same degree. The nitrogen fluxes, on the other hand, are increased to a greater extent than the nitrogen pools. Even in the broad-leafed savanna, phosphorus is not removed from dying leaves to the same extent as nitrogen.

Larger organic inputs lead to more soil organic matter, which results in higher rates of mineralisation. More inorganic nitrogen available for plant uptake permits greater production and higher tissue nitrogen contents. This is prevented from becoming a perpetual motion machine by the gaseous nitrogen losses which occur during nitrification, fires and leaching. It is inferred that the losses would be greater in the fine-leafed than broad-leafed savanna, since the absolute magnitude of the transformations is greater there. Given that the patches are certainly at least one century old, and could be as old as seven centuries, what prevents their nutrient accumulations from 'running down' to the levels sustained in the surrounding broad-leafed savanna? The answer to this question is believed to lie in the nutrient fluxes resulting from large mammal feeding behaviour.

Faunal differences

The ecological differences between the nutrient-rich and nutrient-poor savannas at Nylsvley are maintained at all trophic levels, despite the small scale of the patches relative to the potential foraging range of the organisms involved. Large ungulate herbivores such as cattle, impala and kudu feed in the fine-leafed savanna in preference to broad-leafed savanna in all months of the year, except when the latter has been recently burned (Carr 1976; Monro 1978; both reported in Gandar 1982c; O'Connor 1977; Zimmerman 1978). The degree of preference varies according to the season (Table 12.4), with the highest preferences for fine-leafed savanna occurring at the end of the dry season, and the lowest during the early wet season.

Gandar (1983) reported that the average biomass of phytophagous insects was 0.25 g m^{-2} in the fine-leafed savanna, versus 0.10 g m^{-2} in the broad-leafed savanna. In the fine-leafed savanna, 93% of this was contributed by the Acridoidea (grasshoppers), which made up 76% of the biomass in the broad-leafed savanna. Only the non-acridoidid Orthoptera and Lepidoptera had marginally greater biomasses in the broad-leafed areas; other insect groups were the same in both. The species composition in the two vegetation types differed for all insect groups. The species diversity of grasshoppers was higher in the broad-leafed savanna, because its insect fauna included species found in the fine-leafed savanna, but not vice versa. The grasshopper

Table 12.4. *Habitat preferences by impalas. The broad-leafed savannas make up about 83% of the study site, while the rest is fine-leafed savanna*

	Number of impala (km^{-2})	
	Broad-leafed savanna	Fine-leafed savanna
Seasonal trend (Monro 1980)		
January	16.5	2.4
February	9.2	6.5
March	32.1	59.5
April	29.1	268.5
May	17.4	119.6
June	59.7	302.9
July	33.4	355.1
August	73.9	225.0
September	29.4	365.0
October	26.3	170.6
November	70.8	169.3
December	27.4	67.0
September 1978 (Monro 1978, quoted in Gandar 1982a)		
Pre-fire	6.0	39.0
Post-fire burnt	51.0	2.0
Post-fire unburnt	20.0	0.0

population characteristics also differed: the fine-leafed savannas had more small individuals, and maturation occurred earlier.

Tarboton (1980) noted a higher bird diversity in broad-leafed than fine-leafed savanna (78 vs 63 resident species), but a lower density (5.86 vs 11.85 birds ha^{-1}) and biomass (0.0407 vs 0.0847 g m^{-2}). Both types were dominated by insectivores, but there was a clearly different species composition (Table 12.5). The species found in the fine-leafed savanna had biogeographical affinities with the arid savanna regions of southern Africa, while those of the broad-leafed savanna were typical of the avifauna of the moist, infertile *miombo* savannas of south Central Africa. The Nylsvley avifauna was faithfully reproducing, at a scale of a few hundred metres, their geographical distribution, at a scale of thousands of kilometres. This observation triggered the concept of the fundamental ecological distinction in African savannas being between the arid/dystrophic and the moist/eutrophic types.

Sue Frost (Frost, 1987) compared the feeding ecology of a pair of closely related flycatchers, one of which resides exclusively in the fine-leafed

Table 12.5. *The bird species resident (either seasonally or permanently) in broad-leafed and fine-leafed savanna at Nylsvley, and sporadically observed or absent from the other type. Data from Tarboton (1980)*

	Resident in broad-leafed savanna	Resident in both types	Resident in fine-leafed savanna
Raptors	0	0	2
Insectivores	17	18	34
Granivores	8	4	1
Fructivores	2	1	0
Mixed feeders	4	11	9
Total	31	34	46

savanna (the Marico Flycatcher, *Melaenornis mariquensis*), and the other exclusively in the broad-leafed savanna (the Pallid Flycatcher, *M. pallidus*). Food availability was higher in the fine-leafed savanna in the wet season, but similar for both savannas during the dry season (Table 12.6). The difference in preferred perch height is thought to be related to the difference in the height of the grass layer, which the birds need to scan for their prey. The Marico Flycatcher made more foraging attempts per minute than the Pallid, and waited on a given perch for a shorter period of time before giving up and moving to another perch. Although the mean prey size taken by the Pallid was slightly larger, the overall foraging rate remained higher for the Marico Flycatcher. Frost (1987) suggested that the Marico Flycatcher could therefore outcompete the Pallid within the resource rich patches, but that the more patient feeding behaviour of the Pallid was advantageous in the low-resource areas.

As part of the TSBF study, the soil fauna of the broad-leafed and fine-leafed savannas were compared. This study (Figure 12.5) confirmed the observation of Ferrar (1982a, c) that termites, particularly the soil-eating varieties, were much more prevalent in the broad-leafed than fine-leafed savanna. The pattern for ants was the other way around. The larvae of the Scarabaeinae (dung beetles) were also more prevalent in the fine-leafed savanna, probably reflecting the large amount of dung deposition which occurs there. The soil fauna biomass was higher in the fine-leafed savanna due to the large mean body mass of the organisms found there.

Table 12.6. *Differences in the feeding ecology of the Marico Flycatcher and the Pallid Flycatcher. These closely-related species are confined to the fine-leafed and broad-leafed savannas respectively. Data from Frost (1987)*

	Pallid Flycatcher	Marico Flycatcher
Habitat	Broad-leafed savanna	Fine-leafed savanna
Mass (g)	24	23
Resource availability (mg prey m^{-2})		
Wet season	8.8	77.5
Dry season	5.1	8.4
Grass height (cm)	6–25	2–15
Home range (ha)	10–22	2–3
Mean perch height (m)	2.0	1.5
Foraging rate (attempts min^{-1})	0.2–0.6	0.7–1.7
Mean prey size	Larger	Smaller
Median giving-up time (sec)		
Wet season	28	18
Dry season	46	22
Time spent foraging (%)		
Wet season	90.6	79.1
Dry season	96.7	95.4
Breeding season	Shorter	Longer
Behaviour	Retiring	Aggressive

Differences in the mode of defence against herbivory

The trees in the nutrient-rich sites tend to be armed with thorns, while those in the nutrient-poor sites are not. The leaves of trees in the infertile savanna tend to be unpalatable due to their chemical composition, while those in the fertile savanna are highly sought after by ungulate browsers, despite the thorns. Outbreaks of lepidopteran larvae are a conspicuous feature of the infertile savanna, but never seem to occur in the fertile savanna. How can these different patterns of herbivory and defence against herbivory be explained?

Coley *et al.* (1985) present a model which relates the degree of chemical defence against herbivory to the soil fertility status. The model was developed for boreal forest species, but could equally be applied to savanna trees. They suggest that plants adapted to low-nutrient soils are slow-growing, with inherently low rates of photosynthesis and nutrient uptake. They cannot afford to lose leaf tissue or nutrients by herbivory. The optimum allocation pattern for such plants includes a substantial investment in defensive chemicals, despite the opportunity cost of doing so. Plants growing on fertile substrates, on the other hand, are fast-growing and have high rates of photosynthesis and nutrient uptake. With a plentiful supply of nutrients and carbon, the optimum solution is to sacrifice some tissue to herbivory, rather than invest in energy- and nutrient-demanding defensive compounds. As Cooper & Owen-Smith (1986) showed, the presence of spines functions not to prevent herbivory, but to restrict it to acceptable rates.

The predominance of insect over mammalian herbivory on trees in the infertile savannas and the apparently reversed situation in fertile savannas may be related to the control on insect population dynamics exerted by predators in the two types. The populations of birds and their prey are higher in the fine-leafed savanna, and insect predators such as ants may also play a role in stabilising the impact of insect herbivores. The larger number of insectivore species in the broad-leafed savanna suggests a greater degree of specialisation. This may be because the insects use the plant defensive compounds to render themselves distasteful, requiring adaptations in their avian predators. The insect herbivores usually occur at a low density, which means that the avian predators occur at a low density as well. Outbreaks occur when a local population temporarily overwhelms the ability of their predators to control them: control is re-established when the food source is exhausted or when predator numbers respond following increased breeding success, in-migration, or prey-switching.

Figure 12.5. Soil fauna in fine-leafed (nutrient-rich) and broad-leafed (nutrient-poor) savannas at Nylsvley. These values are the means of five 0.25 × 0.25 m samples in each savanna type, taken to a depth of 0.3 m in May 1988.

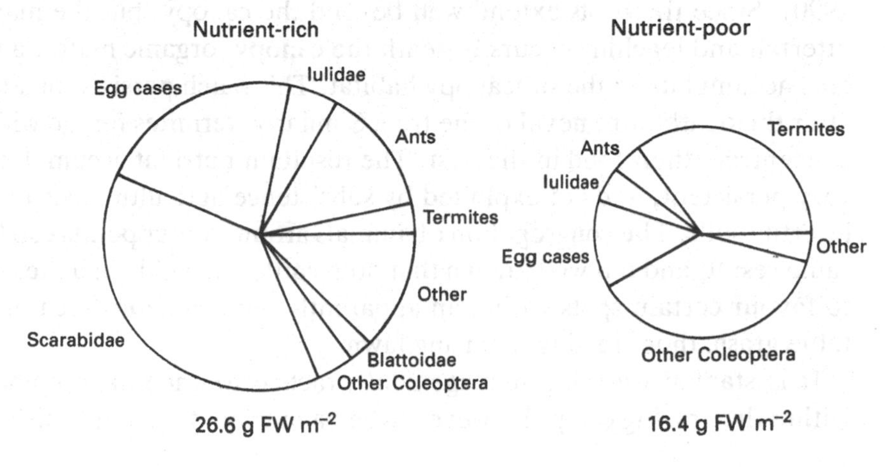

Landscape interactions: why do the fertile patches persist?

It would be intuitively expected that processes of nutrient loss, as well as the action of plants and animals, would rapidly lead to the dispersal of nutrient-rich islands in a nutrient-poor environment. However, in the case of the fine-leafed savanna patches at Nylsvley, the opposite effect seems to occur: there appear to be reinforcing mechanisms which counterbalance the losses due to leaching, erosion and volatilisation. The principle mechanism for nutrient fluxes into the patches appears to be the grazing behaviour of ungulates.

Grazers spend six-fold more time in the fine-leafed patches, although the grass production there is less than twice that in the broad-leafed savanna. The preference for fine-leafed savanna can be explained in terms of the higher quality of the forage there. However, the herbaceous layer productivity in the fine-leafed savanna is insufficient to support the grazer biomass it contains, and the grass is consequently closely cropped. The bulk of the grazer's energy requirements must come from the broad-leafed areas, while their nutrient requirements are satisfied in the fine-leafed areas. Couple this with an assumption that defaecation and urination occur in proportion to the time spent in an area, and the result is a mechanism for continued nutrient enrichment of the nutrient-rich sites.

Nutrient concentration as a significant process in savannas

The nutrient-rich patches at Nylsvley are not an isolated example. Similar vegetation changes are used to detect archaeological sites on the Kalahari fringe (Denbow 1979). Although an anthropogenic origin for such enrichments is probably quite common in Africa, there are many other mechanisms that achieve the same result. At a small scale, a savanna tree operates as a nutrient concentrator (Kellman 1979; Weltzin & Coughenour 1990). Since the roots extend well beyond the canopy, but the majority of litterfall and leaching occurs beneath the canopy, organic matter and nutrients accumulate in the subcanopy habitat. This patch persists for some time after the death or removal of the tree. Similarly, termites forage widely, but concentrate their food in the nest. The resultant nutrient accumulations are very persistent, and are exploited by subsistence agriculturalists in areas of infertile soils. The congregation of animals around water points can have the same result, and it is well known that both cattle and wild ungulates will tend to favour certain spots within an apparently homogeneous area of unpalatable grass, thus creating a grazing lawn.

It is standard grazing management practice to encourage homogeneity within the grazing camp. However, when a nutrient threshold exists, such as the requirement by ruminants for a minimum crude protein content in the forage, then the natural processes whereby the fertility of one patch is subsidised by its surroundings could be a great advantage. Certainly the secondary production at Nylsvley would be substantially lower were it not for the nutrient-rich patches. The formation of locally enriched patches seems to be a universal phenomenon in infertile savannas, and perhaps in nutrient-poor environments in general.

Fertile and infertile savannas in southern Africa

The Nylsvley fertility contrast is a microcosm of a pattern which is widespread in southern Africa, and maybe in Africa as a whole. The broad-leafed savannas at Nylsvley are representative of the extensive savannas that occur on the ancient, highly weathered African Surface, or on more recent surfaces which are nutrient-poor due to their parent material (such as the Kalahari sands or, as at Nylsvley, quartzites). The fine-leafed savanna patches represent the savannas of the younger Post-African surfaces which are nutrient-rich, either due to the active soil formation taking place on those surfaces, or due to their being located on volcanic material of recent volcanic origin. Thus the fine-leafed savannas tend to be located along the lower-lying eastern part of the continent and the valleys of the major rivers (Figure 12.6). These areas also tend to be hotter and drier than the core region of the infertile savannas. Therefore adaptations such as microphylly could well be primarily a response to aridity, and only coincidentally associated with nutrient availability.

Why should the savannas of southern Africa fall into two distinct classes, rather than the fertility continuum that might be expected (Scholes 1990b)? Part of the reason lies in the discreteness of the African and Post-African erosion surfaces. The great age and stability of the African surface has tended to erase the effect of differing parent material on the soils of that surface. The distinction is often less clear on the Post-African surface, where, in addition to clearly nutrient-rich and nutrient-poor variants, there are intermediate mixtures, and some types that are difficult to classify. The savannas growing on the areas of Kalahari sand receiving less than 500 mm of rainfall per annum include many fine-leafed species, despite the infertility of the substrate. This suggests that nutrient sufficiency should be judged relative to the carbon assimilation of the plant: even the infertile sands are able to satisfy the limited plant nutrient demand in arid savannas.

The distinction between nutrient-rich and nutrient-poor savannas in Africa apparently post-dates the breakup of Gondwanaland. As such it should not be expected to be true of savannas world-wide. For this reason the

Responses of Savannas to Stress and Disturbance (RSSD) programme, an international effort to understand the structure and function of savannas, has adopted a plane, with plant available moisture as one axis and plant available nutrients as the other, as the basis for the conceptual organisation of global savannas.

Summary

The savannas of South and East Africa can generally be divided into two types: those growing on nutrient-poor substrates, and those growing on nutrient-rich substrates. Many ecological features of savannas can be predicted on the basis of this classification. The nutrient-poor savannas can be recognised in the field by the absence of thorns and the dominance by the tree families Caesalpiniaceae and Combretaceae, which have leaves several centimetres in diameter. Nutrient-rich savannas are dominated by the Mimosaceae, which are thorny and fine-leafed.

The nutrient-poor savannas occur on the high, ancient African erosional surface. In general, they are cooler and wetter than the nutrient-rich

Figure 12.6. The subcontinental distribution of fertile and infertile savannas.

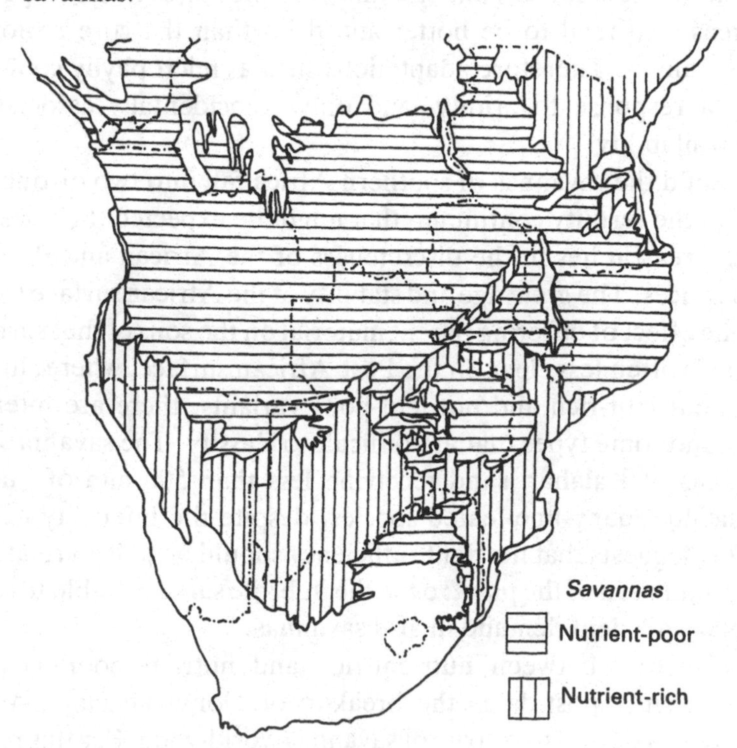

savannas, which occur on the lower, Post-African erosional surfaces and river valleys. Areas of nutrient-poor savanna can be converted to nutrient-rich savanna by the accumulation of nutrients, for instance by the activities associated with human settlements. Nutrient-rich savannas have a higher grass layer productivity, and the grasses growing there are much more acceptable to grazers. Consequently, the grazer carrying capacity is substantially higher in nutrient-rich savannas.

Several ecological processes serve to create patches of nutrient enrichment within a generally nutrient-poor landscape. If these patches are large enough, they become self-sustaining due to changes in herbivore feeding behaviour. We believe that the landscape-level interaction between rich and poor patches is an important feature of the ecology of infertile savannas.

13

Community structure, composition and dynamics

An important lesson learned during the IBP era was that it is not possible to predict all the workings of an ecosystem on the basis of fluxes of material and energy alone. In many crucially important processes, the exchanges of energy and matter are inconsequential: for instance, in the process of pollination. The sensitive techniques of population dynamics and community ecology provide an essential complement to the blunt but powerful instruments of systems ecology.

Community-level studies occurred in two separate phases during the Nylsvley project. Early in the programme, several studies aimed to identify the major communities, and associate them with environmental factors. Ben Coetzee, Roy Lubke, Gillaume Theron and Robert Whittaker made important contributions to this phase. A decade later, Peter and Sue Frost, Dick Yeaton and Graham von Maltitz revived interest in community and population dynamics with a series of manipulative experiments.

Patterns of species distribution

The phytosociological survey of the study site had suggested that the broad-leafed savanna of the study site, named the *Eragrostis pallens–Burkea africana* tree savanna, was not homogeneous. It included two levels of community organisation (Coetzee *et al.* 1976). However, three detailed quantitative studies failed to identify species groupings which were sufficiently consistent to justify treating them as a community (Lubke *et al.* 1976, 1983; Whittaker *et al.* 1984; Theron *et al.* 1984). These studies were carried out in transects of contiguous quadrats, 300–1000 metres long, and located in the mid–slope area of the broad-leafed savanna. The same woody and herbaceous species occurred in all transects, but their density and biomass varied considerably. For the woody plants this variation contained little apparent pattern. There was some clumping of the smaller size classes of *Burkea africana* and *Ochna pulchra* at a scale of 10–40 m, but the mature trees appeared to be randomly distributed (Lubke 1976).

When Detrended Correspondence Analysis was applied to the herbaceous layer data, the major axis of variation in all the studies was found to lead from the subcanopy microhabitats to the open subhabitats. This axis also corresponded to soil depth, and was therefore tentatively linked to water availability. The study of Whittaker *et al.* (1984) associated the second axis of variation with a subtle change in soil fertility between the areas underneath *Burkea africana* canopies (more fertile), and those beneath *Terminalia sericea* canopies (less fertile). There was pattern in the herbaceous layer at a scale of about 30 m, which related to the scale of tree canopies. They observed a tendency for *B. africana* to alternate with *T. sericea*. The third and higher axes were attributed to herbivory and animal disturbance.

The study of Theron *et al.* (1984) associated the second ordination axis with the degree of disturbance, since the perennial species were concentrated at one end, and annual species at the other. On the basis of the scatter of species and the small part of the overall variance removed by the axis, they concluded that the herbaceous layer of the broad-leafed savanna was functionally homogeneous. Based on comparisons with other *Burkea africana* savannas, they concluded that the grass sward at Nylsvley was a subclimax resulting from grazing disturbance. The climax would have less *Eragrostis pallens* and *Digitaria eriantha*, and more *Panicum maximum* and *Setaria perennis*.

In summary, the main factor influencing the pattern of grass species distribution within the core of the broad-leafed savanna area is the distribution of the woody plants. The woody plant distribution varied substantially, but without apparent pattern. This remained the consensus until Yeaton *et al.* (1986) and Yeaton (1988) reorientated the transect and lengthened it slightly, so that it stretched from the centre of the broad-leafed savanna to its edge, instead of remaining in the core area. This meant that it ran for a distance of 2000 m, from the top of the sandstone ridge, down the gentle slope to the *turfvlei* hydromorphic grassland in the valley bottom. According to the Harmse (1977) soil map, this transect would cross four soil series, increasing in clay content from 5% at the ridgetop, to 20% at the bottom. If this soil sequence could be linked to a vegetation change, the combination would constitute a catena, a unit of landscape description that has proved very useful in semi-arid savannas elsewhere in Africa (Milne 1936).

The pattern of dominance of the two main broad-leafed savanna trees does change along this transect (Figure 13.1). Furthermore, although the total density of woody plants increases by 50% between the top and bottom, the total basal area decreases by 2%, implying that the mean tree size increases towards the bottom. A gradient of soil water availability from the ridgetop downwards could be inferred from inspection of the soil profile. The soils at the very bottom of the slope showed the blue-grey hue and yellow/orange mottles typical of profiles which are seasonally wet. This provided a direct gradient by which to elucidate species distribution patterns, rather than the indirect gradient approaches which had been previously employed.

The species composition of the grass layer was also found to change along this catena (Yeaton *et al.* 1986). There was a strong association between a soil series and the plant community that it supported. Furthermore, within the apparently homogeneous Chester Series, which covers the top 1200 m of this catena, there is a gradual replacement of *Eragrostis pallens* and *Digitaria eriantha* by *Elionurus muticus* and *Diheteropogon amplectens* towards the slope bottom. Modifying this gradient, which is attributed to soil water availability, is a secondary gradient between open and subcanopy areas. The picture is further complicated by an overlaid mosaic of disturbance patches of different ages, resulting from mammal burrows and fire (Yeaton *et al.* 1986).

Figure 13.1. Change in the relative proportions of *Burkea africana* and *Terminalia sericea,* the two main tree species of the broad-leafed savanna, as a function of distance down the catena between the crest of the ridge and the valley bottom (redrawn from Yeaton 1988).

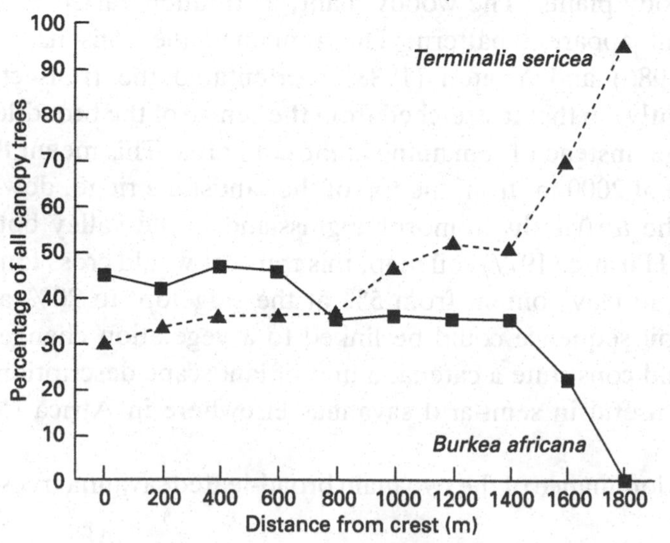

The effects of disturbance

Savannas in nature conservation areas are frequently perceived to be in an undisturbed state. Nylsvley has experienced about 1000 years of significant human disturbance (see Chapter 2), and is not unusual in that regard. Detailed examination of what were previously believed to be stable, 'climax' communities, such as tropical and temperate forests, have shown that ongoing disturbances, often at a relatively small scale, are essential to the maintenance of the apparently stable state. All the studies of the herbaceous layer composition at Nylsvley have revealed disturbance to be an important local modifier of the overall pattern imposed by water and nutrient availability. The main causes of disturbance in the herbaceous layer are the burrowing activities of small and large mammals, the death of trees, and fire. Although fire occurs sufficiently frequently perhaps to be regarded as 'included disturbance', it does have an impact on the grass species composition, productivity, nutrient cycling and woody plant mortality. The size, density and rate of formation of burrows is such that, on average, every part of the broad-leafed savanna should be disturbed once in 300 years. The burrowing disturbances involve the uprooting of plants and the redistribution of soil. They range in size from a feeding scrape 20 cm across, to a warthog burrow complex tens of metres in diameter. The length of time for which the effects of the disturbance persist ranges from a single season, for a small scrape, to decades or centuries for a large burrow complex or a termite mound.

The size of the disturbance has predictable consequences on the rate and pattern of vegetation recovery (von Maltitz 1990). Three processes of re-establishment are involved: the vegetative expansion of neighbouring plants, which occurs at the rate of a few centimetres per year; the formation of daughter plants by tillering and stolon extension, which can colonise a metre per year; and establishment from seed dormant in the soil or recently dispersed. The last process occurs at larger scales, but is critically dependent on the dynamics of the soil seed store and the conditions for establishment. The soil water content is depressed for areas within 0.5 m of the edge of the patch, because of uptake by roots of surrounding grasses. Solar radiation at the soil surface is depressed by up to 90% within 0.25 m of the patch edge, due to shading. The range in soil surface temperature at the centre of a disturbance was higher than in undisturbed vegetation. These factors together result in a stimulation of forb seedling establishment in the centre of disturbance patches, and a depression near the edge. Large patches therefore have more seedlings than small (Figure 13.2). The pattern of grass establishment from seed is fairly even across the patch, but small patches have a higher number of stolon-derived colonists. Species richness was unaffected by disturbance patch size.

The rate of plant establishment increased by an order of magnitude in the second season year after the disturbance. This was partly attributable to differences in the amount and timing of rainfall. The timing of the disturbance in relation to the growing season has a large influence on the rate and pattern of subsequent regrowth. The longevity of seeds of different species in the soil varies greatly, as does their sensitivity to germination triggers. An early-wet season disturbance favours those species with short-lived, easily germinated seeds. Disturbance in the middle of the growing season favours species with long-lived, tardily germinating seeds, since the majority of short-

Figure 13.2. The density of forb seedlings is influenced by the size of the disturbed patch, in both the first and second season after disturbance (redrawn from von Maltitz 1990).

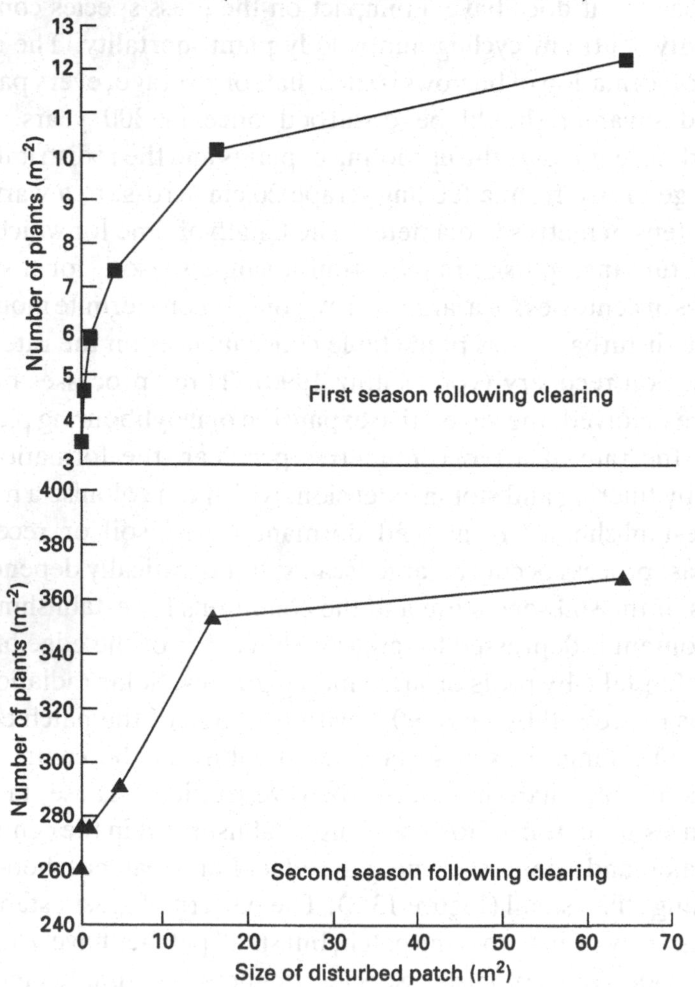

lived seeds in the seed bank have already germinated and died before the disturbance occurs. Since the seed bank of long-lived seeds is much smaller than that of short-lived seeds, disturbances later in the growing season result in progressively lower densities of seedlings.

In summary, both the size and the timing of disturbances of the herbaceous layer influence the recovery trajectory. The first of the three main models of plant succession (facilitation, inhibition and tolerance: Connel & Slatyer 1977), does not appear to apply to the herbaceous layer at Nylsvley. Plants representative of the undisturbed state are able to establish during the early post-disturbance phase. Mature grass tufts suppress the germination and establishment of grass and forb seedlings, suggesting that the inhibition model may be the most appropriate one.

Seed banks

The total number of seeds per square metre for the dominant trees in the broad-leafed savanna is about 38 for *Burkea africana* and 134 for *Terminalia sericea* (Janse van Rensburg 1982). However, 50% of buried *B. africana* seeds are germinable, whereas 70% of *T. sericea* seeds are parasit ised, and only 16% of the unparasitised seeds could be germinated in the laboratory. Thus the viable seed stores are about 19 and 6 per m^2. *B. africana* seeds are distributed slightly deeper in the soil than *T. sericea* (85% vs 95% in the top 20 mm). The dominant grass, *Eragrostis pallens*, had 883 seeds m^{-2}, of which 16.6% were germinable, giving a viable store of 146 m^{-2}. These values are in good agreement with the study conducted by Venter (1976) 5 years earlier, suggesting that the seed bank dynamics for these species are relatively stable. For all three species the highest numbers of seeds were found in the immediate vicinity of the parent plants.

The germination percentage in *T. sericea* and *B. africana* was significantly reduced at temperatures below 17 °C. The daily mean temperature at Nylsvley is below this level from April to September. Daily maximum temperatures up to 37 °C had no effect on germination. There was no significant effect of temperature on germination of *E. pallens*, down to a temperature of 12 °C. Germination under bright light was a quarter of that of seeds kept in the dark. The germination success in the laboratory averaged 25% for 4-month old seeds, increased to a peak of 52% after 8 months, and then declined to 41% at 15 months. Exposure to light had no effect on the germination of tree species, but scarification of the hard testa of *Burkea africana* increased germination four-fold. There was no significant change in viability over a 15-month period. The highest germination success in *T. sericea* was obtained with 6-month old seeds, with a slow decline thereafter.

Woody plant establishment and growth

There are 400 to 700 seedlings per hectare of *Burkea africana* (Henderson 1979; Janse van Rensburg 1982), of which 31% survived one year in the field. The seedlings are more common beneath the canopies of trees than between canopies, but the sapling growth rate is slower beneath a *B. africana* canopy than beneath the canopy of a *Terminalia sericea* or a *Dombeya rotundifolia* (Yeaton 1988). *T. sericea* had 240 seedlings ha^{-1} (Janse van Rensburg 1982), of which 32% survived a year. Seedlings of *T. sericea* are seldom found under the canopy of any tree. *D. rotundifolia* saplings are more commonly found beneath *B. africana* than beneath *T. sericea* or *D. rotundifolia* itself. This suggests a successional sequence whereby *T. sericea* is first to establish in a gap. Individuals of *B. africana* and *D. rotundifolia* establish beneath its canopy, and in time grow through it and shade it out. On the lower part of the slope, *T. sericea* is able to dominate on soils with seasonally waterlogged deep horizons due to its superficial rooting system. Complete domination of the upper catenal positions by the latter two species is precluded by their high susceptibility to porcupine and fire damage as mature plants (Yeaton 1988).

The size structure of dominant trees of the broad-leafed savanna has been reported by Walker *et al.* (1986). Growth ring studies indicated that the average annual radial increment for *B. africana* was 2.21 mm, independent of the size of the individual. Therefore the largest individuals of this species at Nylsvley are approximately 120 years old (Henderson 1979).

Summary

The main environmental factor shaping the community species composition of the broad-leafed savanna at Nylsvley is the moisture and nutrient gradient which stretches from the crest of the interfluve down to the bottomlands. The density and species composition of grasses and trees respond to this gradient. For the herbaceous layer, there is a smaller-scale gradient superimposed on the first gradient, between sites beneath tree canopies, and sites between canopies.

A variety of disturbances, including fire, grazing and burrowing animals complicates this basic pattern, by creating transient microhabitats for a wide range of species. The species succession on a disturbed patch is influenced not only by the type of disturbance, but is also critically dependent on its seasonal timing, size and neighbouring vegetation. These disturbances are essential for the maintenance of the structure, species composition and function of the broad-leafed savanna.

14

Tree–grass interactions

If the distinguishing feature of savannas is the co-dominance by trees and grasses, then the central question in savanna ecology must be the mechanism of their coexistence. The interaction between the woody component and the grass layer is a feature absent from either pure grasslands or closed woodlands. All biomes include a range of plants types; but savannas are unusual in that two functional types ('trees' and 'grass' for short), share the primary production in a more or less equitable fashion. Plants, lacking the mobility of animals, all use essentially the same resources: radiant energy, water and nutrients. Simple competition theory predicts that when competitors use the same resource, one should be superior to the other, and should become dominant. Trees and grasses have evolved very dissimilar patterns of environmental interaction. In large areas of the world either one or the other is dominant. What is special about the savanna environment that allows trees and grasses to coexist? This is the core of what Sarmiento (1984) calls the 'savanna problem'.

The proportions of trees and grasses in a given savanna are not always predictable from environmental conditions, nor are they stable over time. On the same substrate and under the same climate there may be local variations in the proportions. Nevertheless, the large-scale, long-term pattern is one of a remarkably stable coexistence. An increase in woody plant density usually follows sustained heavy grazing. This phenomenon, called 'bush encroachment', has been observed in savanna regions all over the world. Since the main economic uses of savannas are based on grass-eating animals, and the degree of woodiness has a substantial impact on the productivity of the grass layer, a change in woodiness is of direct economic consequence.

The investigations into interactions between woody plants and grasses at

Nylsvley were initiated and led by Brian Walker, with important contributions by Wendy Knoop, Tom Smith and Dick Yeaton.

The effect of trees on grasses and vice versa

It has been repeatedly demonstrated that the removal of woody plants from savannas results in an increase in herbaceous layer production. The reverse experiment (the effect of grass removal on woody plant productivity) has seldom been attempted. Knoop & Walker (1984) performed both manipulations in the broad-leafed savanna at Nylsvley. There was an increase in the height and basal area of herbaceous plants after the removal of competition by woody plants, but no stimulation of tree shoot extension or radial increment following grass removal. In the fine-leafed savanna, on the other hand, there was a small stimulation in tree shoot extension when grasses were removed from their surrounds. The effect of trees on the grass layer was not measured in the fine-leafed savanna. This experiment demonstrates two important points: first, competition between mature trees and grasses is strongly biased in favour of the trees; and secondly, the outcome of the tree–grass interaction is influenced by factors such as the soil conditions

In general, the relationship between measures of the woodiness of a savanna (such as tree biomass, density, basal area, or leaf area) and grass production is not linear: a halving of the woodiness usually results in less than a doubling of herbaceous layer production. A tree-thinning experiment to determine the shape of this curve was initiated at Nylsvley, but was abandoned as a result of the inefficacity of the arboricide used.

The presence of trees imposes a complex mosaic of microenvironments on the grass layer. The net radiation in the subcanopy habitat is 50 to 75% less than that in areas between trees (Harrison 1984). Although the area-averaged LAI of the tree layer seldom exceeds a value of 1.0, beneath the tree canopy it can reach 3.0. Rainfall is reduced by up to 15% due to interception losses (de Villiers 1981), except for the area immediately adjacent to the stem, which gains from the stemflow. Since the water status of plants reflects the balance between water supply and water demand, the grasses in the subcanopy habitat have a more favourable water regime than those in the open. This is demonstrated by their remaining green and unwilted for longer into dry periods than the grasses from open areas. Furthermore, the grass community beneath the tree canopy contains more species with a mesic distribution, such as *Panicum maximum*. The evaporative demand beneath the tree canopy is not reduced proportionally to the reduction in net radiation, since a large part of the subcanopy energy budget is contributed by advection of heat from the surrounding unshaded areas (Harrison 1984).

The subcanopy habitat is also richer in nutrients than the between-canopy areas (Kellman 1979; Olsvig-Whittaker & Morris 1982) and has higher organic carbon, total nitrogen and microbial activity (Bezuidenhout 1978; S. J. McKean, unpublished data). This is a consequence of the nutrient pump effect of trees. The tree roots extend considerably beyond the canopy (Rutherford 1983), but leaf litterfall and nutrient leaching occurs mainly beneath the canopy. Trees may also bring nutrients up from soil layers not accessible to grass roots, although this is not thought to be an important factor at Nylsvley. The consequence of enhancement of nutrient availability on the one hand, and reduction of photosynthetic capacity on the other (due to the reduced irradiance), is to increase the nutrient concentration of the grasses from the subcanopy habitat. The productivity of the subcanopy grasses is boosted by the improved water and nutrient status, but suppressed by the low irradiance, and by competition between tree and grass roots for the belowground resources. The net result can be an increase or decrease in productivity relative to a treeless area, depending on the interaction of those four factors (Stuart-Hill, Tainton & Barnard 1987). For an average rainfall year in the broad-leafed savanna at Nylsvley, Grossman *et al.* (1980) estimated that the grass layer productivity in the open areas between tree canopies was 44% higher than grass layer productivity beneath tree canopies. This is unusual for dry savannas. The situation is certainly the reverse in the fine-leafed savanna, although it has never been properly quantified at Nylsvley. Bosch & van Wyk (1970), working 200 km west of Nylsvley, reported increased productivity of *Panicum maximum* beneath tree canopies.

The characteristic shift in grass layer species composition beneath the tree canopy could be caused by any one of the altered environmental factors, but is probably a complex expression of their interaction. The classic subcanopy species of southern African savannas, *Panicum maximum*, has been shown to have a higher productivity in full sunlight than in shade, provided it is well supplied with water (Scholes 1987). Soil type had little effect on productivity in either the sun or the shade. Kennard & Walker (1973) found that establishment of *P. maximum* seedlings could occur only in a shaded environment.

The impact of a tree on its surrounds must decline as some function of the distance from the tree stem. Rutherford (1983) showed that tree roots extend into the area beyond the canopy: as a rule of thumb, up to seven times the canopy radius. There should be no part of the Nylsvley savanna free of tree root influence. However, the density of tree roots probably declines with distance from the tree. The combined tree and grass root biomass shows an inverse relationship to an index of woody plant influence (R. J. Scholes, unpublished data). The woodiness index of a point is defined as the sum of the squared inverse distances from the point to all neighbouring trees, multiplied by each tree basal area. When a geostatistical analysis (Robertson 1987) is applied to root biomass samples collected in a coordinate system, the semivariogram has a range of about 12 m. In other words, root biomass samples collected less than 12 m from one another show a degree of correlation, indicating an organising process at about that scale. This is approximately the scale of influence of an individual tree.

In summary, the herbaceous layer in a savanna can be considered to consist of a mosaic of microenvironments, dependent on the spatial relationship of the tree canopies and rooting zones (Figure 14.1). Most savanna grass production studies average over this patch-variability. The overall productivity and composition of the herbaceous layer depends on the relative extent and properties of the component subhabitats. These depend, in turn, on the density, size distribution, morphology and spatial pattern of the trees. A

Figure 14.1. A conceptual model of the grass layer subhabitats in a savanna. Subhabitat A is directly below a tree canopy. B is not below a tree canopy, but is within the tree rooting radius. C is beyond both the canopy and the roots of the trees. The productivity, species composition and nutritional value of grasses in the different subhabitats will vary according to the balance of irradiance, nutrients, water and competition with trees. The area-averaged properties of the grass layer are therefore influenced by the spatial arrangement and morphology of the trees, as well as by total tree biomass, rainfall and soil fertility.

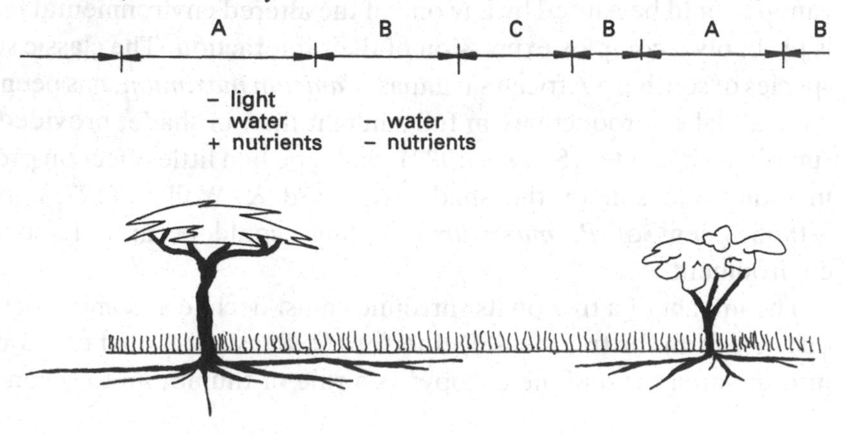

Table 14.1. *The survival and growth rate after 24 weeks of seedlings of* Burkea africana *grown in pots with and without tufts of* Eragrostis pallens, *and with and without 19 mm week^{-1} of water in addition to the natural rainfall. Four pots were used per treatment. The pots were 0.79 m^2 in area and 35 cm deep, and were planted with eight seeds of* B. africana *each. The tufts of* E. pallens *had 24 to 28 tillers each. Data are from Knoop (1982)*

	Without *E. pallens* competition		With *E. pallens* competition	
	Rain only	Watered	Rain only	Watered
Biomass (mg)	263	321	130	82
Stem diameter (cm)	1.1	1.2	0.8	0.7
Survivors	14	12	14	3

model of tree–grass interactions which takes into account these spatial factors could go a long way to explaining the non-linearity of the tree density vs grass production curve, as well as reconciling the different forms which have been reported for this curve. Such models are under development for savannas (Penridge & Walker 1986).

While Knoop & Walker (1984) showed that grass biomass had a minimal effect on the growth rate of mature trees, they also showed, in a series of pot trials, that grass biomass strongly influenced both the survival and growth rate of woody plant seedlings (Table 14.1). Supplementation of the water supply to the pots did not significantly improve tree seedling survival. This finding was supported by a field experiment in which seeds were planted in plots which contained no grass, dead grass (previously killed using a herbicide) or live grass, all either with tree roots excluded by means of barriers, or with roots from surrounding trees present. The presence or absence of woody plant roots had a significant effect on the growth of seedlings, but only when live grass was absent. When live grass was present, it had a strong effect on both the survival and growth of the woody plant seedlings. The presence of dead grass had an intermediate effect, suggesting that competition for light could be one of the mechanisms of growth suppression. Whatever the cause, the presence of a dense herbaceous layer exerts a strong control on the rate of recruitment of mature trees.

Fire is another mechanism whereby the grass layer influences the tree layer. As indicated in Chapter 8, the susceptible parts of woody plants (buds and leaves) are usually killed if they occur within the flame zone of the fire.

This acts as a grass-dependent recruitment control for woody plants, since the frequency and intensity of fires depends on the fuel load provided by the herbaceous layer. The main cause of mature tree mortality at Nylsvley appears to be the porcupine–fire–wind interaction described in Chapter 8. Therefore, although fire is seldom the sole agent in killing mature savanna trees, it is instrumental in controlling their biomass and density in interaction with other factors, such as herbivory.

Competition between trees

A patch of broad-leafed savanna on the farm Mosdene, adjacent to Nylsvley, has been protected from fires for three decades, and has developed a basal area double that typical for similar communities on Nylsvley. The annual radial increment of this stand is 4% (Rutherford & Kelly 1978), by comparison with the 6–10% recorded on Nylsvley. The implication is that competition between trees has an important effect on tree growth. Interplant competition has frequently been inferred from spatial pattern within the community. Clumped distributions of individuals suggest that competition is not an important force in structuring the community, while evenly spaced plants could result from the competitive interactions. Grant (1983) found that in two study areas within the broad-leafed savanna, the two dominant trees (*Burkea africana* and *Terminalia sericea*) were significantly clumped ($p < 0.01$) when considered together, although on one site both species were significantly regularly spaced ($p < 0.05$) with respect to other individuals of their own species.

Another technique used to infer competitive interactions is the relationship between the distance between neighbouring plants and their combined canopy cover. Applying this technique, Grant (1983) found strong evidence ($p < 0.004$) for competitive interactions between neighbouring trees of the same species, and weaker evidence for competition between neighbours of different species ($p < 0.017$).

Competition between grasses

The dominant grass of the between-canopy areas is *Eragrostis pallens*, with *Digitaria eriantha* as a subdominant. Beneath the canopies, *Panicum maximum* and *D. eriantha* dominate, with *E. pallens* virtually absent. When grown in pots, either alone or in mixtures, *D. eriantha* outproduces *E. pallens*, even under suboptimal watering regimes and severe defoliation (Kemper 1984). This was particularly true for belowground production. Nearest-neighbour analyses using these two species, conceptually identical to the analysis described above for the dominant trees, indicated strong intra- and interspecific competition between grass tufts in the field (Yeaton *et al.* 1986), especially on sites with an intermediate fire frequency.

The 'radius of influence' of individual grass tufts at Nylsvley is at least 0.38 m (von Maltitz 1990), which is considerably more than the mean intertuft spacing. This value was determined by isolating grass tufts in the field within steel cylinders of various sizes, hammered into the ground to a depth of 30 cm. The four species of grass and one forb studied all responded positively to increasing available space up to the maximum area used, 0.45 m^2 (Figure 14.2). There was also considerable variation between species in the number of seedlings that successfully established in the same cylinder as the isolated tuft, indicating differential ability to suppress the establishment of potential competitors.

Mechanisms of coexistence

The coexistence of apparently competing species is the rule rather than the exception in plant communities. There are many theoretical mechanisms which could allow this. They can be divided into two groups: the equilibrium and disequilibrium mechanisms. Coexistence in equilibrium models is attributable to special features of the interaction: niche separation and differential herbivory are examples. Disequilibrium models would tend to domination by one competitor, but disturbances from outside the system keep restarting the race.

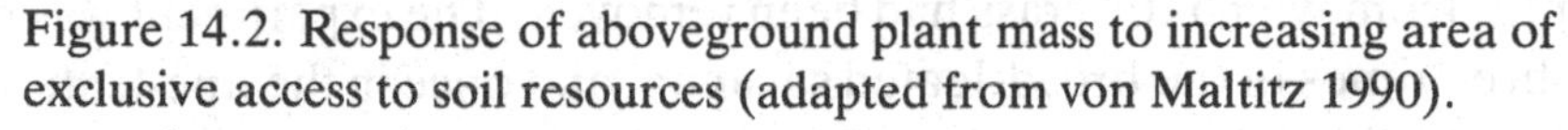

Figure 14.2. Response of aboveground plant mass to increasing area of exclusive access to soil resources (adapted from von Maltitz 1990).

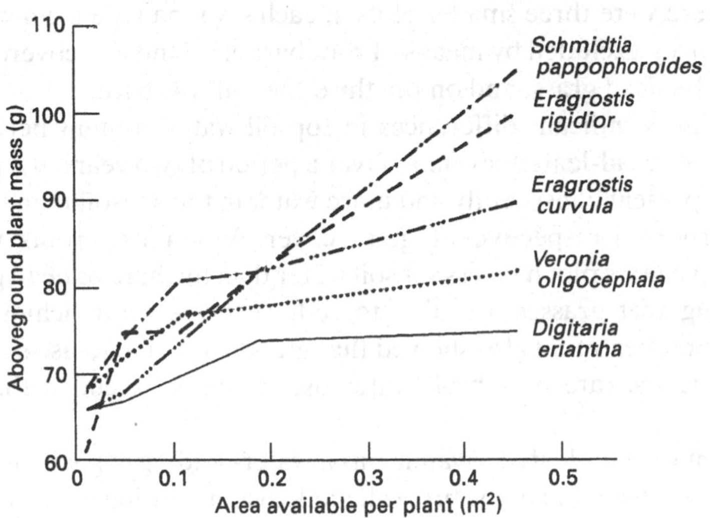

The classic model for coexistence in savannas is an equilibrium model, proposed by Walter (1971). The 'Walter Hypothesis' contends that water is the limiting factor in semi-arid savannas, and that grasses are superior competitors for water in the surface horizons of the soil. Trees, however, have exclusive access to water in the deeper horizons, as well as access to the topsoil water. This model does not need absolute root depth separation to provide an equilibrium solution, but a degree of separation proportional to the difference in water use efficiency is necessary. The model is able to explain many observed spatial and temporal trends in savannas (Walker *et al.* 1981; Walker & Noy-Meir 1982; Walker 1985).

Knoop & Walker (1984) and Scholes (unpublished data) measured the vertical distribution of fine roots of trees and grass in the fine-leafed and broad-leafed savannas at Nylsvley, using the profile count technique (Böhm 1979). Both studies show that tree roots and grass roots exploit essentially the same soil volume (Figure 14.3). Tree roots peak in density at a slightly greater depth than grass roots, but in absolute root length terms, grass roots are dominant to a depth of one metre, which is the average soil depth at Nylsvley. When these data are taken in conjunction with the storm depth distribution at Nylsvley and the soil water retention characteristics (Chapter 6), it is apparent that water is seldom present in the deeper soil horizons where trees have their putative advantage.

Knoop & Walker (1984) monitored the soil water content in the topsoil (0–30 cm) and subsoil (30–130 cm) in three treatments: a control containing both trees and grass; a plot from which all woody plants had been removed; and a plot from which the grass had been removed. The experiment was replicated three times in the broad-leafed savanna, and once in the fine-leafed savanna. In addition, there were three smaller plots in each savanna type from which tree roots had been excluded by means of root barriers. One was covered by live grass, one by dead grass, and on the third the soil was bare.

There were no significant differences in topsoil water content between treatments in the broad-leafed savanna. Over a period of two years, the plots with tree roots present consistently had more water in the subsoil than those without tree roots, irrespective of grass cover. When tree roots were excluded, the live grass plot had less subsoil water than the bare or dead grass plots, indicating that grasses are able to reduce the water reaching the subsoil. The same treatment also showed that grasses are able to use subsoil water; however, the rate of subsoil water use by trees exceeded that of grasses.

The topsoil in the fine-leafed savanna has a water-holding capacity nearly three times greater than that of the broad-leafed savanna, owing to the higher organic carbon content. The subsoil was virtually dry for the entire duration of the experiment. The topsoil and subsoil water contents in the grassless treatments were higher than in the treatments with live grass cover. The treatment with tree root exclusion and dead grass cover had the highest topsoil and subsoil water content. More water penetrated to the subsoil in the

Figure 14.3. Depth distribution of fine roots of woody plants and grasses in the broad-leafed and fine-leafed savannas. Solid symbols are tree roots, open symbols are grass roots. The bars represent ± 1 SE. The root lengths have been calculated from the formula $L_R = N$, where N is the number of cut roots intercepted by a plane of unit area. The values are the means of eight profiles 0.5 m wide by 1.0 m in each vegetation type, while the bars indicate the standard error of the estimate.

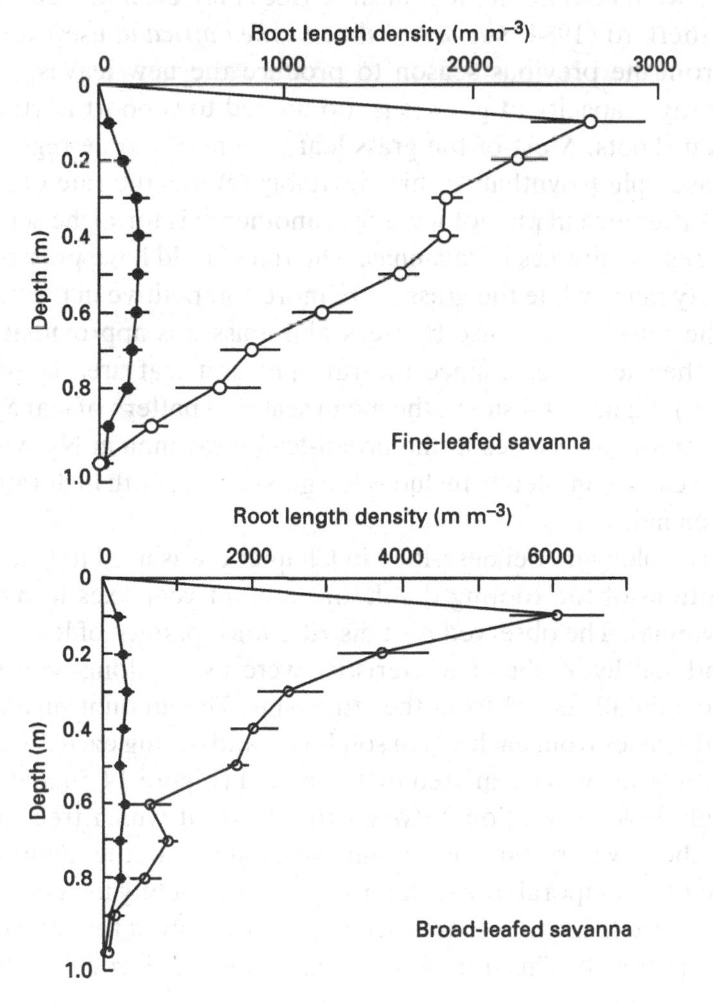

absence of tree roots than in their presence. Infiltration was reduced, and evaporation increased, on the plots with bare soil. This experiment indicates the ability of the grass layer to reduce the amount of water reaching the subsoil, particularly in soils with a high water-holding capacity, but also that the water used by grasses is partly offset by the reduction in transpiration and increase in infiltration which they bring about. As in the broad-leafed savanna, grass roots were able to use subsoil water as effectively as tree roots, and tree roots dried out the topsoil nearly as fast (and completely) as grass roots alone.

The predominance of savannas in climates of high rainfall seasonality would suggest that temporal rather than spatial niche separation may be the key to coexistence in savannas. It has frequently been observed that woody plants achieve full leaf expansion before the grasses do; if the autumn rains in the previous year were favourable, leaf flush in trees may even precede the spring rains. Rutherford (1984) suggested that *Burkea africana* uses stored carbohydrates from the previous season to produce the new leaves. The carbohydrate storage capacity of grasses is too limited to support anything but the first green shoots. Most of the grass leaf area needs to be regrown using current season photosynthates: this inevitably retards the rate of leaf expansion. This difference in phenology offers another axis for niche separation between trees and grasses in savannas. The trees could have preferential use of the early rains, while the grasses are more competitive in the mid-rainy season. The rate of water use by trees and grasses is approximately proportional to their leaf areas, since the rate per unit leaf area is quite similar (Chapter 6). Figure 14.4 shows the mean seasonal pattern of leaf area development for trees and grasses in the broad-leafed savanna at Nylsvley. Note that the tree curve completely includes the grass curve, both in duration and peak leaf area index.

The savanna hydrology model described in Chapter 6 was used to test the relative contributions of the rooting depth and time of year axes to niche separation in savannas. The observed root distributions, pattern of leaf area development and soil hydraulic characteristics were used, along with the observed 16-year rainfall record from the study site. The amount of water used by trees and grasses from each 10 cm soil layer, and during each 2-week period through the year, was calculated by the model (Figure 14.5*a* and *b*). There is strikingly little separation between the depth at which trees and grasses extract their water, but significant differences in the temporal pattern. Note that the temporal tree water use niche completely includes the grass temporal niche at the whole-season scale, and probably at the scale of a single wetting event as well. This situation can lead to an equilibrium solution only if the competitor with the included niche has a much higher resource use efficiency than the competitor with the broader niche. The data in Chapter 6 suggest that there is little difference between trees and grasses with respect to water use efficiency.

In summary, there seems little experimental evidence to support the Walter hypothesis in its purest form. The water use niche separation between trees and grasses in savannas is predominantly on the time-of-year axis. Neither axis suggests the possibility of an equilibrium solution. Trees undoubtedly do have deeper roots than grasses, on average, but this is of little assistance in obtaining large amounts of water. Water enters the soil from the top, and most storms only wet the surface horizons. Trees and grasses are forced to use the same water source for the bulk of their needs. The deep-rootedness of trees is advantageous in assuring survival in times of drought, and thereby allows the rapid response by trees to the beginning of a new wetting event.

The observed patterns are to a small degree dependent on the assumptions made in the model regarding root water uptake. The model assumes that uptake per species is proportional to root length and water content in each layer, to a maximum constrained by the evaporative demand, maximum transpiration rate per unit leaf area, and LAI. These errors are likely to be small.

Figure 14.4. The pattern of leaf area development in trees and grasses of the broad-leafed savanna at Nylsvley. The tree curve is the mean of 10 years of phenological observations on thirty trees (B. J. Huntley and E. Grei, unpublished data), while the grass curve is the mean of 3 years of clipping data reported by Grunow *et al.* (1980).

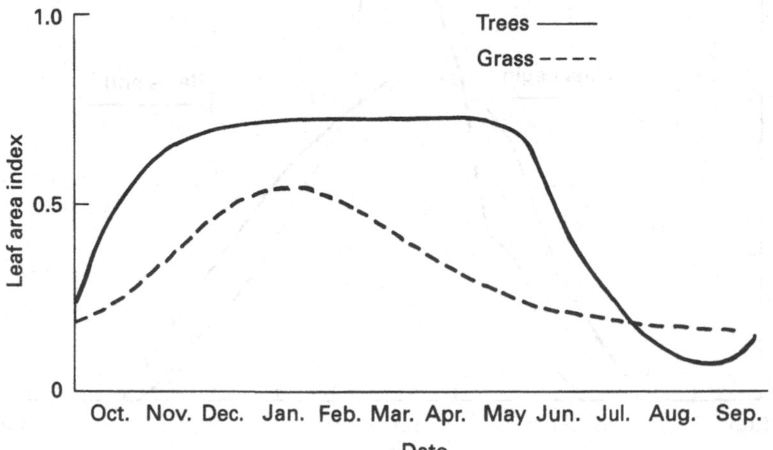

Competition for what?

For competition to occur, two organisms need to share a resource that is essential to their growth, and in limited supply. It has been implicit in the debate about competitive interactions in savannas that this common limited resource is water. One piece of evidence used to justify this choice is the linear relationship between annual rainfall and grass production in

Figure 14.5. Spatial (*a*) and temporal (*b*) patterns of water use by trees and grasses in the broad-leafed savanna at Nylsvley, as predicted by the savanna hydrology model described in Chapter 6. The curves have been scaled such that the area under the tree curve and the grass curve are the same. The predicted total water use by trees was twice that by grasses.

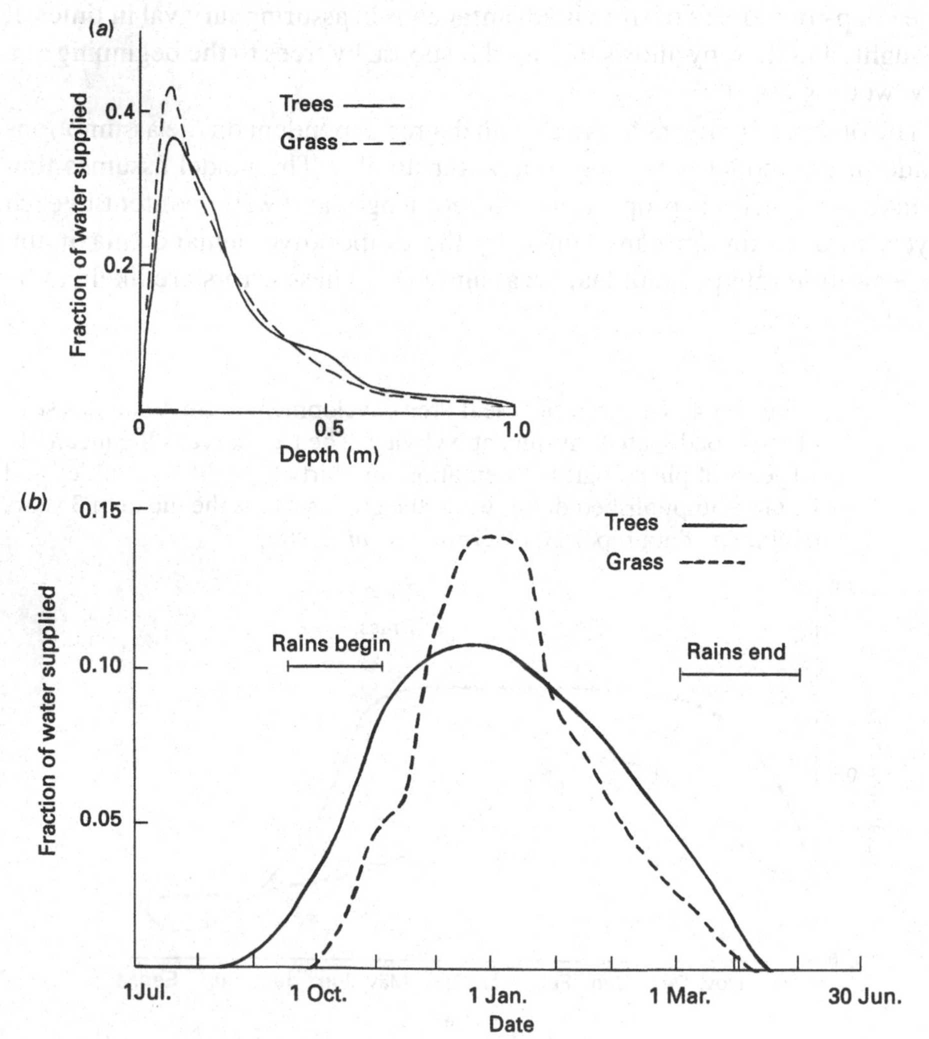

savannas. It has also been implicit that the mechanism of water limitation is via the control on carbon assimilation through stomatal closure. Evidence presented in this book suggests that neither of these assumptions is strictly true.

In Chapters 6 and 7 we have argued as follows:

1. Water acts as an on–off switch to savanna processes, and therefore controls the duration of growth, not the rate.
2. Production in savanna plants is not constrained by their carbon metabolism, but by their nitrogen metabolism.
3. Nitrogen assimilation is controlled by the nitrogen mineralisation rate, which is in turn controlled by water availability.

Therefore, it is likely that the immediate limiting resource in the broad-leafed savanna at Nylsvley is nitrogen, and the key effect of water is to control the availability of inorganic nitrogen, not the control of photosynthesis. This may be true of many other savannas too. This hypothesis would go a long way towards explaining the non-linear relationship between grass production and tree biomass typical of savannas. It would also explain why the degree of non-linearity of this relationship increases as the productive potential of the site declines.

The following sections will discuss classical theories of tree–grass interaction in savannas, which are based on the assumption of water limitation. The *caveat* expressed here should be kept in mind while reading them.

Equilibrium and disequilibrium in savannas

An equilibrium solution for a competitive system requires either niche separation, or competitive superiority by the species with the included niche. The preceding section has shown that there is little evidence for niche separation of trees and grass in either the depth of rooting or the season of uptake. The tree water use niche includes the grass niche in both these respects, and grass does not have a substantially higher water use efficiency than trees do. This suggests that savannas do not represent a stable mixture of trees and grasses, as has been suggested in the past, but an inherently unstable mixture which persists owing to disturbances such as fire, herbivory and fluctuating rainfall.

Work done at Nylsvley and elsewhere has shown that there is strong interaction between the tree and grass layers in savannas. The interaction is usually competitive, but under some combinations of water and nutrient supply, tree morphology and spatial pattern, the interaction could be beneficial, at least to the grasses. The competitive interaction between mature trees and the grass layer is strongly asymmetrical, with trees having a much larger effect on grasses than vice versa. However, while tree seedlings are within the grass layer, this asymmetry is reversed. There is also strong competition between trees and other trees, and between tufts within the grass layer.

A relatively treeless grassland is able to remain that way by suppression of mature tree recruitment through competition, browsing and fire. When the integrity of the grass layer is disrupted, for instance by drought or sustained heavy grazing, tree seedlings can establish. These seedlings persist in the grass sward despite their slow growth rate, possibly because of access to resources in the deep soil layers and a phenological pattern different from that of grasses. If the period between fires is sufficiently long, or the fires are too cool to kill the young trees, and in the absence of intense browsing, the saplings grow beyond the influence of fire or grass competition, and begin to suppress grass growth. It is now less likely that they can be killed by fire, and the woody biomass increases until it is limited by competition with other trees. At this stage (a closed woodland) grass production is greatly suppressed, but grasses persist owing to their ability to exploit opportunities such as exceptionally wet years or treefall gaps.

It appears that many, if not most, savannas have a historical tree density and biomass considerably below the upper limit imposed by intertree competition. This is why the phenomenon of bush encroachment is so widespread. The disequilibrium tree density is maintained by the episodic mortality of mature trees through various agents of disturbance. Fires, herbivory (from Lepidoptera through porcupines to elephants), droughts, frost, lightning, wind and wood harvesting may individually have a small effect, but together, and particularly when they coincide, can significantly influence the tree density. Alteration of the disturbance regime leads to an increase in woody plant density or biomass. The density and biomass will not necessarily revert to their former levels if the disturbance is reintroduced: at least, not in the short term.

Encroachment occurs rapidly, but because of the long lifespan of woody plants, is slow to reverse without direct intervention. If the encroachment occurred in a single episode, the result is often a dense, low-growing thicket of even-aged trees. This situation can return spontaneously to more open grassland when the encroaching trees more or less simultaneously reach the end of their lifespan. More typically, self-thinning and succession will lead in time to a savanna with mixed species and age classes, taller but less dense than the thicket.

Summary

The classical view of the co-dominance of trees and grass in savannas is that it represents a competitive equilibrium, in which the trees use the deep soil water, and grass uses the surface soil water. There is little field evidence to support this idea, and several lines of argument that undermine it.

The alternative view is that the tree–grass mixture in savannas is inherently unstable. In the absence of disturbances which control woody plant biomass (fire in particular), savannas increase in woodiness until tree growth is limited by competition with other trees. Evidence from Nylsvley supports this view.

Trees have a well-known suppressive effect on grass production, but grasses have little direct competitive effect on mature trees. They do have a strong suppressive effect on tree recruitment, both through competition with tree seedlings, and through the effect of fire on small trees. When the grass layer is reduced by heavy and continuous grazing, tree seedlings can grow beyond the control of grass competition and fire, and the result is bush encroachment.

15

Plant–herbivore interactions

The plants and herbivores of African savannas have coexisted for millions of years. This, and the rich variety of organisms occurring in savannas, makes them an ideal laboratory for examining plant–animal interactions. Trees, forbs and grasses have evolved different modes of defence against herbivory, and browsers and grazers have different approaches to overcoming them. Mammalian and insect herbivory have elicited different responses in plants, and there are many variations within each group. There are characteristic differences in the type and degree of herbivory on nutrient-rich versus nutrient-poor sites, and the method of defence adjusts accordingly. The Nylsvley study provided opportunities for examining all these issues.

This chapter should be read as a companion to Chapter 9, on herbivory. That chapter took an ecosystem-level view of the degree and type of herbivory, and its impact on savanna function. This chapter concentrates on individual species, and asks the question: why are some plants eaten, and others not?

This question has major implications for the livestock industry. In Chapter 9 it has been shown that only a small portion of the aboveground primary production is converted into secondary production. This is largely because much of the plant material on offer is unacceptable to animals. As the livestock industry moves towards more complete use of the savanna resource, especially through the incorporation of browsers in the production system and the trend towards multispecies stocking, it is crucial to know what controls the plant–herbivore interaction.

This chapter is based on the work of Norman Owen-Smith, Sue Cooper and John Bryant.

Adaptations which reduce herbivory

Several classes of plant adaptation can have the effect of reducing the impact of herbivory.

1. *Structural adaptations*. The accessibility of the susceptible plant parts to the herbivore is reduced.
2. *Chemical adaptations*. Compounds which are distasteful, toxic or otherwise disadvantageous to the herbivore are accumulated in the plant tissue.
3. *Phenological adaptations*. The susceptible parts are exposed to the risk of herbivory for a reduced period.

Adaptations which reduce herbivory need not have evolved for that purpose originally. A given adaptation can have multiple consequences, and reflects the interaction of a whole suite of environmental factors. The following example from Nylsvley illustrates this point.

For plants growing in nutrient-poor savannas, substantial loss of nutrients is highly disadvantageous, whether it occurs through fire or herbivory. One defence against both is to store a large proportion of the nutrients in belowground organs. A large investment in belowground organs may also have other advantages, such as increasing the ability of the plant to extract nutrients and water from the soil, and to store them for periods when they are less available. Thus there are several possible explanations for the unusually high root:shoot biomass ratios in nutrient-poor savannas. There is a class of woody plants, widespread in nutrient-poor savannas, and especially in those on deep sands, which take this trend to its extreme. Only the leaves are exposed at ground level; a large 'lignotuber' is concealed belowground. In some places they may form an 'underground forest' less than 0.6 m tall (White 1976). There are several such species at Nylsvley, including *Paranari capensis* and *Dichapetalum cymosum*. The consequence of such a subterranean growth form, however, is strong competition for light within the herbaceous layer. One way to overcome this is for the underground trees to grow at a time of the year when the grasses are inactive. The possession of water and nutrients, stored in the lignotuber, permits them to produce green leaf during midwinter. This is well before the beginning of the growing season for grasses, which must await the onset of the rain. This 'window of opportunity' is especially beneficial following early winter burning, which removes the layer of dead grass. However, having the only green leaf during the dry season exposes the underground trees to a high risk of herbivory. They are therefore invariably heavily chemically defended at this time of the year. *Dichapetalum cymosum* accumulates fluoroacetate, a powerful cardiac poison, in its leaves during the vulnerable period. A few mouthfuls are sufficient to kill a cow. The presence of *D. cymosum* is the main reason why the broad-leafed savannas around Nylsvley are not used for cattle grazing in the winter; this is an important constraint on cattle farming in the region. Indigenous herbivores rarely eat *D. cymosum* during its toxic phase.

This example illustrates the difficulty of attributing a particular adaptation to nutrient status, drought, herbivory or fire alone, since all four factors contribute to the environment in which it evolved. It also shows how structural, phenological and chemical strategies are often interdependent.

Structural defence

Structural deterrents in grasses

The complementarity between the grass growth habit and the morphology of grazing ungulates has been suggested as an example of coevolution (Stebbins 1981). A key factor in the success of grasses is the location of the meristem which produces new leaf close to ground level, instead of at the branch tip as in dicotyledonous plants. Loss of leaf through grazing therefore does not mean the loss of opportunity to produce future leaves; this makes moderate grazing losses tolerable. With a few notable exceptions, grasses do not have large spines or distasteful chemicals; this contributes to their relatively high growth rates (at Nylsvley, about $0.03\ g\ g^{-1}\ d^{-1}$, compared with about $0.002\ g\ g^{-1}\ d^{-1}$ for woody plants). Grasses do have siliceous bodies within their epidermal cells, which have the effect of wearing down the teeth of grazers. Among the mammals, only those with high-crowned teeth are able to have a diet consisting mainly of grass. When the teeth wear out, the animal dies.

Some of the grass species most preferred by mammal grazers, such as *Brachiaria* sp., are ignored by grasshoppers (Gandar 1982b). This may be attributable to the presence of hairs on the leaf surface, which are an insignificant deterrent to mammals, but pose a structural barrier to grasshoppers, analogous to a row of thorns.

Structural defence by woody plants

The woody plants growing on nutrient-rich sites at Nylsvley are conspicuously thorny and small-leafed, while the dominant vegetation on the nutrient poor sites is virtually thornless, and large-leafed. Thorns come in many designs, including the long straight variety, the short, backward-curving type and the small epidermal prickle. Thorns are too large to have any effect on insect herbivores; therefore it is assumed that their main purpose is to deter browsing ungulates. Hairy leaves may be the equivalent of thorns to insects, but may also serve to increase the leaf boundary layer resistance to water loss. The shrub *Grewia flavescens* is low in chemical deterrents and is highly acceptable to mammalian browsers, but is not consumed to a large degree by caterpillars, possibly due to the dense hairs on its leaf surface (Owen-Smith & Cooper 1987a).

The amount of food which a browser consumes per unit time is a product of the bite rate and the mean amount of leaf obtained per bite. It is possible that small leaves reduce the bite size, although there are many other factors, such as water relations and energy balance, which could also favour small leaves.

Cooper & Owen-Smith (1986) conducted an experiment at Nylsvley to test the effect of thorns on bite rate and size. Observations on tame animals showed that for kudu, impala and goats the bite rate was little affected by the presence or absence of thorns, but the bite size was substantially reduced in all three browsers when they ate thorny plants (Figure 15.1). The animals slightly increased the average time which they spent feeding on one plant to compensate for the lower intake rate, but the overall consequence of having thorns was to reduce the amount of leaf lost per meal to browsers. The hooked thorns were especially effective against goats and impalas (in particular, they caught in the long floppy ears of the goats), while the long straight thorns prevented the larger browsers such as kudu and giraffe from simply biting off the entire tip of the branch. The browse intake rate was significantly increased in three out of five plant species when the thorns were pruned off the branches before they were exposed to the animals. This was a result of an increase in both mean bite size and bite rate.

Chemical defence and attraction

A large range of plant chemicals can be involved in the selection or rejection of plants by herbivores (Box 15.1). In general, toxic or digestion-inhibiting chemicals predominate in the woody plants and forbs at Nylsvley,

Figure 15.1. The influence of spinescence on bite size and bite rate by three browsers. The diagonal lines indicate the intake rate at various combinations of bite size and rate. Solid symbols represent spiny plants; open symbols, spineless plants (adapted from Cooper & Owen-Smith 1986).

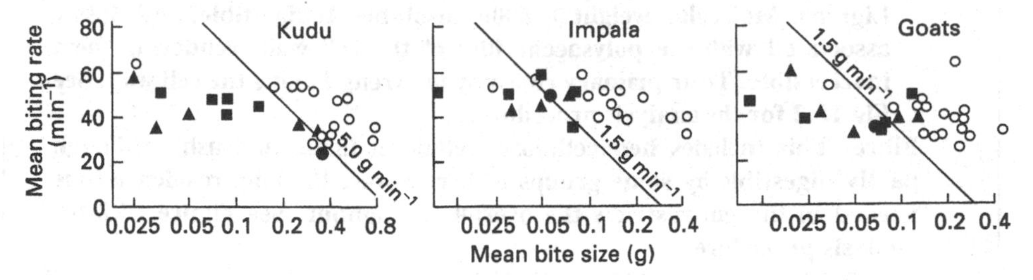

BOX 15.1. A brief guide to the categories of plant compounds involved in mediating plant-herbivore interactions (Rosenthal & Jansen 1979). Since most categories contain a wide range of individual chemicals, and different extraction methods recover them to different degrees, the extraction procedure is a necessary part of the definition of the classes. This is not a complete list, but covers the compounds mentioned in the text.

Nutrients. These compounds are expected to act as attractants to herbivores.

Proteins. (N) Estimated by multiplication of the inorganic nitrogen in a Kjeldahl extract by 6.25.

Minerals. (P, Ca, Mg, Na etc) Determined by spectroscopy following digestion in acid.

Carbohydrates. Make up most of the plant mass. Sugars and starches provide energy to most organisms, but cellulose and hemicellulose can only be digested by specialist bacteria found in the guts of ungulates and termites. To other organisms they are digestion retardants.

Sugars and starches. Extracted in boiling water, converted to glucose for assay.

Hemicellulose.

Cellulose. See Figure 15.2 for assay procedure.

Digestibility reducers. Although not toxic themselves, these compounds reduce the value of a particular food resource to the herbivore, and should therefore be negatively correlated with acceptability.

Phenolic polymers. Immobile and expensive to make, because of the large number of carbon atoms, but require no maintenance. Present in large quantities (up to 40% of leaf dry mass), and therefore called 'quantitative' defensive compounds. 'Total phenolic' content can be indexed by extraction in 75% dimethylformamide.

Tannins. Molecular weight 300–3000, mostly soluble. A diverse group of compounds united by their ability to form insoluble complexes with protein and carbohydrates, thereby rendering them indigestible.

Hydrolysable tannins. Isolated in vacuoles within the cell, and thought to be primarily insect deterrents. Deactivated by mammalian saliva, and digestible by microbes. Extracted in aqueous acetone and assayed by their protein-precipitating power.

Condensed tannins. Located in the cell wall. Probable main function is to protect the living leaf against microbial attack. Also deterrent to large mammal browsers. Estimated by extraction in 70% acetone and assaying for a component molecule, proanthocyanidin.

Lignins. Molecular weight > 5000, insoluble. Indigestible, and closely associated with the polysaccharides of the cell wall, rendering them inaccessible. Their primary role may be strengthening the cell wall. See Fig 15.2 for the analysis procedure.

Fibre. This includes hemicellulose, cellulose, lignin and ash. Although partly digestible by some groups of herbivores, the long residence time needed in the gut restricts the overall throughput. See Figure 15.2 for analysis procedure.

BOX 15.1. (*Cont.*)

Ash. Consists of insoluble minerals, and in the case of grasses, mostly silica. Determined as the residue after combustion at 550 °C.

Saponins. Have detergent properties which interfere with digestion. Could be regarded as toxins, especially to insects, where they inhibit development. Detected by a foaming test.

Toxins. Should be avoided by all herbivores except those specialised in detoxifying particular compounds. Mostly mobile, low molecular weight nitrogen-containing compounds, present in small absolute quantities and therefore known as 'qualitative' defensive compounds.

Alkaloids. Heterocyclic nitrogen-containing compounds. Very many types, each with a specific action. Important in defence against insect herbivory.

Terpenes. Aromatic oils, uncommon in savannas except in some grasses. Extracted in ether, and characterised by Thin-Layer Chromatography. Some are toxic.

Cyanogenic glycosides. Generate cyanide when the tissue is ruptured. Detected by the darkening of picrate paper.

Lectins. A type of amino acid, frequently found in seeds.

but are virtually absent from grasses. Grasses show a range of acceptability according to their fibre and protein content (Figure 15.2 illustrates the analytical definition of fibre used here). Nitrogen is a herbivory attractant, in the sense that protein content is positively correlated with plant acceptability by insect and mammal herbivores at Nylsvley. Cattle and impala select for the species, plant parts and growth stages of grass which have higher than average protein contents (Grundlehner 1989). Sodium is also attractive, since it is required by animals but not plants, and is therefore usually deficient in plant tissues. There is strong selection by large mammal browsers for the plants growing on the sodium-rich soils near the Nyl river. Bailey (1990) showed that these plants had higher leaf sodium contents than plants of the same and other species growing on nearby low-sodium soils.

Chemical deterrence by woody plants

The large mammal browsers at Nylsvley do not consume leaves of different species in the same proportion as they are encountered. Owen-Smith & Cooper (1987b) compared several indices of the acceptability of a range of woody plants to kudu, and concluded that 'Site-based Acceptability' (SA) was the easiest to measure and interpret. They defined the SA as the ratio of the number of 30 minute observation periods during which a particular species was observed to be eaten, to the number of times that it was available within a 10 metre radius of the browser during that period. Three fairly clear categories of acceptability emerged: favoured (SA > 0.45), intermediate and neglected (SA < 0.1; Figure 15.3). These categories were also reflected in the mean feeding duration on a particular species. The intermediate plants were mostly species which changed dramatically in acceptability during the year. The unpalatable category accounts for nearly 80% of the available browse in the broad-leafed savanna.

The SA scores for the most palatable species were fairly constant between times of the year, but three seasonal patterns were discernible among the less palatable plants: those that were always unacceptable; those in which the acceptability increased during the dry season; and those which were acceptable in spring, but declined in acceptability as the summer progressed (Owen-Smith & Cooper 1987a).

Figure 15.2. A fractionation scheme for quantifying the cell wall constituents, known as 'fibre'. The soluble sugars and starches which occur within the cell are digestible by all animals, but the complex carbohydrate polymers which make up the cell wall can only be digested by certain symbiotic microorganisms, found in the gut of termites and ungulates. Fibre is a major source of metabolic energy for ungulates, but it is at the same time a digestion-retarding factor. Efficient digestion of high-fibre requires a long residence time in the gut, which limits the amount that the ungulate can eat. Lignins are indigestible, and may protect some cellulose from digestion.

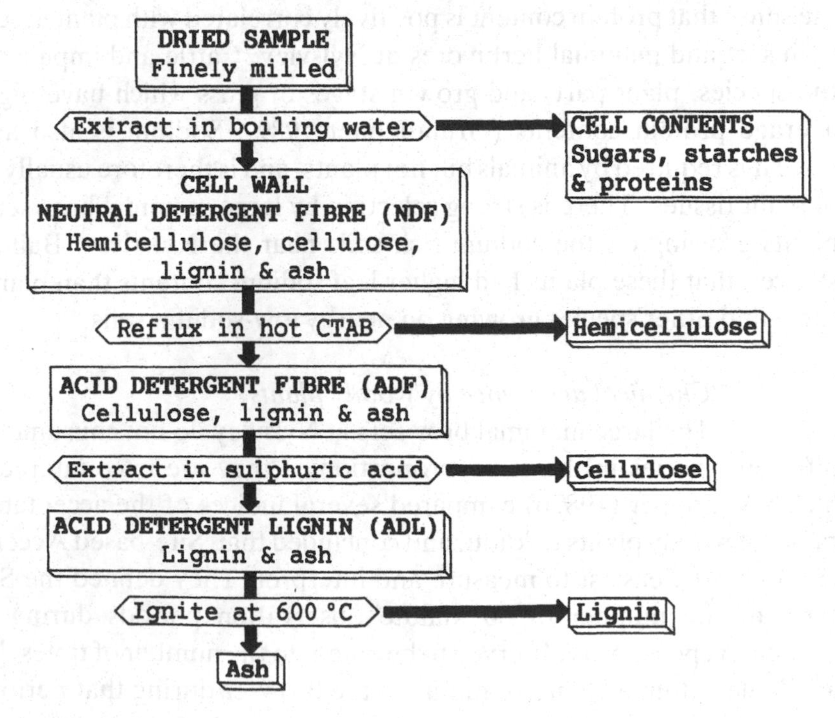

The relative acceptability of woody plants was found to be broadly similar in the three species of mammalian browser studied (kudu, impala and goats). For this comparison, 'Plant-based Acceptability' was used, defined as the ratio of plants of a given species eaten to the number encountered within 0.5 m of the animal's path through the savanna. Kudu was the only species which was entirely dependent on woody plants for food, and showed the clearest discrimination between favoured and disfavoured food plants. Goats and impala, consuming 57% and 19% of their diet as woody plants, respectively, were much less discriminating (Figure 15.4). There was no correlation between acceptability to mammalian herbivores and acceptability to insect herbivores. One of the species most palatable to mammals, *Grewia flavescens*, is virtually untouched by insects, perhaps because of the stellate hairs on the leaf surface. Similarly, the species which support caterpillar outbreaks all have low mammal palatability.

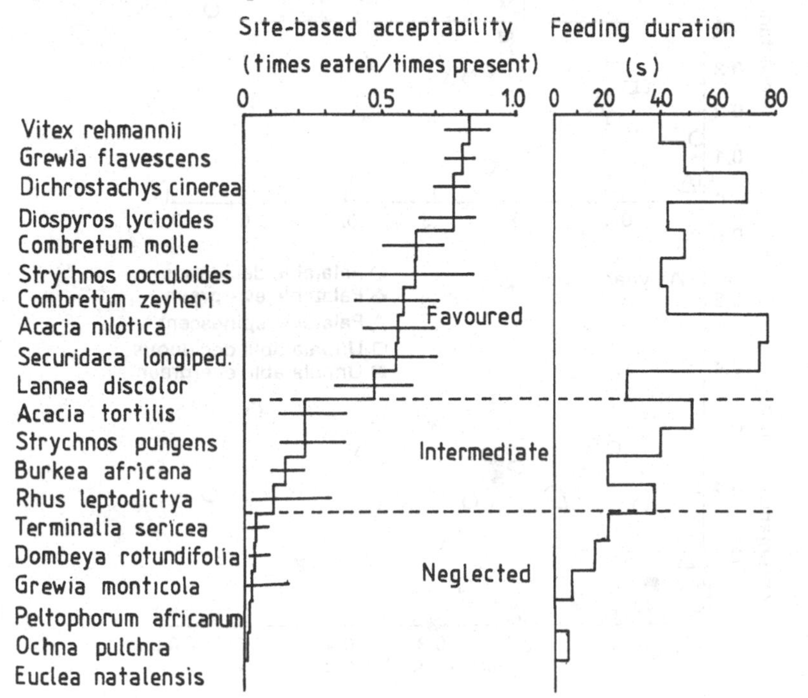

Figure 15.3. Site-based Acceptance scores for woody plants by kudu (*Tragelaphus strepsiceros*) at Nylsvley. Acceptance is defined as the number of 30-minute intervals during which a given species was observed to be eaten divided by the number of intervals when that species was within visible range (10 m) of the browser (redrawn from Owen-Smith & Cooper 1987b).

Associating the browse acceptability with specific aspects of leaf chemistry proved to be much more difficult than expected. Individual chemical factors were poorly correlated with acceptance. Phosphorus and magnesium were significantly ($p < 0.05$) positively correlated with SA at some times of the year, while total fibre, cellulose and condensed tannins were negatively correlated with SA at others. The picture was no clearer using multivariate techniques. The first axis of a principle component analysis of a matrix of chemical attributes by species accounted for 54% of the variation in leaf chemistry, but only about 15% of the variation in kudu and impala SA scores. The second axis removed a further 26% of the chemical variation, but only another 9% of the SA variation (Cooper, Owen-Smith & Bryant 1988).

Figure 15.4. Plant-based acceptances by kudu in relation to those by impala and goats. Plant-based acceptances are defined as the number of plants of a given species eaten divided by the number encountered within 0.5 m of the browser (data from Owen-Smith & Cooper 1987a).

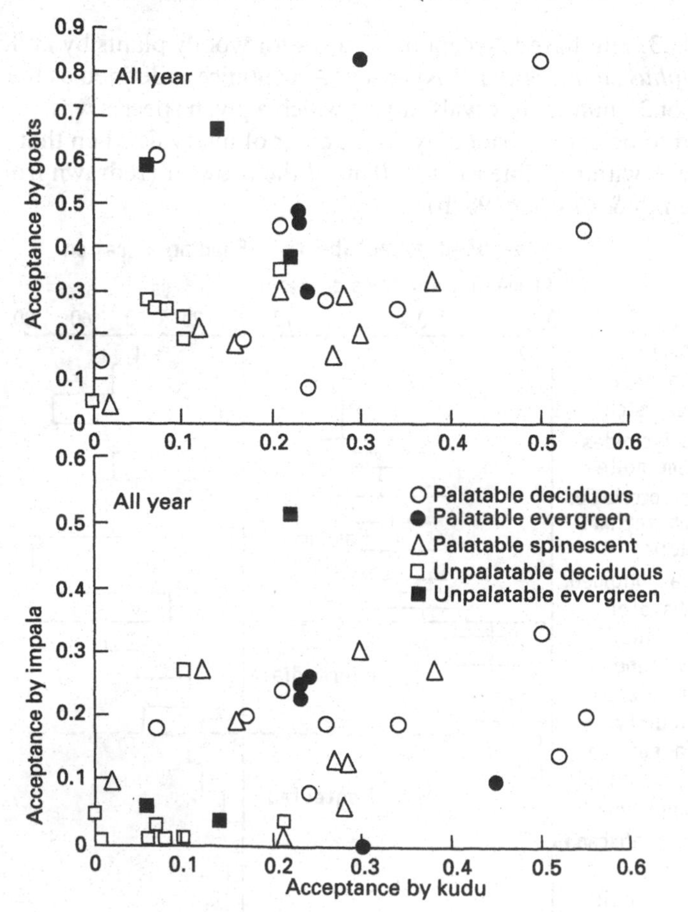

Correlative techniques such as those applied above assume a linear relationship between the variables, whereas palatability can be a strongly non-linear response. A scatter-plot of the late wet season acceptance values versus condensed tannin concentration in the leaves suggests a threshold effect: high acceptabilities (> 0.25) occur only when the condensed tannin content is less than about 3% (Cooper & Owen-Smith 1985). Total polyphenolic content, which includes the hydrolysable tannins as well as the condensed tannins, shows no such pattern. When data from all seasons are considered, however, this simple rule fails. Instead, excellent discrimination between palatable and unpalatable leaves can be obtained using an index including both protein and condensed tannin content (Figure 15.5; Cooper *et al.* 1988). High acceptabilities occur at values above 90 of leaf protein content minus half of the leaf condensed tannin content, both expressed in mg kg^{-1}.

Another approach to the problem of palatability assessment is via the concept of digestibility. This is based on the assumption that evolution will have resulted in the animals selecting mainly those plants which they are able to digest. A study of *in vitro* digestibility conducted by Boomker & van Hoven (1983) produced digestibility estimates which correlate well with the SA values reported in the Owen-Smith & Cooper (1987a) study ($r^2 = 0.74$, $n = 7$). A particularly low digestibility was recorded for *Peltophorum africanum*, which contains large amounts of saponins and is ignored by browsers despite a low condensed tannin content. A complementary study (van Hoven & Boomker 1983) showed marked differences between browser species in the degree to which their rumen fluid was able to digest a standard substrate (antelope cubes). Furthermore, the fermentation rate was substantially lower when the assay was conducted with rumen fluid collected during the winter months.

The impala and kudu used in feeding studies at Nylsvley were tame animals, raised in captivity from the age of 2–5 days. This provided an opportunity to study the process whereby animals learn their feeding habits. The findings support the hypothesis that food choice is partly innate, and partly learned. Frost (1981) presented eight young animals with freshly cut branches of 48 woody plant species. One fifth were immediately accepted, half were smelled and then rejected, and the remainder were tasted and rejected. The first two classes were fairly consistent between experimental animals, while most species in the third class were only tasted by 1–3 animals, and rejected on smell by the remainder. Once tasted and rejected, the same species were later rejected on smell alone by all the animals. The young impala rejected 10 species out of the 16 recorded by Monro (1979) as being eaten by adult animals; however, all the plants that they accepted are listed by Owen Smith & Cooper (1987a) as palatable. The feeding habits of the captive impala changed after they were released into a large pen with a variety of woody plants. They were soon eating plants which they had previously rejected.

Figure 15.5. Discrimination of palatable from unpalatable browse species on the basis of an index incorporating the effect of condensed tannins on proteins in the leaf. The derivation of the canonical axes used in the lower two diagrams is illustrated in the top diagram, where it appears as the broken line. It is a linear function of the leaf protein and tannin contents (data from Cooper *et al.* 1988).

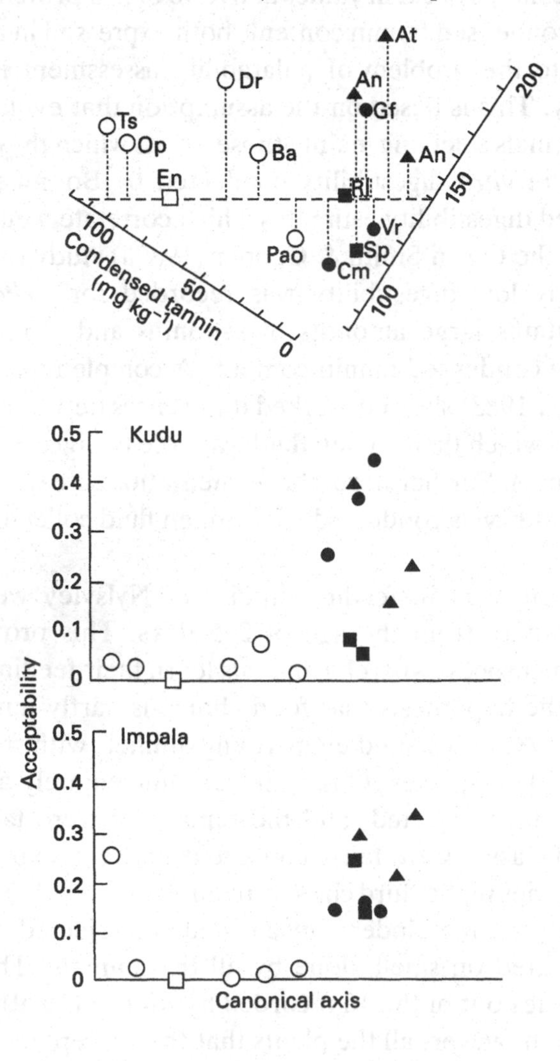

The cost/benefit metaphor for plant defence

Plant defence and animal foraging theory is built largely on concepts of marginal value borrowed from economics (Pyke, Pulliam & Charnov, 1977). Defence incurs a direct or indirect cost to the plant, either in terms of the energy, carbon, nitrogen or other nutrients used in constructing and maintaining the defence, or in the 'opportunity cost' of not having grown productive structures instead. The benefits of defence lie in preventing the loss, due to herbivory, of energy, carbon, nutrients or the opportunity to produce or reproduce. Consumption of a leaf represents not only an immediate loss of carbon and nutrients, but the loss of the carbon which that leaf would have assimilated in the future. The theory predicts that defence will occur when the benefits exceed the costs. The type and degree of defence will depend on risk of herbivory of various types, and the relative availability of resources to the plant. Because the pattern of growth differs between plant species, and the pattern of herbivory and resource availability varies with time and species, there is usually more than one way of reducing herbivory, even in one environment. Usually several species exhibit similar adaptations to herbivory, either through common ancestry, or through convergent evolution: this is known, somewhat loosely, as a 'defence strategy'.

Competing theories: apparency versus resource availability

There are currently two main theoretical frameworks for explaining the observed patterns of chemical plant defence (Howe & Westley 1988). 'Apparency theory' suggests that plants which are easily found by animals ('apparent') will have large quantities of broad-spectrum herbivore deterrents, whereas 'unapparent' plants need only small quantities of simple, highly herbivore-specific toxins (Feeny 1976; Rhoades & Cates 1976). This theory views the plant–herbivore interaction essentially from the herbivore side.

'Resource availability theory', on the other hand, takes a plant-centred view, in that it proposes that the degree and type of defence is constrained by the amount of various resources available to the plant. It makes three predictions. Firstly, the degree of defence is related to plant growth rate. Fast-growing plants in resource-rich environments invest little in defence, because lost tissue can easily be replaced. Plants growing in resource-poor situations are inherently slow-growing and therefore need to be heavily defended. Fixed defences, such as tannins and lignins, are initially expensive but maintenance free, while mobile defences, such as alkaloids, are initially cheap because they are made in small quantities, but require continuous resynthesis and are therefore expensive in the long run. The second prediction is therefore that the type of defensive chemical is related to the longevity of the organ: long-lived plant parts such as the leaves of evergreen trees should have predominantly fixed defensive chemicals. The third prediction is that the defensive chemicals will reflect the relative availability of carbon and nitrogen to the plant. Where carbon is restricted, for instance in shaded habitats, nitrogen-based chemicals such as alkaloids and cyanogens will be favoured; while where nitrogen is limited, for example on infertile soils, carbon-based polymers such as lignin and tannin will predominate.

Resource availability theory was originally developed to explain patterns of defence in Alaska, but important supporting evidence has been provided from Nylsvley (Bryant *et al.* 1989). For instance, plants with low acceptability to kudu and impala have low inherent growth rates.

Plant defence in nutrient-rich versus nutrient-poor sites

The broad-leafed savanna at Nylsvley is relatively low in plant-available nutrients. About one sixth of the study site is occupied by fine-leafed savannas, which are much richer in a whole range of plant nutrients. The origin of these patches, and the many ecological differences associated with them, are discussed in Chapter 12. This section will enlarge upon those aspects that relate to plant defence and herbivory.

The prevalence of spinescence and tiny leaves in the fertile savannas, and large leaves and no spines in the infertile savannas, has already been mentioned. Both browsers and grazers show a strong preference for the fine-leafed savannas. They are found there up to six times more frequently than would be expected from a random distribution model. The dominant trees of the fertile patches are characterised as palatable to browsers (Owen-Smith & Cooper 1989), while those of the infertile savannas are unpalatable. On the other hand, the broad-leafed savannas have much higher levels of tree defoliation by lepidopteran larvae, characteristically concentrated into massive outbreaks every few years.

The apparency theory is not supported for woody plants at Nylsvley, since tree leaves in the fine-leafed and broad-leafed savannas are equally apparent, but differently defended. The general pattern among woody plants is a reliance on structural deterrents (thorns) in the fertile savannas, and chemical deterrents in the broad-leafed savanna. Neither savanna is considered to be constrained by carbon assimilation, so the defences are carbon-based in both cases.

In broad terms, the defence allocation pattern conforms with the resource availability theory of Coley *et al.* (1985). The woody plants in the infertile broad-leafed savanna have net photosynthetic rates about 10% lower than those in the fertile savannas, and this translates into lower litterfall per unit tree basal area. The growth rate of seedings of infertile savanna trees is three to five times lower than that of seedlings from fertile savannas, even in the presence of abundant light, water and nutrients (Bryant *et al.* 1989).

The slow-growing trees of the infertile savannas have a large defensive investment. Non-lignin polyphenols contribute up to 30% of the leaf dry mass, of which up to 12% consists of condensed tannins. The fibre content is up to 52%, with lignin contributing up to 22%. The total defence investment in the fertile savanna is much less: thorns make up only a small percentage of the plant mass.

There are some unexplained features, however. One is the caterpillar outbreaks in the broad-leafed savanna. The fine-leafed trees of the fertile savannas can have total polyphenol contents as high as those recorded in the broad-leafed savanna, although the condensed tannin and lignin contents are lower. This may be why insect herbivory is low on the fine-leafed savanna trees. Why does *Grewia flavescens* have high growth rates, despite being found in the infertile savannas? One theory is that it favours locally disturbed sites with high nutrient availability, such as warthog burrows. *Terminalia sericea* also has high inherent growth rates, but is hardly touched by mammalian browsers.

The Coley *et al.* (1985) model predicts lower leaf protein contents in the nitrogen-poor site. This is true for the grasses at Nylsvley, but is less valid for the dominant trees, which are relatively high in protein in both cases. The leaf protein content appears to be related more to the tree family (legume or non-legume) than to the fertility status. Where does the nitrogen for this protein come from, given that none of the leguminous trees at Nylsvley has been demonstrated to fix nitrogen in the field ? At present, the best explanation which can be given is the high degree of internal nitrogen cycling which takes place in the trees, but not the grasses.

Browser feeding patterns

There is a predictable gradation in the body size of coexisting browsers in savannas: each species is 2.0–2.5 times larger than the next (Owen-Smith 1985b). Body size has a major impact on the feeding behaviour of browsers, not only in the maximum height at which they can feed, but also in the type of material which they select. The size of the mouthparts and the gut compartments influences ingestion and throughput rates, which determine the efficiency of digestion. Body size also controls the metabolic energy requirements of the animal. Owen-Smith (1985b) developed an optimisation model based on these considerations to predict that smaller browsers required progressively more protein in their diet. Thus the smallest browsers (such as duiker) are forced to pick out only the highest-quality plant parts, while large browsers, such as black rhinoceros, can afford to be less selective.

Cooper (1982) compared the feeding behaviour of an indigenous mixed feeder (impala) with that of an 'alien' mixed feeder (goat) of similar body mass. Impala spend a higher proportion of their time eating grass: 79% in summer, and 45% in autumn; while goats grazed for 27% in summer and 11% in autumn. Fallen leaves and pods were an important part of the diet of both during the dry months. Goats ate a wider range of species than impala did: eight plant species were consistently eaten by both, five were eaten by goats but rejected by impala, and one was never eaten by goats but sometimes eaten by impala.

There is a fundamental difference between the resource limitation experienced by grazers and that experienced by browsers, as has been pointed out in Chapter 9. The feeding behaviour of grazers can be interpreted as promoting the intake of higher quality forage than is generally on offer, while that of browsers promotes the maximisation of energy gains (Owen-Smith & Novellie 1982).

Browsers in the broad-leafed savanna have a further problem in that most of the plants contain digestion-retarding substances. While the diet is dominated by the leaves of palatable woody plants and forbs during the wet season, food scarcity in the dry season forces them to eat increasing quantities of unpalatable species (Owen-Smith & Cooper 1989). However, by not browsing on a single species for longer than a few minutes, even when plenty of leaf tissue remains available on that plant, the kudu appear to be able to keep their intake of any one plant chemical to sub-critical levels. The total polyphenol content of the diet stays below 6% throughout the year, while condensed tannins stay below 3%.

Grazer feeding behaviour at Nylsvley

Differing browsing heights and tolerance to plant chemicals can explain the diversity of browsers, but what about grazers, where the diversity is just as high? All grass grows at essentially the same height, and is therefore available to all grazers. Few grasses have chemical defences. Some niche differentiation is possible on the basis of selection for different species or plant parts within the sward, enabled by variations in mouth structure and digestive efficiency (Owen-Smith 1985b). The overall high efficiency of the foregut fermentation system found in the bovids allows small differences in the quality of forage intake to have large consequences. There is considerably less speciation in the equids, which have a less efficient hindgut fermentation system. The categorisation of grazers into 'coarse' and 'fine' grazers is popular among African wildlife managers. The suggestion is that fine grazers are more selective about their intake than coarse grazers are. In reality, all grazers are probably as selective as the size of their mouthparts allows them to be.

Cattle have relatively large mouths, and so tend to bite off the entire tuft, rather than selecting the more nutritious plant parts within the tuft. Antelope such as impala have smaller mouths, and can therefore be more selective. Grundlehner (1989) compared the feeding patterns of impala and cattle, with particular reference to their impact on the grass plant. At the same grazing intensity, impala and cattle selected slightly different proportions of the grass species on offer. Impala ate mostly leaf tissue, while cattle ate both leaf and stem. Grass tufts clipped to simulate impala-like feeding showed no difference in the amount of regrowth from tufts clipped in a cattle-like pattern, although in the case of *Eragrostis pallens*, the latter resulted in greater tiller production.

Since Africander steers grazing in the Nylsvley study site consumed only 0.1% of the available grass, Zimmerman (1980b) did not consider lack of forage to be a constraint on the rate of intake, but the time spent searching for the better quality resources may have been. He related live-weight gain of fistulated Africander steers grazing in broad-leafed savanna to the intake of both digestible organic matter (DOM) and its crude protein content (Zimmerman 1980b). The animals showed good weight gain (greater than 7 g LM kg $LM^{-0.75}$ d^{-1}) from October to March (the wet season), and poor weight gain or even weight loss for the rest of the year. Forage intake varied by about 20% between the wet and dry season, but was not strongly related to either the *in vitro* digestibility or crude protein content of the forage. The digestible crude protein content of the forage intake was the best predictor of live weight gain (r^2 = 0.89). The maintenance requirement (zero weight gain) was 53% digestible organic matter in the forage (44 g DOM kg $LM^{-0.75}$ d^{-1}), and a crude protein content of 5.6%. The seasonal pattern of crude protein intake is consistently higher than the average protein content of the dominant grass (Figure 15.6), indicating that the animals were successfully selecting the more nutritious grass species. The steers foraged, on average, for 9 hours a day.

Quantification of digestibility and crude protein in the forage intake through the use of fistulated animals is an unpleasant procedure. Zimmerman (1980a) found that the nitrogen and moisture content in the dung correlated very well with the protein content and digestibility of the forage, thus offering a much simpler way of collecting this information. This approach could be used only on animals feeding almost entirely on grass.

Summary

Savanna plants show many adaptations which serve to reduce herbivory, although they may serve other functions as well. The pattern of anti-herbivore defence in woody plants supports the idea that the degree and type of defence is related to the availability of resources to the plant, rather than how easily the plant is located by herbivores. For savanna trees growing on nutrient-poor soils, the main deterrent to mammalian herbivory is the quantity of condensed tannins in relation to the protein content of the leaf. Hydrolysable tannins are also found in these leaves, and are probably more related to defence against insect herbivory. Savanna trees on nutrient-rich substrates are defended with thorns, which do not prevent browsing, but slow it down to a tolerable level.

The plant adaptations have resulted in considerable specialisation on the part of the indigenous herbivores, which is not matched in imported domestic herbivores such as cattle and goats. Indigenous browsers feed from a single plant for less than a minute, and move on even if leaf material remains accessible on the plant. This is thought to keep the concentration of any one

Figure 15.6. The protein content of grass eaten by Africander steers, in relation to the average protein content of dominant grasses in the broad-leafed savanna. The intake data are from Zimmermann (1980b). The grass data were collected in a different year (M. C. Scholes, unpublished data), but serve to illustrate that the animal must be selecting for the parts of the sward with above average protein content.

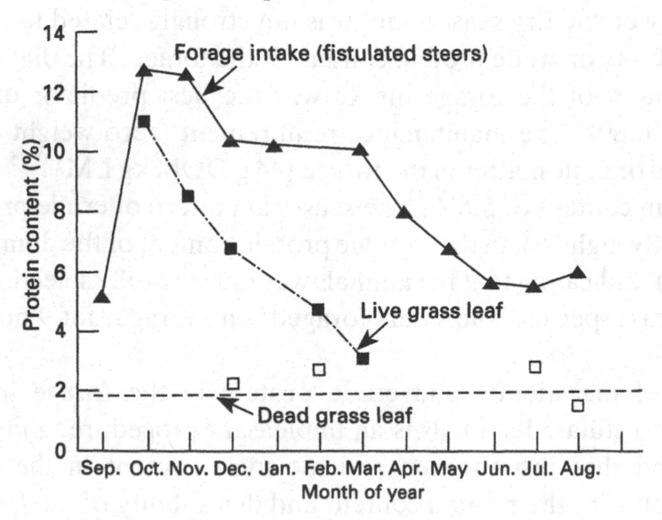

plant chemical in the forage intake below the toxic threshold. Indigenous grazers avoid plants in the herbaceous layer which are fatal to cattle.

This process of adaptation and counter-adaptation is an important contributory factor to the diversity of both plants and animals in savannas.

Part V

Lessons from Nylsvley

16

An overview of savanna ecology

A great deal has been learned about savannas during the past two decades, both as a consequence of the South African Savanna Biome Programme and of the increased interest in savannas in the world scientific community. Compare the state-of-the-art review of savanna ecology before the Nylsvley study began (Bourlière & Hadley 1970) with the range of texts now available, of which this book is just one example (Huntley & Walker 1982; Bourlière 1983; Sarmiento 1984; Tothill & Mott 1985; Cole 1986; Walker 1987; Werner 1991). Over this period there have been significant conceptual shifts within the broad science of ecology as well. It is often difficult to say exactly where and when these ideas originate, but within the South African context, most of them have been introduced through the Ecosystem Programmes.

When the Nylsvley study began, the prevailing view of ecosystems was one of reasonably stable entities, when left to their own devices. Twenty years on, they are seen as frequently being far from equilibrium, with a structure and function largely driven by disturbances. Thus although there are more answers now than twenty years ago, many of the questions have changed as well.

Determinants of savanna structure and function

When the programme began, the only predictive model of savanna structure and function was the two-layer soil hypothesis proposed by Walter (1971). This model predicts that the proportion of trees in the savannas is a function of rainfall amount and distribution in relation to the physical properties of the soil, in particular the water-holding capacity. An important contribution of the Nylsvley work was to emphasise the pivotal role of soil fertility, a soil chemical property, in defining savanna structure and function.

The Walter Hypothesis, in a modified form, still persists as a model of tree–

grass interaction (see below), but the current conceptual model of savanna composition and structure is based on a four-determinant scheme which originated in the Savanna Biome Programme (Frost *et al.* 1986). It is not a mechanistic model in the sense that the two-layer model was, but is an empirical scheme for classifying savannas which emphasises continuous gradients rather than discrete classes.

The four-determinant model gives water availability and nutrient availability equal status in establishing the range of possible forms a savanna can assume, and in constraining functional attributes such as primary production. Fire and herbivory then determine the actual form and function within that range. The predictions of the model, such as tendency to higher woody biomass with increasing water availability and decreasing nutrient supply, are currently being tested.

Water, nutrients and primary production

In savannas receiving less than about 900 mm rainfall per year, the aboveground production of the herbaceous layer is linearly related to the annual rainfall. The slope of this line is controlled by nutrient availability, and the *x*-axis intercept by the soil physical characteristics (for which clay content is an index) in relation to the climate. The number, size, species and spatial distribution of woody plants also has an impact on the position of this line, particularly on the *x*-axis intercept. Woody plant production is also related to rainfall, but less strongly. There is some carry-over of the effect of rainfall in the previous season, causing tree production to be less variable than either annual rainfall or grass production.

The broad-leafed savanna at Nylsvley is towards the dry and infertile end of the savanna spectrum, and has a total annual productivity of about 950 g $m^{-2} y^{-1}$, and a peak phytomass of about 4 kg m^{-2}. This is split almost equally between trees and grasses, and just over half is belowground production. Primary production in the more fertile, fine-leafed savanna at Nylsvley is about 50% higher, with a smaller proportion of the production and biomass belowground. High belowground biomass is typical of broad-leafed savannas, and is thought to be a response to fire and herbivory in a low-nutrient environment.

Fire in savannas

We believe that fire is axiomatic in African savannas. The evolution of the savanna biome was the consequence of the establishment of summer-wet winter-dry tropical climates, which permitted frequent, hot fires. The Nylsvley study, and many others, provide evidence of increases in woody biomass in savannas when the fire frequency or intensity is reduced. A salutary lesson from Nylsvley is the subtlety by which fire-related processes may operate. Death of the dominant tree species, *Burkea africana*, is largely a consequence of repeated fires, over a long period, acting on a small scar resulting from feeding by a porcupine when the tree was young. The death of the fire-weakened tree is ultimately caused by a strong wind, which snaps it off at the base. On the basis of their contribution to herbivore biomass, energy cycling and nutrient flux, porcupines are an insignificant part of the savanna ecosystem, yet they are a key species in terms of control of the woody biomass and composition.

There are strong interactions between fire and the other savanna determinants. The combination of plant available water and nutrients determines the fuel load, which in turn determines fire intensity. Water availability defines the duration of the main fire season. Fire also has a transient impact on the hydrology of savannas, and a substantial effect on nutrient cycling, particularly of nitrogen. Occasional fires accelerate the mineralisation of nitrogen, and therefore boost the grass layer production and protein content in the following year. Annual fires can reduce the carbon and nitrogen pool in the savanna, and therefore reduce its average production and quality.

Herbivory and savanna structure

Herbivory influences savanna structure either directly, or by reducing the fuel load available to fires. The direct effects are obvious in the case of large herbivores such as elephants or giraffe, but even relatively minor herbivores, evidenced by the porcupine example above, can have an important effect on tree density. The impact of browsers also interacts strongly with fire: fire keeps the browse palatable and within the reach of browsers, while browsers keep woody plants within the flame zone. The indirect effects of herbivory, operating via fire, are just as important as the direct impacts, particularly in fertile savannas. Grazing ungulates can consume a large fraction of the aboveground herbaceous standing crop, even if it is dead, thereby reducing the fire intensity. Commercial farmers are loath to burn in fertile savannas, knowing that cattle will eat the grass, and therefore the fire frequency decreases as well. The consequence is that woody plant saplings are able to escape from the herbaceous layer. As the tree and shrub cover increases, so the grass production is suppressed.

Patterns of herbivory and plant defence

A substantial portion of the herbivory in African savannas is attributable to insects. Both insects and mammals are attracted to plants with a high protein content, but they are deterred by different factors. The content of condensed tannins in relation to the protein content is strongly related to the acceptability of leaves by several species of browsing mammals. The factors controlling the acceptability of savanna tree and grass leaves to herbivorous insects have not been demonstrated and may be species-specific, but there is circumstantial evidence to suggest that they include the hydrolysable tannin content and structural features of the leaf surface, such as hairs.

The pattern of anti-herbivore defences in nutrient-poor savannas is different to that in nutrient-rich savannas. The former have leaves with a high condensed tannin and lignin content, while the latter have thorns to slow down, but not prevent, mammalian browsing. With reference to defence against mammalian browsers, the observed patterns in the degree and type of plant defence support the hypothesis that plant defence mechanisms are controlled by the availability of carbon and nutrients to the plant, rather than by the probability of the plant being eaten.

The outcome of the different woody plant defence patterns characteristic of the contrasting types of savannas is to reduce the feeding rate in nutrient-rich savannas, and reduce the feeding duration in nutrient-poor savannas. Both result in reduced loss of leaf through herbivory: only about 10% of leaf production is consumed. Mammalian browsers feed for an average of less than a minute on a single plant, even if it is palatable and still has available browse. The feeding duration on unpalatable plants is less than 25 seconds. This feeding behaviour results in varied diet, and keeps the dosage of any one type of plant chemical below the threshold at which it could interfere with the herbivore metabolism. There are few deterrent substances in the grasses found at Nylsvley, although the grasses of the infertile savannas are generally high in digestion-resistant compounds. Feeding behaviour in grazers is related to selection for the plant parts, microhabitats and landscape patches which offer forage with higher than average protein content.

The limits of herbivory

Some of the heat has gone from the debate about whether or not herbivores are limited by food supply, but the mechanism of herbivore limitation remains an issue of economic importance. The Nylsvley data illustrate a widespread paradox: herbivores can be food-limited despite consuming only a small portion of the primary production. This can occur for a variety of reasons. In the case of mammalian browsers, the food supply is restricted at a particular time of the year, the late dry season after the leaves have fallen from the trees. The phenology of browse plants in relation to the climate determines the duration and intensity of this period. Alternatively, the limit may be set by the abundance of a single key resource, such as the palatable shrub *Grewia flavescens*, which forms a minor component within the generally unpalatable broad-leafed savanna. The amount of forage within feeding height is crucial to browsers. Grazers within nutrient-rich savannas are able to consume a high proportion of the aboveground production, and may in fact be directly limited by primary production. The persistence of their food plants is thanks to the remarkable adaptations of the grass plant under these conditions to sustained high removal rates. In infertile savannas the low protein and high fibre content of the grasses during the dry months constrains their acceptability to ungulates, particularly those with a fore-gut fermentation digestive system.

Two broad generalisations can be made regarding the limitation of herbivory in savannas. The first is that secondary production in infertile savannas is controlled by the fraction of the primary production which is acceptable to herbivores, rather than by the total primary production; while in fertile savannas the primary production may itself impose the limit. The second is that secondary production in grazing mammals is constrained by the quality of the food they can ingest during the dry season (and possibly also by its quantity in heavily stocked, fertile savannas), whereas browsing mammals are constrained by the quantity of food they can locate in the period following leaf fall.

A model of the tree–grass interaction

The interaction between trees and grasses lies at the very core of what makes savannas different from forests or grasslands. Water and nutrient availability determine the primary productivity of the savanna as a whole, but it is the degree of woodiness of the savanna which determines what fraction of that production is provided by grasses. The two-layer soil model (Walter 1971) was proposed to account for the coexistence of trees and grasses in savannas, and the way in which they partitioned the available water. Theoretical studies (Walker *et al.* 1981; Walker & Noy-Meir 1982) demonstrated that this model could reproduce some of the observed structural and dynamic features of savannas, and subsequent tests of the model (Knoop & Walker 1984) showed that its application was highly dependent on the particular combination of rainfall and soil type. Measurements of water and root distribution at Nylsvley and elsewhere (Scholes 1987) revealed that the roots of both grasses and trees extend into the subsoil layers (contrary to the Walter Hypothesis) and in soils shallower than 1 m they overlap almost completely in their vertical distribution. On the infertile soils at Nylsvley, water-holding capacity is low and a significant portion of the rain penetrates through to the deeper layers where woody plant roots are relatively more abundant. On the fertile patches the water-holding capacity is higher (due to the higher soil organic matter content) and water gets through to the subsoil less frequently. Most uptake therefore occurs in the upper 25 cm, where grass roots are relatively more abundant. So, while the vertical separation hypothesis does distinguish different savanna types, the original Walter Hypothesis assumptions are not fully supported. The largest degree of niche separation between trees and grass exists in the time at which their roots are active: trees develop a full leaf canopy, and presumably a root system to match, within weeks of the first rains, while grasses reach their peak after only a few months. The early growth of trees, which frequently precedes the rains, is permitted by a carry-over of carbohydrates, nutrients and water from the previous season. This pattern gives trees prior access not only to water, but to nutrients mineralised during the early part of the wet season. The necessary price is a low stomatal conductance, which results in a conservative water use pattern, but constrains the peak photosynthetic rate to below that of grasses. Grasses have a much denser root network, which permits freer use of water when it is present. Grass production is controlled by the duration of the periods for which water is available in the soil, and by the nutrient supply during these periods. The early-season nutrient access, and the capacity to store and internally recycle nutrients, reduces nutrient constraints on trees to some extent.

There is a third axis of niche separation between trees and grass, in the horizontal spatial dimension. Tree roots are not evenly distributed over the aerial extent of the savanna, and neither are the shaded habitats caused by tree canopies, nor the nutrient-enriched areas beneath their canopies. Therefore, from a grass perspective, a savanna is a mosaic of subhabitats with different irradiance, nutrient and water supply characteristics. As the density of woody plants in a savanna increases, so the proportions of the various subhabitats alter, and with them the grass quantity, quality and species composition. This is believed to be responsible for the non-linearity of the grass production versus woody plant density relationship typically found in savannas.

In all three of these axes, the grass niche is entirely within the tree niche. This, and the ability of trees to shade grass, results in the interaction between established trees and grasses being strongly asymmetric in the favour of trees. Therefore, the tree–grass mixture does not represent a competitive balance, but has an inherent tendency to become more woody. This trend is periodically offset by fires, and is limited at the upper end by tree-on-tree competition. The proportion of woody and grass plants in a savanna at a given moment is mostly quite different from what an equilibrium analysis would predict. It is determined more by the recent history of episodic establishment and mortality events relating to the woody plants than by competitive interactions between trees and grass.

The competitive situation between seedling trees and established grasses is completely different. The asymmetry in this case is in favour of the grasses, and the young trees are very susceptible to fire and browsing. Therefore it is possible to maintain a virtually treeless savanna with regular burning, provided that the integrity of the grass layer is not breached by disturbance or excessive grazing. This is a quasi-stable state, in that if fire is withheld or reduced in intensity by grazing of the fuel load, the trees will escape from the grass layer. As the woody biomass increases, grass production (and therefore fire intensity) decreases, leading to a new quasi-stable state at a higher tree density, determined by competition between trees (Smith & Walker 1983).

The 'multi-dimensional asymmetric' model of the tree–grass interaction lacks the simple elegance of the two-layer model. The additional complexity is necessary to account for the observed structural patterns and responses to disturbance. A minimum model must include competition between trees and trees, as well as trees and grass, for light, water and nutrients. It must also consider the interactive effects of fire and herbivory.

The stability of savannas

The savanna model discussed above is an equilibrium model, with the potential for multiple stable states with sudden and relatively irreversible transitions between them, as well as neutrally stable and unstable behaviour. It replaces a notion of a single equilibrium state, determined by climate and soil. The present view of savannas is, however, one of disequilibrium rather than equilibrium, with disturbance as a major factor determining the form and composition of African savannas as we know them: in other words, a discrete-state probabilistic view rather than a continuous, deterministic view. Savannas are thought to be largely driven by external events. The equilibrium model is imbedded within a stochastic savanna environment. In particular, certain combinations of circumstances, individually weak and perhaps quite commonplace occurrences, are collectively powerful and rare, and determine the form and direction of savanna dynamics for long periods of time. Between these contingent events, savannas are in a state of continuous internal adjustment within defined limits, but may be relatively insensitive to the same driving factors, acting independently.

Nutrient cycling in savannas

Nylsvley's nutrient budgets show that savannas, like other ecosystems, have cycles within cycles, operating at different scales and rates, and intersecting at certain points. For instance, the internal recycling of nitrogen which occurs before a tree drops its leaves accounts for a large part of the nitrogen requirements of the new leaves. As a consequence of this storage and the slow decay rate of tree leaves in savannas, the tree–soil cycle is much slower than the grass–soil cycle, although at Nylsvley it accounts for a greater proportion of the total nutrient flux. Similarly, the difference between nutrient-rich and nutrient-poor savannas lies not only in the total quantity of nutrients present, but in the rate at which they are turned over. Faster nutrient turnover in a small patch of infertile savanna may be induced by repeated grazing on the same spot, creating a grazing lawn. Another example of fast cycles nested within slow cycles is the alternate pathways to mineralisation: rapidly via fire; intermediate by passage through an animal gut; and slowly by microbial decomposition in the litter layer or soil. Despite dominating the soil fauna, termites play a surprisingly small role in the comminution of litter in the nutrient-poor broad-leafed savanna at Nylsvley, perhaps because the soil is too sandy for nest construction by litter-feeding genera such as *Macrotermes*. They are even less important in the nutrient-rich fine-leafed savanna, where the soil fauna is dominated by ants and dung beetles. Grazing mammals recycle a greater fraction of aboveground primary production in the fine-leafed savanna than the broad-leafed savanna.

The savanna nutrient cycles do not operate in a smooth and continuous fashion. The rate of chemical and biological processes is strongly controlled by water availability, and weakly controlled by temperature. Therefore key steps in the cycles, such as mineralisation and uptake, occur in a pulsed fashion.

The decomposition rate of litter at Nylsvley is extremely slow. This is attributable to the relative aridity of the site, and also to the chemistry of the litter. Grass litter disappears in about one year, but tree leaves persist for 3–5 years. The decomposition rate is positively related to the nitrogen content of the leaves, and negatively related to the lignin and condensed tannin content. The leaves of the dominant trees in the broad-leafed savanna are high in lignin and condensed tannin, for reasons related to the aridity and infertility of the site and protection against herbivory and microbial pathogens. The consequence of the low decay rate is a buildup of leaf litter, which protects the soil surface against erosion and suppresses evaporation, but intercepts a significant fraction of the incoming rainfall. The accumulated litter is partially burned during fires, but the interval between fires means that only a minor portion of the nutrients in the leaf litter is mineralised in this way.

Multiple limitation in savannas

No single nutrient has an overriding limiting effect on the primary production of the broad-leafed savanna at Nylsvley. Production responds to fertilisation with water, nitrogen, and phosphorus. The role of water is central, since it controls the duration of the period for which growth and uptake of other nutrients can occur. Although it also regulates the rate of those processes to a degree during the wetting and drying phases, the discrete pattern of rainfall events in dry savannas make it convenient to treat water availability as an on–off switch operating on many other aspects of savanna function. Savannas therefore alternate between water limitation and nutrient limitation.

Nitrogen has a key role because of its impact on forage quality, anti-herbivore defences and the decay rate of litter. There is a significant degree of exchange between the terrestrial part of the nitrogen cycle and the atmosphere, amounting to about 10% of the annual mineralisation flux, and 0.1% of the nitrogen standing stock. The main losses are pyrogenic volatilisation and probably denitrification as a leakage from the nitrification step, but are closely balanced by inputs through wet and dry deposition, and to a small extent through biological nitrogen fixation. The dominant trees of the fertile and infertile savannas are legumes, with a relatively high leaf nitrogen content. In the infertile savannas they belong to the Caesalpiniaceae, which mostly do not support symbiotic nitrogen-fixing bacteria, while in the fertile savannas they belong to the Mimosaceae, many members of which have been shown to fix nitrogen. At Nylsvley, however, there is no evidence that they do so. The small amount of biological nitrogen fixation is attributed to leguminous forbs growing within the grass layer.

Ammonium in the soil is rapidly converted to nitrate and taken up by plants. There is virtually no loss as leachate, owing to the small volume of deep-draining water and its low nitrogen content. Because of the atmospheric link, the nitrogen stock in long-established savannas accumulates until the losses balance the inputs, or until another nutrient, such as phosphorus, which has a much weaker atmospheric link, becomes limiting. It should therefore not be unexpected that more than one element should be close to limiting production; and when fertilised with one nutrient, the others should soon impose limits. A further consequence of the near-limiting status of a number of nutrients, coupled with the fact that the intact savanna is very efficient with respect to nutrient recycling, is that disturbance to the soil results in nutrient loss through leaching, which may be replaced very gradually. Many of the nutrient cycles are closely tied to one another, because of their mutual link to the water and carbon cycles. There is a limited degree of flexibility in the C:N and C:P ratios required for tissue growth, so savannas can respond to fertilisation with one element alone, but the greatest production increases will come from mixed fertilisers.

Spatial patterns in nutrient distribution

Nutrients are not evenly distributed in the savanna landscape. There are several mechanisms which lead to nutrient concentration, and once such a 'hot spot' has formed, it can be very persistent due to the feedback mechanisms which ensure a continued influx of nutrients to the enriched site. Human settlement was the mechanism which produced the high-fertility patches at Nylsvley, but geomorphological processes such as catena formation, and biological processes such as foraging patterns in termites or root distribution in trees can have the same consequence. Large herbivores make use of this spatial inequity to satisfy their nutrient requirements and, in so doing, perpetuate it. Nutrient patchiness may be a universal and necessary feature of the ecology of infertile environments.

Biodiversity in savannas

The public awareness of African savannas, such as it is, is based on the diversity of large mammals found there. The high faunal diversity is associated with a high plant diversity. The diversity of plants and animals has its origin in a long evolutionary history, uninterrupted by glaciation. The pre-Pleistocene megafauna of Europe and America was equally rich, but was subject to extinctions which seem to be linked to the expansion of *Homo sapiens* in those continents. The African savanna fauna adapted to, and evolved with, Man.

Savannas, and in particular the infertile forms, have a high number of species on a small-plot basis (so-called alpha diversity). The environmental gradients in savannas tend to be gradual, and therefore the beta diversity (species turnover on gradients) tends to be lower than, for instance, the *fynbos* vegetation of the southern Cape. The landscape-level (gamma) diversity of savannas is variable. At Nylsvley it is high, which accounts for the exceptionally high diversity of plants and animals found there. At one stage, a team of Nylsvley researchers held the world record for the largest number of bird species observed in one location (within walking distance) in a 24-hour period.

The maintenance of biodiversity in savannas is dependent on the temporal and spatial variability of the main savanna determinants (rainfall, soil nutrients, fire and herbivory) and physical disturbance. For at least a million years, Man has been an increasingly influential component of this disturbance regime. The ability of large mammals to exploit this variability by migration is an important factor in maintaining their diversity and biomass. The large-scale seasonal migrations that occur in savanna ecosystems such as the central Kalahari and the Serengeti, and the small-scale movements of mammals between different habitats at Nylsvley are manifestations of the same process: the spatial exploitation of seasonal variation. Reduction in the degree or scale of spatial variability, by limiting the agents of disturbance or by reducing the size of management areas, will increase susceptibility to temporal variations (which are extreme in savannas). Diversity is also reduced by restricting animal movements with fencing or water provision.

Future directions in savanna research

There are three main trends in savanna research at an international level. These themes are expressed in the publications of the Responses of Savannas to Stress and Disturbance programme, a joint initiative of the International Union of Biological Sciences and Unesco's Man and the Biosphere programme. The first is a concern with the functional classification of the savanna biome. The driving force behind this theme is the necessity to gain a predictive understanding of the responses of savannas at large scales to a changing global environment: altered rainfall, temperature, nutrition, herbivory and fire. The second theme is motivated by the same concern, but at the scale of individual plants: the interaction of the life-history characteristics of savanna plants with environmental cues and the characteristics of other plants. The third trend is towards an explicit consideration of the role of spatial pattern in savanna ecosystem function, at all scales from that of the individual plant up to the region.

An emerging fourth theme has to do with the high species diversity in savannas and relations between biodiversity and ecosystem function. The key issue is determining the amount of functional redundancy that exists between species; in other words, how many species can be lost before ecosystem function is materially altered? The reciprocal question relates to the degree of biodiversity change that can be expected for a given degree of functional change, which could be due to local changes in the fire or grazing regimes, or global changes in the climate or atmospheric composition.

17

Managing savannas

The Savanna Biome Programme, from its outset, was intended to be strategic research aimed at ultimately improving savanna management. The Department of Agriculture, which has the responsibility for the tactical research needs of the cattle industry, was a prime motivator for the programme. Likewise, the Division of Nature Conservation of the Transvaal Provincial Administration, which is responsible for wildlife on state and private land in the Transvaal, and which owns and manages the Nylsvley study site, actively participated in the research there in the initial phases. These bodies, and the National Parks Board, were represented on the steering committee. Sadly, all of these user-orientated organisations became less involved in the programme with time, and the crucial task of transferring and interpreting the information from the Nylsvley studies into a management framework has occurred largely by chance, if at all.

This chapter aims to give agricultural and conservation extension officers some indication of the practical implications of the basic research carried out in the Savanna Biome Programme. It is their mandate to translate the findings into management practice as implemented by farmers and game rangers. A number of general management-orientated reviews have originated from the Savanna Biome Programme, notably *Management of Semi-Arid Ecosystems* (Walker 1979), *The Management of Large Mammals in African Conservation Areas* (Owen-Smith 1983), and *Management of semi-arid savannas* (Barnes 1982). There are many other papers on specific topics which have a management slant, which will be referred to below.

Management actions in savannas

The basic savanna manager's toolkit consists, figuratively speaking, of a box of matches, a rifle, a water pump, a pair of fencing pliers and an axe. The low net income per hectare restricts the use of intensive techniques such as fertilisation, irrigation and pasture replacement to small patches and special circumstances. The two main factors which control the plant growth and animal production in savannas, rainfall and soil type, are beyond the control of management. The farmer or conservator has access only to the second-level controls on the way savannas look and work: fire and herbivory. Therefore the degree of change which the manager can bring about is limited. It is usually much less work to influence controls on savanna form and function, than to change the form and function directly. For instance, it is cheaper to adjust the fire regime than to chop out trees. Sometimes the controls will not work, or take too long to act, in which case direct intervention, for instance by bush clearing or grass mowing, may be an option.

The manager can control the fire regime by deciding where, when (time of day and time of year) and how big a fire should be. These decisions will influence the frequency, intensity, season and type of the fire, which will in turn determine its ecological consequences. The manager can also control the number and type of large mammals on the land by culling, refraining from culling and reintroductions. The distribution of the animals can be altered with fences and the strategic location of water points, salt licks and burned patches. The balance of trees and grass can be tipped in favour of grasses by burning, reduced grazing or bush clearing; or in favour of trees by excluding fire and increasing grazing. Finally, limited remedial actions such as the rehabilitation of bare, crusted or eroded patches can be undertaken. Savanna management, like politics, is the art of the possible: this chapter is organised to reflect the things that a manager can actually do.

The importance of timing and coincidence

During the course of the research at Nylsvley, the world-wide trend in ecology, and amongst savanna ecologists more so than most, has changed from thinking of Nature as being in balance, to a view which has Nature mostly out of balance, even under 'natural' conditions. This is especially true for ecosystems such as southern African savannas, where the main driving factor (rainfall) is determined outside the system, and is not much influenced by what goes on inside the system. The state of the ecosystem is always reacting to the most recent event. It may be adjusting towards a balance, but it is generally pushed around with sufficient frequency and force to keep out of balance for most of the time. Along with this conceptual shift goes a belief that change in savannas is seldom gradual and continuous. Rather, it occurs in fits and starts, linked to specific events.

Major changes often depend on the coincidence of two or more events, which then set the pattern for that savanna for a long period, until another event sequence resets it. For example, many southern African savannas regularly experience below-freezing temperatures in winter, with no lasting consequence. If the cold period happens to coincide with out-of-season rainfall, such as occurred in 1968, the result can be extensive tree deaths. The traces of that die-off are detectable in many parts of Zimbabwe two decades later.

The sequence of events leading to the death of savanna trees, described in Chapter 8, is another good example of how individually small and insignificant events can combine to exert great leverage. Fires would not kill mature trees at Nylsvley if it were not for the presence of porcupines and strong winds.

The first consequence of this view for management is that once change has occurred, it will not necessarily, spontaneously, revert to its former state if the factor causing the change is removed. The former state was unlikely to be a balance point, or if it was, the ecosystem may now be in the grip of another, different balance. A second consequence is that there are only brief periods during which substantial change occurs, or can be induced or prevented. What happens in between these 'windows of opportunity' has little effect. The trick in management therefore lies in recognising the moment, and seizing it.

The role of patchiness, size and disturbances

A related issue is the role of disturbance. We previously viewed Nature as being undisturbed before we came along. We now believe disturbance to be the engine which keeps most ecosystems working; and in African savannas many of those essential disturbances have a human origin. Over a period of centuries (which is not long, given the lifespan of savanna trees), every bit of the Nylsvley savanna is churned up by burrowing warthogs, aardvarks, porcupines, gerbils and dung beetles. Soil from deep in the profile is brought to the surface, and a characteristic group of plants occupies the newly created habitat. Management policies which seek to exclude disturbance will result in changes in the biological diversity, species composition, landscape pattern and productivity of the ecosystem, and may reduce the capacity of the ecosystem to survive similar disturbances in the future.

An example of a disturbance which is regarded as highly beneficial by today's managers is the nutrient-rich patches at Nylsvley, resulting from human settlements some centuries ago (Blackmore *et al.* 1990). This nutrient concentration greatly enhances the diversity of plants and animals at

Nylsvley and increases animal production, not only of the site itself, but of the adjacent nutrient-poor areas too. At a smaller scale, a termite heap, dung midden or tree performs the same function: gathering nutrients over a wide area, and depositing them in a smaller area. They also create a nutrient hot-spot, but one which is much less persistent, because it is smaller. The maintenance of the old village sites depends on a continuing inflow of nutrients due to the preferential feeding of herbivores on the nutrient-rich areas. There must be a maximum proportion of the area that can be converted to self-perpetuating nutrient-rich patches, beyond which the nutrient-poor matrix is insufficient to maintain the subsidy. It is not only the degree of alteration and aerial proportion of disturbance patches which matters, but also the size of individual patches (von Maltitz 1990). If they were bigger than the feeding range of the animals, the nutrient influx could not occur, and the patches would disappear.

A fixed management pattern will favour some species over others. We do not know the exact growth requirements of every savanna species, so we cannot predict with any confidence which will be advantaged and disadvantaged under any given regime. In an agricultural situation this seldom matters, since management is aimed at optimising the performance of just a few species. For conservation management, on the other hand, it is crucial. Management for diversity should not aim at a fixed pattern, but at a general range of possibilities, based if possible on the historical pattern, with variation around it. This variation allows the persistence of a wider range of species. For example, late dry season burning every three years may be the target, but an occasional summer burn, or two burns in successive years, is not disadvantageous.

The same argument can be extended to spatial pattern. Current practice is to divide the land into fixed management blocks, within which the treatment is consistent. It would be better, in conservation areas, to allow those management blocks to migrate over the landscape, setting up a complex patchwork of different treatment sequences. In summary, it is the temporal and spatial heterogeneity of savannas which enables the persistence of their high biodiversity and which gives them the characteristic features for which they are valued. Yet there is a widespread urge in savanna managers to even things out in time and space. It is fortunate that they have thus far not been very successful.

Savanna assessment and classification

When assessing the productive potential of an area of savanna, two fundamental features must be considered: effective rainfall and soil fertility.

The strong influence of rainfall on grass production in savannas has been recognised for a long time, but the modifying influence of soil fertility, and the way that it interacts with water supply, has become explicit only recently. Water supply and nutrient supply alternate in controlling plant production in savannas: when there is no water, even a fertile savanna cannot grow; but when there is water present, a fertile savanna grows faster than an infertile savanna. Soil fertility has a profound effect on savanna ecology, since it determines not only the plant production, but also what fraction of it is edible (Chapter 15) and what species of plants and animals will be present.

Mean annual rainfall has some value as an index of water supply, but can be much improved by considering the seasonal distribution of evaporative demand in relation to rainfall and the effect of soil texture, stoniness and depth on water availability to plants. This can best be achieved by modelling the soil water balance (as was done in Chapter 6). A simple 'bucket model' using monthly weather data is usually adequate for this purpose. More detailed modelling is possible, but is seldom supported by sufficient data. Based on the idea that water acts as an on–off switch in savannas, an intuitively obvious way of expressing water availability is to add up the fraction of time during which the switch was 'on'; in other words, when sufficient water was available to permit plant growth. This can be expressed as a number of 'growth days' per year: in savannas it ranges from about 60 to 180.

The traditional soil tests, such as analysis for basic cations and the related concepts of base saturation and cation exchange capacity, are of limited value in assessing the fertility of savannas. This is because the nutrients which control production and forage quality are more typically nitrogen and phosphorus, which have complex cycles not easily reduced to a single analytical value (Chapter 7). They are generally positively related to the concentration of bases, owing to a mutual dependence on the clay mineral fraction. In time, nutrient cycle models and new sorts of soil tests, presently only used in research contexts, will become available to quantify soil fertility in savannas on a routine basis. For the present, the vegetation, underlying geology, soil type and conventional soil analyses provide clues as to the fertility status (Chapter 12). Soils derived from wind-blown sands, sandstone, acid igneous rocks, or deeply weathered basement materials support infertile savannas, while shale, mudstones, dolomite, and basic igneous rocks support fertile savannas. The tree layer of a fertile savanna is dominated by the family Mimosaceae (usually *Acacia* species), with tiny leaflets and prominent thorns. Infertile savannas are dominated by the Caesalpiniaceae and Combretaceae, and have larger leaves and few thorns. Fertile savannas tend to be on clayey soils, while infertile savannas tend to be on sandier soils.

The African savannas can be plotted on a graph which has these two determining factors, water and nutrients, as its axes (Figure 17.1). This serves to identify the main sorts of savannas, while not obscuring the fact that they exist on a continuum. The broad species composition, productivity and

Figure 17.1. Translating the position of savanna samples in geographical space into their position in environmental space. The circles are spinescent, fine-leafed savannas, and the squares are broad-leafed savannas. The environmental space is defined by Plant Available Moisture (PAM) and Available Nutrients (AN). PAM is defined as the number of days in the year on which there is water in the soil available to plants, as calculated by a soil water balance model. Available Nutrients is expressed as the growth rate of *Panicum maximum* in soil from the site, in a growth chamber with plentiful water supply.

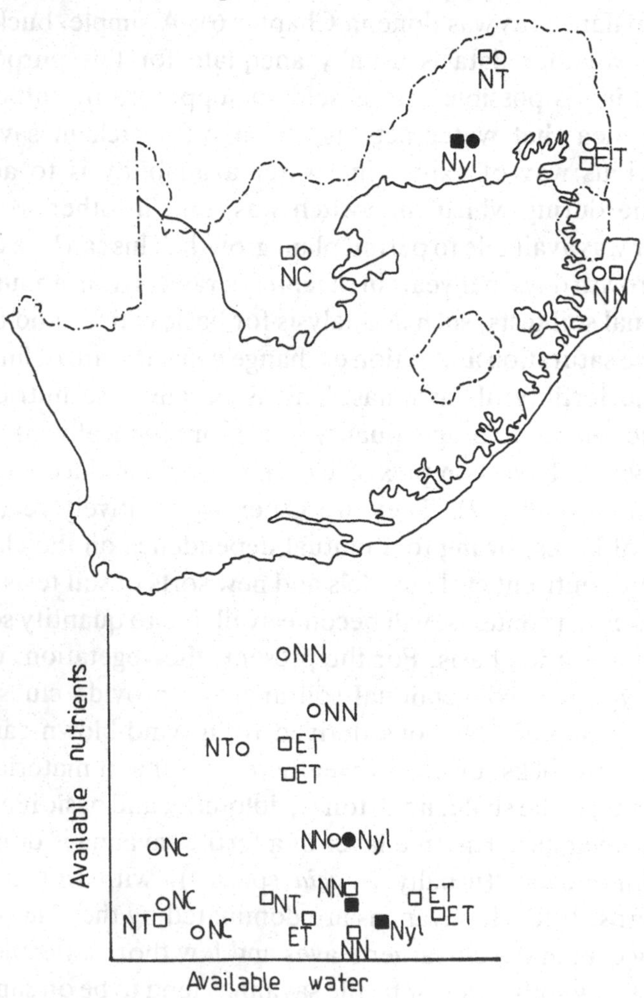

degree of woody cover is partly predictable from the position of a given savanna in this space. The main virtue of this system is that it identifies which savannas are ecologically similar, despite having very different species composition. This allows research results obtained in one savanna to be transferred to others.

Stocking rate and herbivore species mix

The number of grazing animals which can be supported on a given area of fertile savanna (sweetveld, in South African terms) is controlled by the amount of grass produced in a year, which is in turn related to the rainfall and soil fertility. On an infertile savanna, on the other hand, the stocking rate depends to a much smaller degree on the amount of grass, and to a greater degree on what fraction of that grass is acceptable to grazers. The quality of the forage is reflected by its protein content, particularly during the dry season; and the protein content is related to the nitrogen cycle (i.e. soil fertility) of the site. Many African savannas consist of a mosaic of infertile and fertile patches, resulting from soil-forming processes, human settlement, or the activities of animals themselves. Herbivores balance their nutritional requirements between these patches, resulting in an enhanced overall stocking rate.

There are currently no guidelines for setting the stocking rate of wild browsing animals. The risk of causing long-term damage by overstocking with obligate browsers is small, since they starve to death before they kill their resource; and unlike grazing, heavy browsing does not expose the soil surface to erosion. Browsers with an alternate food source, such as goats and impalas, can have a major impact on woody plants. Browsers can be an important part of a bush control strategy, but if there is insufficient browse present, mixed feeders such as goats will consume grass instead. The Nylsvley studies show that browsers are limited by food supply, rather than some factor such as territoriality. Secondly, the amount of browse available in midsummer is largely irrelevant. It is the browse supply during the late dry season which imposes the stocking limit. If this limit could be relieved, for instance by supplemental feeding at this time, the year-round browse usage could be increased.

Kudu and giraffe are ideal candidates for combined browser–grazer systems (Owen-Smith 1985a), since they eat virtually no grass. The diet of impalas, on the other hand, consists of about 80% grass when both grass and browse are plentiful.

The current major economic use of southern African savannas is cattle ranching. Cattle eat mainly grass, leaving the woody plants as an unutilised resource and a management problem. A promising strategy to improve animal production in savannas, and at the same time reduce bush encroachment, is to include browsers in the production system. In order to manage such systems effectively, it is necessary to know what determines food choice by browsing ungulates. In Chapter 15 we argued that an understanding of the factors which make one plant acceptable to herbivores, while others are not, is the key to controlling secondary production in infertile savannas. The ecologically efficient use of the diverse plant and animal resource offered by savannas will require farming systems considerably more complex than the single-species ranching which presently dominates the industry. The challenge for the future is to narrow the gap between what is ecologically optimal and what is economically feasible.

Burning policy

There is no such thing as a savanna which does not burn: the only management choice is between planned fires and unplanned fires. Even under a no-burn policy, it is only a matter of time before an unplanned fire occurs, or the savanna changes into a thicket, woodland or forest. Mature, undamaged savanna trees are not killed by fires of normal intensity. However, the woody biomass is reduced by burning, and plants entirely within the flame range may be killed or burned down to ground level. There is an important interaction between burning and browsing, which enhances the effectiveness of both in woody plant control. Fires promote coppicing and lower the mean height of the tree canopy. This increases the amount of leaf available to browsers, and their impact on the growth rate of woody plants. In turn, the suppression of tree growth by browsing keeps the apical buds within the flame zone, where fire can kill them. Trees with dead-wood scars at the base resulting from previous damage, for instance by porcupine browsing, will eventually be hollowed out and toppled by repeated burning.

Burning does result in the loss of some nutrients from the ecosystem in the form of volatile gases and smoke (particularly nitrogen), but on a large scale these losses are largely balanced by nutrient inputs from rain and dust. Burning also releases carbon dioxide, carbon monoxide and oxides of nitrogen and sulphur, which are important 'greenhouse gases', into the atmosphere. Most of the carbon dioxide would have been released anyway through litter decomposition, and is trapped again in the flush of regrowth following the fire. In the long term, therefore, savanna burning has no net effect of the global carbon balance: however, there is considerable scope for adjusting the amount of carbon stored in savannas, largely by altering the burning regime. The impact of savanna burning on ozone production and greenhouse gases is currently an important research area, and managers may in the future come under pressure to modify their burning practices.

Managing the proportions of trees, shrubs and grass

This topic is a crucial savanna management issue, because the tree–grass mixture is inherently unstable, and when it changes, many other aspects of savanna ecology are affected. Woody plants tend to increase in savannas at the expense of grasses when the grazing pressure is increased and the frequency and intensity of burning is reduced (Chapters 14 and 16). This tendency can be halted, but is not easy to reverse, even with reduced grazing and increased burning. Mechanical removal or chemical poisoning of the trees is generally uneconomical unless some value can be recovered from the trees, but ignoring the problem also results in a declining economic return.

There have been many trials, carried out in southern African savannas and elsewhere, that show an immediate increase in grass production following the removal of woody plants. Partial reviews of these trials are given by Barnes (1979) and O'Connor (1985). The grass production increases are frequently substantial, and since management systems in savannas have been geared towards grazers rather than browsers, many farmers and wildlife managers have been bush clearing on a large scale. Much applied research has gone into the best method of killing trees: chemical, mechanical, biological and various combinations. Most of these experiments are compromised by three factors: they are short-term, take place on small plots, and seldom measure any response variables other than grass production and tree mortality. Whereas the short-term response to bush clearing is increased grass growth, the long-term results may be different. Part of the initial increase is due to the flush of nutrients resulting from decomposition of the clearing residues (especially the belowground organs, which the Nylsvley data show to be substantial). With time, because the total plant productivity has decreased, and particularly since the input of slowly decomposing tree litter has ceased, the soil organic matter content declines. The tree canopy-scale mosaic of nutrient-enriched sites disappears and the protective surface mulch of leaf litter decreases. The decline in soil organic matter is particularly deleterious on coarse-textured, infertile soils, where the major part of the ion exchange capacity of the soil is provided by the soil organic colloids. The increase in grass production following clearing is much less on infertile than fertile sites, since it is constrained by nutrient availability. In the long term the quality and quantity of grass production on sandy soils cleared of trees is likely to decline as a result of nutrient loss. This trend has not been demonstrated since the published trials are short-term, or else the long-term trials (such as the excellent study reported by Dye & Spear (1982)) are conducted on small plots which receive litter inputs from surrounding uncleared areas.

Deep, infertile, sandy soils favour a high tree biomass relative to grasses, and it is therefore difficult to keep them in a treeless state. This observation, coupled with the considerations discussed above, suggests that bush clearing is seldom, if ever, a viable management option on sandy soils.

Nearly all of the southern African bush clearing trials compare complete clearing versus no clearing. The experiment conducted at Nylsvley to examine the effect of partial clearing was unfortunately never completed. The factors which influence the degree of response of grasses to partial clearing must therefore be inferred from theoretical considerations and experiments conducted elsewhere. The optimum density of woody vegetation in southern African savannas, calculated in terms of the cost of clearing and the net value to livestock (Norton & Walker 1985), is currently unknown.

One of the contributions of the Nylsvley study has been to focus attention on the woody plants in savannas. Currently, the main management approach to increasing the grass and animal production in savannas is to remove the trees. However, the trees contribute half of the total plant production at Nylsvley, and two-thirds of the aboveground production. The total plant productivity in a savanna from which the trees have been removed is usually less than that of the original tree-filled savanna, despite the increase in grass growth. An alternative approach to enhanced savanna productivity would be to look into agricultural systems which use the trees as part of the productive resource: for instance, by promoting mixed grazer–browser systems, or harvesting the trees for biomass energy. Some of the traditional systems of savanna use may be more sustainable than bush clearing. For instance, in the 'chitemene' system used in *miombo* woodlands in Zambia, the trees are lopped off at chest height to provide browse for livestock, while at the same time reducing competition with grasses. It does not lead to any apparent long-term decline in the productivity of the system.

The equations developed at Nylsvley for estimating the amount of wood and leaf for a given tree (Rutherford 1979) have allowed the size of the wood resource to be estimated. For Nylsvley, which is reasonably typical of moderate-rainfall savannas on sandy soils, it is about 12 tons ha^{-1}, including branches down to about 4 cm in diameter. Given the low net profit per hectare from cattle ranching, the value of the timber and fuelwood on many savannas may exceed the grazing value.

Annual growth of wood is about 900 kg ha^{-1}. Therefore, one hectare of

Nylsvley broad-leafed savanna could supply the firewood requirements of a typical peasant family for three years, or four hectares could support their needs on a sustainable basis. The removal of wood does represent a small nutrient drain on the system, but even in a nutrient-poor system such as the broad-leafed savanna at Nylsvley, less than 1% of the nitrogen and 2% of the phosphorus is contained in aboveground woody tissues.

Rehabilitation

One square metre of savanna grassland contains tens of thousands of grass seeds in the top few centimetres of soil, but only about a hundred grass plants. Each plant produces thousands of grass seeds every year, which have a soil life of a few months to a few years. Established grass plants are at a huge competitive advantage relative to grass seedlings. Therefore there is little point in overseeding a savanna which already has some grass cover, or has had in the past two years. The existing seed bank and seed production are more than adequate to ensure recovery. Where an area is bare and has received (or retained) no new seeds for several years, reseeding may be called for, provided the soil surface is suitable for germination and establishment. A hard, compacted or crusted surface is an inhospitable environment for a germinating grass plant. The grasses of all but the driest savannas are mostly perennial, and reproduce and occupy new openings by forming tillers rather than by new seedlings. Therefore, even if regular grazing and burning reduces grass seed production, the grasses will not disappear. The ecological function of grass seeds is to colonise distant patches and to survive short periods of adversity. Even so, most of the seeds fall very close to the parent plant. Therefore a bare patch needs to be only a few metres across before it is starved of seed input. Where there has been a prolonged history of selectively heavy grazing, the situation arises where palatable perennial species may be locally extinct. Under these circumstances it may be necessary to reintroduce viable seeds of these species. Getting them to establish in the face of competition from the existing sward depends on being able to identify and seize one of the windows of management opportunity, discussed earlier. In this case it will generally consist of a period of good and sustained rainfall following a prolonged drought.

Soil erosion is a natural process resulting in the redistribution of soil and water in the landscape. The surface of a healthy savanna consists of a repeating sequence of erosion, transport and deposition zones. Soil lost from one small patch is not lost from the ecosystem, but is deposited in a drift just a few metres away. This drift comprises a new habitat, rich in nutrients and seeds, which provides a favourable environment for new plants to germinate. This small-scale soil movement is a necessary process, not to be confused with the massive net soil losses from slopes denuded of vegetation.

The manageable factor which controls the erosion rate most strongly in savannas is the degree to which the soil is protected from direct raindrop impact. Even in a relatively well-grassed and woody savanna like Nylsvley, the tree and grass leaf cover is quite small, and much of the soil protection is provided by leaf litter. Tree leaf litter is more effective than grass litter in this role, since it decays much more slowly, providing another reason for maintaining some trees in the savanna system.

Transferring knowledge into practice

One of the disappointing aspects of the Savanna Biome Programme was the weakness of the interaction between research workers and savanna managers. Perhaps it is too soon to expect the Nylsvley research findings to have made a significant impact on the way that savannas are managed, but this transfer of information will not occur unless both parties make an effort to communicate.

18

Reflections on ecosystem studies

Ecosystem-level studies, such as the one described in this book, are expensive and time-consuming. Critics have suggested that they are not the most efficient way of conducting ecological research. There are certain circumstances, however, where the alternative approaches have little chance of success. Candidates for collaborative studies include those that require inter- or multi-disciplinary teams; are too big, expensive or extensive for single organisations; and need new, rather than conventional approaches. The scale and complexity of the environmental issues facing South Africa and the world is such that large cooperative studies will continue to be a prominent feature of research organisation in the future. Within the 16-year duration of the Savanna Biome Programme at Nylsvley, the style and structure of environmental research management in South Africa changed substantially. The aim of this chapter is to discuss the organisational experience of the Nylsvley programme, in order to guide future undertakings of this kind.

The substance of the chapter is based on a review paper by Brian Huntley, the manager of Ecosystem Programmes at the Foundation for Research Development for most of the duration of the Nylsvley study (Huntley 1987). The chapter also draws on the assessment conducted by Paul Risser, following the termination of the programme in 1989 (Risser 1989).

Historical background

Ecology has long been a strong component of South African biological research, thanks to a rich fauna and flora, a developed conservation awareness and an agricultural economy based largely on unimproved rangeland. However, by the late 1960s, rangeland science in South Africa had got into a rut. The prevailing conceptual models and techniques no longer seemed adequate to deal with the problems of the time. Issues such as bush encroachment, overgrazing, soil erosion and alien plant invasion had been on the research agenda for decades, but little progress had been made towards understanding the processes which underlay them. The contemporary issues were much more complex and cross-disciplinary than those which had been addressed in the past. There were relatively few South African ecologists at that time, and they were mostly involved in descriptive vegetation studies. No single organisation had a sufficient range of expertise to tackle the new challenges alone.

The emerging field of systems ecology, and the associated idea of collaborative research, offered a promising new approach. The International Biological Programme (IBP) provided both the impetus and the initial model for a South African Savanna Ecosystem Project (SASEP). Nylsvley was chosen as the study site, largely because it was available and convenient to the major research institutes and universities. The choice of Nylsvley as the main site was criticised on the basis that it is an example of a floristic type which is neither of great extent in South Africa, nor of significant economic importance due to its low agricultural value. It turned out to be a lucky choice in several respects, since it inadvertently included ecological features which were of great significance and stimulated some very productive research directions. The representativeness of the site turned out to be less important than its suitability as a research venue. The characteristics of the site generated research opportunities which could not have been foreseen at the outset.

The SASEP was the first of several cooperative ecosystem studies in South Africa. Experience gained in the savanna programme strongly influenced the planning and organisation of the studies which followed. The Council for Scientific and Industrial Research (CSIR) was responsible for programme management and coordination. The section which undertook this task became part of the Foundation for Research Development (FRD), which ultimately became a parastatal body independent of the CSIR. In the process of this transition, the FRD changed its mission to manpower development, and terminated national scientific programmes such as SASEP. Participants in the research programme came from government departments, research institutes and conservation bodies, museums and universities. Initially, most brought their funds with them, but in the last decade the FRD acted as broker for most of the grants.

The programme was intended to last ten years, and to have three phases (Anon. 1978). The first phase involved baseline descriptions of the soils and vegetation, and was completed by 1976 with very few problems, since this was the sort of task in which South African scientists were accomplished. Phase two consisted of process-level studies, and was initially influenced by the IBP emphasis on ecosystem energetics. By the late 1970s, however, some of the blind spots of this approach were becoming apparent, and the emphasis shifted to identifying and understanding the key processes in the ecosystem. This phase overlapped into the third phase, which began in the mid-1980s. It had the objective of extending the ideas generated at Nylsvley to other southern African savannas, and savannas in general.

The SASEP (by this stage known as the Savanna Biome Programme, SBP) was terminated in 1989. Nylsvley remains a nature reserve, and some research continues there under a variety of other programmes.

Costs and products

The cost of the SBP in South African currency (Rands) rose steadily between 1974 and 1985, but the inflation-adjusted costs were fairly steady at around R 500 000 per annum in 1989 terms (about US$ 175 000). After 1985 the cost declined in both absolute and real terms, to about R 250 000 per year. This does not include self-funded research, university overheads and the time of the research leaders. The real costs to the taxpayer were probably about twice this amount. The source of the funds disbursed by the FRD was approximately 70% from the Department of the Environment and the remainder from the FRD's own parliamentary grant. Approximately 66% of the money was spent on salaries for researchers, 27% on running costs and 7% on capital items.

The measurement of research productivity is not a simple matter. The tangible products of research are publications and degrees awarded. The reference section of this book includes 114 papers in refereed journals, conference proceedings or books originating from the SBP. This represents most, but not all, of the published work. Two books have been published synthesising the Nylsvley study results (including this one), and a further five have resulted from initiatives originating in the SBP. Twenty-one reports were published by the FRD, mostly in the *South African National Scientific Programmes Report* series. Twenty-two masters degrees and eight doctoral degrees were awarded for work done at Nylsvley, with three masters and one doctorate still in preparation. The indirect benefits include the improvements in the state of ecological science in the nation, the transfer of knowledge which occurs when people involved in the programme move into other activities, and the relationship between South African and international science.

The evolution of management structures and style

The SBP was steered by a committee of ten to fifteen people. Its task was to direct the research to where it was most needed, and to this end it was comprised mostly of senior representatives of organisations which might need savanna information. However, a strongly hierarchical management structure was not found to be very effective in identifying any but the most obvious research directions. The top-down approach resulted in rigid research priorities and narrowly defined projects, which tended to yield lots of data but little understanding. The attempts at bridging gaps between disciplines by retraining scientists, already specialised in one field, were unsuccessful.

In time, a more decentralised, opportunistic approach developed. The steering committee maintained the overall direction by defining general fields, but most of the ideas for projects were generated by small working groups and workshops, including many relatively junior scientists. The projects in the later phases tended to be targeted at hypothesis-testing through experimentation or comparative studies.

Strengths and weaknesses of the Savanna Biome Programme

The SBP attained most of its scientific objectives (Risser 1989). Although the savanna management problems which instigated the programme still persist, there is a vastly improved understanding of their causes and mechanisms, and therefore of how to go about solving them. South African scientists are recognised among the world leaders in savanna ecology, and work performed at Nylsvley is regularly cited in the international literature. The ecosystem programmes, led by the savanna programme, revolutionised South African ecology. A cohort of young scientists were trained in new ways of thinking about ecological problems, and the number of savanna specialists increased from none to a score or more.

On the negative side, a greater published output should have been achieved. In particular, the research findings have been poorly synthesised. This book is one attempt to address that problem. Several projects yielded no publications, and much of the thesis work was not published in any other form. The most prominent shortcoming was the failure to transfer the research findings between the scientific and management realms: Nylsvley research has had no directly attributable impact on savanna management practice. There has been an indirect impact, particularly in conservation management circles, which is due to the employment of graduates of the ecosystem programmes in the conservation agencies. In hindsight the main mistake was not having management agency scientists and extension workers involved in the research from the beginning. We simply did not put enough effort into this after the rather lukewarm reception to our first overtures. Had we persisted and been successful, the technology transfer would have happened automatically.

The SBP steering committee, chosen to represent the user, funding and research communities, was not an effective mechanism for information transfer in either direction, since it connected the organisations at too high a level. In later ecosystem programmes strong low-level links were formed through encouraging the active participation in research projects by members of the user groups. The Savanna Biome Programme was not truly cooperative, except in that it involved many people from several institutions, working at one place. Activities tended to be funded on a project-by-project basis, rather than according to the synergistic benefits that could be had by relating projects to one another. This partly reflects the absence of a common conceptual model at the onset of the study. There was poor data transfer between projects, despite attempts at creating a common database. The database failed because the data it included were too detailed, and therefore the data volume and management load was excessive. Since it was stored on a remote mainframe computer, it was not easily accessible to researchers. Consequently the records of what experiments were performed where in the study site are inadequate. Advances in desktop computers and user-friendly database programmes should alleviate this problem in future programmes, but it is important to recognise the principle that data transfer between projects seldom occurs at the level of raw field data. It is much easier to store, understand and transfer in a semi-worked state.

Principles for managing an ecosystem study

1. A rigidly-phased experimental programme is unnecessarily restrictive, but a thorough planning phase is essential. During the planning phase, there should be extensive consultations between researchers and user groups regarding the programme objectives, approaches, products and responsibilities. Research sites must be identified, bibliographies and reviews of existing work compiled, and conceptual models developed before field research begins.

2. During the experimental phase, the SBP applied the following principles for allocating support to various activities:

   a. Within a broadly defined field, fund any project that proposes good science on an appropriate topic, rather than sticking to a rigid project priority and narrow subject definition. Internal and external peer review is a key element in identifying 'good science'.

b. Actively encourage the best available worker (who may be relatively inexperienced) to become involved in a particular topic, rather than waiting for unsolicited proposals.

c. Concentrate support on the outstanding scientists, rather than distributing it completely equitably among all participants; but beware of exhausting a few individuals by making them responsible for too many activities. This principle should not be pursued to the extent that competent but unspectacular scientists are disillusioned and unproductive owing to their inability to attract funds.

d. Invest a significant proportion of the available funds in activities which encourage communication, idea generation and synthesis (workshops, synthesis reports, modelling activities). Synthesis does not happen automatically: it has to be encouraged.

3. A programme which attracts fewer than 12 researchers who interact frequently is unlikely to sustain momentum. Equally important is that there should be more than one (preferably at least three) 'idea generating' people in the group. If there is only one such person, the programme is vulnerable to the loss of that person and can become easily sidetracked. The incessant demands on a lone 'idea generator' prevents them from pursuing their own ideas, and isolates them from the source of their inspiration. A programme which has too many thinkers and not enough doers can become lost in theoretical debate.

4. The link between the science and its applications has to be structured into the programme from the bottom up, rather than from the top down. This can be achieved by ensuring that the research team consists of people with both an academic and practical bent, and by ensuring that they interact with one another and the outside world effectively. The important rule is to ensure that the research is done in collaboration with the user agencies, rather than in the form of a consultancy for them, and independently of them. The latter approach will always have the subsequent problem of trying to sell the results to groups who are committed to their own ideas and therefore inevitably resist suggestions which are at variance with them.

In conclusion

Collaborative ecosystem studies are not to be undertaken lightly, and once in progress require sensitive and flexible management if they are to operate efficiently. The benefits can be substantial, and it is hard to see how many of the environmental problems addressed by this type of research could be tackled in any other way.

It is the belief of the authors that the Savanna Biome Programme conducted at Nylsvley was a milestone in South African ecology, and a significant contribution to the global understanding of this important, but neglected, ecosystem type. It is our hope that this synthesis will bring the findings of the programme to a wider audience, and stimulate interest in the field of savanna ecology. Many pressing problems remain to be solved.

BIBLIOGRAPHY

Acocks, J. P. H. (1953). Veld types of South Africa. *Memoirs of the Botanical Survey of South Africa* 28.

Anderson, R. J. (1976). A preliminary study of rates of CO_2 efflux from the soils of an *Eragrostis-Burkea* savanna community in the Nylsvley Provincial Nature Reserve. BSc Honours dissertation, University of the Witwatersrand, Johannesburg.

Anderson, J. M. & Ingram, J. S. I. (1989). *Tropical Soil Biology and Fertility: A Handbook of Methods*. Wallingford, UK: CAB International.

Anonymous (1978). Nylsvley – a South African savanna ecosystem project: objectives, organisation and research programme. *South African National Scientific Programmes Report No*. 27. CSIR, Pretoria.

Ariovich, D. & Cresswell, C. F. (1983). The effect of nitrogen and phosphorus on starch accumulation and net photosynthesis in two variants of *Panicum maximum* Jacq. *Plant, Cell and Environment* **6**, 657–64.

Ariovich, D., Vinograd, P. & Cresswell, C. F. (1981). The regulation of starch accumulation in *Panicum maximum* Jacq. by nitrogen. *Proceedings of the Grassland Society of southern Africa* **16**, 151–4.

Atjay, G. L., Ketner, P. & Duvigneaud, P. (1987). Terrestrial primary production and phytomass. In *The Global Carbon Cycle*, ed. B. Bolin, *et al.*, pp. 129–81. SCOPE 13. New York: John Wiley.

Bailey, C. L. (1990). Nutrient content in vegetation growing on sodic soil. BSc Honours report, Botany Dept, University of the Witwatersrand, Johannesburg.

Baines, K. A. (1989). The water use, growth and phenology of four savanna grass species in response to changing soil moisture availability. MSc thesis, University of the Witwatersrand, Johannesburg.

Barnes, D. L. (1979). Cattle ranching in semi-arid savannas in East and southern Africa. In *Management of Semi-arid Ecosystems*, ed. B. H. Walker. Amsterdam: Elsevier Scientific.

Barnes, D. L. (1982). Management strategies for the utilisation of southern African savannas. In *Ecology of Tropical Savannas*, ed. B. J. Huntley and B. H. Walker *Ecological Studies* 42. Berlin: Springer-Verlag.

Bate, G. C., Furniss, P. R. & Pendle, P. G. (1982). Water relations of southern African savannas. In *Ecology of tropical savannas*, ed. B. J. Huntley and B. H. Walker *Ecological Studies* 42. Berlin: Springer-Verlag.

Bate, G. C. & Gunton, C. (1982). Nitrogen in the *Burkea* savanna. In *Ecology of Tropical*

Savannas, ed. B. J. Huntley & B. H. Walker, pp. 498–513. *Ecological Studies* 42. Berlin: Springer-Verlag.

Baylis, G. T. S. (1972). Minimum levels of available phosphorus for non-mycorrhizal plants. *Plant and Soil* **36**, 233–4.

Beard, J. S. (1962). Rainfall interception by grass. *Journal of the South African Forestry Association* **42**, 12–15.

Bell, R. H. V. (1982). The effect of soil nutrient availability on community structure in African ecosystems. In *Ecology of Tropical Savannas*, ed. B. J. Huntley & B. H. Walker, pp. 193–216. *Ecological Studies* 42. Berlin: Springer-Verlag.

Bezuidenhout, J. J. (1978). Die aktiviteit van microörganismes in die grond van die savanne-ekosisteem by Nylsvley. MSc thesis, University of Pretoria.

Bezuidenhout, J. J. (1980). 'n Ekologiese studie van die ontbinding van bogrondse plantreste in die Nylsvleysavanne-ekosisteem. PhD thesis, University of Pretoria. Pretoria.

Bezuidenhout, J. J. & Morris, J. W. (1978). A preliminary study of leaf litter input and decomposition at Nylsvley. Report to the National Programme for Environmental Sciences, CSIR, Pretoria. 16 pp.

Bicchieri, M. G. (1972). *Hunters and Gatherers Today*. New York: Holt, Rinehart & Winston.

Bigalke, R. C. (1978). Mammals. In *The Biogeography and Ecology of Southern Africa*, ed. M. J. A. Werger, pp. 983–1049. The Hague: Junk.

Blackmore, A. C. (1992). The functional classification of South African savanna plants based on their ecophysiological characteristics. MSc thesis, University of the Witwatersrand, Johannesburg.

Blackmore, A. C., Mentis, M. T. & Scholes, R. J. (1990). The origin and extent of nutrient-enriched patches within a nutrient-poor savanna in South Africa. *Journal of Biogeography* **17**, 463–70.

Bodot, P. (1967). Etudes écologiques des termites des savannes de Basse Côte d'Ivoire. *Insectes Soc.* **14**, 229–58.

Böhm, W. (1979). *Methods of Studying Root Systems. Ecological Studies* 33. Berlin: Springer-Verlag.

Booysen, P. de V. & Tainton, N. M.(ed.) (1984). *Ecological Effects of Fire in South African Ecosystems. Ecological Studies* 48. Berlin: Springer-Verlag.

Boomker, E. A. & van Hoven, W. (1983). *In vitro* digestibility of plants normally consumed by the kudu, *Tragelaphus strepsiceros. South African Journal of Animal Science* **13**, 206–07.

Bosch, O. J. H. & van Wyk, J. J. P. (1970). Die invloed van bosveldbome op die produktiwiteit van *Panicum maximum. Proceedings of the Grassland Society of southern Africa* **5**, 69–74.

Bourlière, F. (ed.) (1983). *Tropical savannas. Ecosystems of the World* 13. Amsterdam: Elsevier.

Bourlière, F. & Hadley, M. (1970). The ecology of tropical savannas. *Annual Review of Ecology and Systematics* **1**, 125–52.

Brain, C. K. (1981). *The Hunters or the Hunted.* Chicago: University of Chicago Press.

Brain, C. K. & Sillen, A. (1988). Evidence from the Swartkrans cave for the earliest use of fire. *Nature* **336**, 464–6.

Brookes, P. C., Powlson, D. S. & Jenkinson, D. S. (1982). Measurement of microbial biomass phosphorus in soils. *Soil Biology and Biochemistry* **14**, 319–29.

Brown, G., Mitchell, D. T. & Stock, W. D. (1984). Atmospheric deposition of phosphorus in a coastal fynbos ecosystem of the south-western Cape, South Africa. *Journal of Ecology* **72**, 547–51.

Bryant, J. P., Kuropat, P. J., Cooper, S. M., Frisby, K. & Owen-Smith, N. (1989). Resource availability hypothesis of plant antiherbivore defence tested in a South African savanna ecosystem. *Nature* **340**, 227–9.

Buechner, H. K. & Golley, F. B. (1967). Preliminary estimation of energy flow in Uganda Kob (*Adenota kob thomasi* Neumann). In *Secondary Productivity of Terrestrial Ecosystems*, ed. K. Petrusewicz, pp. 243–56. Poland: Panstwowe Wydawnictwo Naukowe.

Campbell, G. S. (1985). *Soil Physics with* BASIC. *Developments in Soil Science* 14. Amsterdam: Elsevier.

Carr, R. D. (1976). Progress report on habitat preferences of impala at Nylsvley. Report to the National Programme for Environmental Sciences. Typescript, 21 pp.

Cass, A., Savage, M. J. & Wallis, F. M. (1984). The effect of fire on soil and microclimate. In: *Ecological Effects of Fire in South African Ecosystems*, ed. P. de V. Booysen and N. M. Tainton, pp. 311–25. *Ecological Studies* 48. Berlin: Springer-Verlag.

Clayton, W. D. (1981). Evolution and distribution of grasses. *Annals of the Missouri Botanical Garden* **68**, 5–14.

Coe, M. J., Cumming, D. H. & Phillipson, J. (1976). Biomass and production of large African herbivores in relation to rainfall and primary production. *Oecologia* **22**, 341–54.

Coetzee, B. J. (1983). *Phytosociology, Vegetation Structure and Landscapes of the Central District, Kruger National Park, South Africa. Dissertationes Botanicae*. Vaduz: J. Cramer.

Coetzee, B. J., van der Meulen, F., Zwanziger, S., Gonsalves, P. & Weisser, P. J. (1976). A Phytosociological classification of the Nylsvley Nature Reserve. *Bothalia* **12**, 137–60. (Also published as *South African National Scientific Programmes Report* 20. CSIR, Pretoria).

Cofer, W. R. III, Levine, J. S., Winstead, E. L. & Stocks, B. J. (1991). New estimates of nitrous oxide emissions from biomass burning. *Nature* **349**, 689–91.

Cole, M. M. (1986). *The Savannas: Biogeography and Geobotany*. London: Academic Press.

Coleman, C. D. (1973). Soil carbon balance in a successional grassland. *Oikos* **24**, 195–99.

Coley, P. D., Bryant, J. P. & Chapin, F. S. III (1985). Resource availability and plant anti-herbivore defence. *Science* **230**, 895–9.

Collins, N. M. & Wood, T. G. (1984). Termites and atmospheric gas production. *Science* **224**, 84–5.

Collinson, R. F. H. & Goodman, P. S. (1982). An assessment of range condition and large-herbivore carrying capacity of the Pilanesberg Game Reserve, with guidelines and recommendations for management. *Inkwe* **1**, 1–54.

Connel, J. H. & Slatyer, R. O. (1977). Mechanisms of succession in natural communities and their role in community stability and organisation. *American Naturalist* **111**, 1119–44.

Cooper, S. M. (1982). The comparative feeding behaviour of goats and impalas. *Proceedings of the Grassland Society of southern Africa* **17**, 117–21.

Cooper, S. M. (1985). Factors influencing the utilisation of woody plants and forbs by ungulates. PhD thesis, University of the Witwatersrand, Johannesburg.

Cooper, S. M. & Owen-Smith, N. (1985). Condensed tannins deter feeding by browsing ruminants in a South African savanna. *Oecologia* **67**, 142–6.

Cooper, S. M. & Owen-Smith, N. (1986). Effects of plant spinescence on large mammalian herbivores. *Oecologia* **68**, 446–55.

Cooper, S. M., Owen-Smith, N. & Bryant, J. P. (1988). Foliage acceptability to browsing ruminants in relation to seasonal changes in the leaf chemistry of woody plants in a South African savanna. *Oecologia* **75**, 336–42.

Corbet, E. S. & Crouse, R. P. (1968). Rainfall interception by annual grass and chaparral. US Forestry Service Research Paper PSW-48. 12 pp.

Coughenour, M. B., Ellis, J. E. & Popp, R. G. (1990). Morphometric relationships and developmental patterns of *Acacia tartilis* and *Acacia recifiens* in southern Turkana, Kenya. *Bulletin of the Tomey Botanical Club*. **117**, 8–17.

Cresswell, C. F., Ferrar, P., Grunow, J. O., Grossman, D., Rutherford M. C. & van Wyk, J. J. P. (1982). Phytomass, seasonal phenology and photosynthetic studies. In *Ecology of Tropical Savannas*, ed. B. J. Huntley & B. H. Walker, pp. 476–97. *Ecological Studies* 42. Berlin: Springer-Verlag.

Crutzen, P. J. & Andreae, M. O. (1990). Biomass burning in the tropics: Impact on atmospheric chemistry and Biogeochemical cycles. *Science* **250**, 1669–78.

Crutzen, P. J., Heidt, L. E., Krasnec, J. P. & Pollock, W. H. (1979). Biomass burning as a source of atmospheric gases CO, H_2, N_2O, NO, CH_3Cl and COS. *Nature* **282**, 253–6.

Cumming, D. H. M. (1982). The influence of large mammals on savanna structure in Africa. In *Ecology of Tropical Savannas*, ed. B. J. Huntley & B. H. Walker, pp. 217–45. *Ecological Studies* 42. Berlin: Springer-Verlag.

Dasmann, R. F. (1964). *African Game Ranching*. London: Pergamon.

Daubenmire, R. (1968). Ecology of fire in grasslands. *Advances in Ecology* **5**, 209–66.

Dean, W. R. J. (1987). Birds associating with fire at Nylsvley Nature Reserve, Transvaal. *Ostrich* **58**, 103–6.

de Jager, J. & Harrison, T. D. (1982). Towards the development of an energy budget for a savanna ecosystem. In *Ecology of Tropical Savannas*, ed. B. J. Huntley & B. H. Walker, pp. 456–75. *Ecological Studies* 42. Berlin: Springer-Verlag.

Denbow, J. R. (1979). *Cenchrus ciliaris*: an ecological indicator of Iron Age middens using aerial photography in Eastern Botswana. *South African Journal of Science* **75**, 405–8.

de Villiers, G. du T. (1977). Netto reënval onderskeppingsverlies in 'n savannabedekking. *South African Geographer* **6**, 465–75.

de Villiers, G. du T. (1981). Net rainfall and interception losses in a *Burkea africana–Ochna pulchra* tree savanna. *Water SA* **7**, 4–25.

Drent, R. H. & Prins, H. H. T. (1987). The herbivore as prisoner of its food supply. In *Disturbance in Grasslands*, ed. J. van Andel *et al.*, pp. 131–47. Dordrecht: Dr. W. Junk.

du Preez, D. R., Gunton, C. & Bate, G. C. (1983). The distribution of macronutrients in a broad leaf woody savanna. *South African Journal of Botany* **2**, 236–42.

Dye, P. J. (1983). Prediction of variation in grass growth in a semi-arid induced grassland. PhD thesis, University of the Witwatersrand, Johannesburg.

Dye, P. J. & Spear, P. T. (1982). The effects of bush clearing and rainfall variability on grass yield and composition in south-west Zimbabwe. *Zimbabwe Journal of Agricultural Research* **20**, 103–18.

Dye, P. J. & Walker, B. H. (1980). Vegetation–environment relations on sodic soils of Zimbabwe Rhodesia. *Journal of Ecology* **68**, 589–606.

Dyer, C. (1980). The development of stomata in *Ochna pulchra* Burch ex DC., *Terminalia sericea* Hook. and *Burkea africana* Hook. BSc Honours report, Botany Department, University of the Witwatersrand, Johannesburg.

East, R. (1984). Rainfall, soil nutrient status and biomass of large African savanna mammals. *African Journal of Ecology* **22**, 245–70.

Eberhard, D. (1987). Energy consumption patterns and alternative energy supply strategies for underdeveloped areas of Bophutatswana. Report to Agricor, Mafekeng.

Edwards, D. (1983). A broad-scale structural classification of vegetation for practical purposes. *Bothalia* **14** (3 & 4), 705–12.

Edwards, D. (1984). Fire regimes in the biomes of South Africa. In *Ecological Effects of Fire in South African Ecosystems*, ed. P. de V. Booysen & N. M. Tainton, pp. 19–38. *Ecological Studies* 48. Berlin: Springer-Verlag.

Eller, B. M. (1984). Variation of the optical properties during development of *Ochna pulchra*. *South African Journal of Botany* **3**, 134–5.

Eller, B. M., van Rooyen, N., Theron, G. K. & Grobbelaar, N. (1984). Optical properties of some plant species of the sourish Mixed Bushveld. *South African Journal of Botany* **3**, 43–9.

Ellis, J. E. & Swift, D. M. (1988). Stability of African pastoral ecosystems: alternate paradigms and implications for development. *Journal of Range Management* **41**, 450–9.

Endean, F. (1967). The productivity of 'miombo' woodland in Zambia. *Forestry Research Bulletin* 14, Lusaka: Government Printer.

Endrödy-Younga, S. (1979). The scarabaeid dung beetle fauna of Nylsvley. An ecological and biological study. Final report to the Cooperative Scientific Programmes, Council for Scientific and Industrial Research, Pretoria. 78 pp.

Endrödy-Younga, S. (1982). An annotated checklist of dung-associated beetles of the Savanna Ecosystem Project study area, Nylsvley. *South African National Scientific Programmes Report* 59. 30 pp.

Epstein, H. (1971). *The Origin of the Domestic Animals of Africa*, Vols 1 & 2. New York: Africana Publishing Corporation.

Evers, T. M. (1975). Recent Iron Age research in the Eastern Transvaal, South Africa. *South African Archaeological Bulletin* **30**, 71–83.

Fairley, R. I. & Alexander, I. J. (1985). Methods of calculating fine root production in forests. In *Ecological Interactions in Soil*, ed. A. H. Fitter, pp. 37–42. Oxford: Blackwell Scientific.

FAO/Unesco (1974). *Soil map of the world* 1:5 000 000. Vol 1. Legend. Paris: Unesco.

Farquhar, G. D. & von Caemmerer, S. (1984). Modelling photosynthetic response to environmental conditions. In *Physiological Plant Ecology II. Encyclopedia of Plant Physiology* (New Series), Vol. 12B, pp. 550–87. Berlin: Springer-Verlag.

Feeny, P. (1976). Plant apparency and chemical defence. In *Biochemical Interactions between Plants and Insects*, ed. J. W. Wallace & R. L. Mansell, pp. 168–213. New York: Plenum.

Ferrar, P. (1980). Environmental control of gas exchange in some savanna woody species. I. Controlled environment studies of *Terminalia sericea* and *Grewia flavescens*. *Oecologia* **47**, 204–12.

Ferrar, P. (1982a). Termites of a South African savanna. I List of species and subhabitat preferences. *Oecologia* **52**, 125–32.

Ferrar, P. (1982b). Termites of a South African savanna. II Densities and populations of smaller mounds and seasonality of breeding. *Oecologia* **52**, 133–8.

Ferrar, P. (1982c). Termites of a South African savanna. III Comparative attack on toilet roll baits in subhabitats. *Oecologia* **52**, 139–46.

Ferrar, P. (1982d). Termites of a South African savanna. IV. Subterranean populations, mass determinations and biomass estimations. *Oecologia* **52**, 147–51.

Ferrar, P. (1982e). The termites of the Savanna Ecosystem Project study area, Nylsvley. *South African National Scientific Programmes Report* 60. 37 pp.

Fichardt, J. (1957). Prehistoric cultural materials from Wellington Estate, Settlers, Springbok Flats, Transvaal. *South African Archaeological Bulletin* **12**, 50–61.

Ford, J. (1960). The influence of tsetse fly on the distribution of African cattle. *Proceedings of the 1st Federal Science Congress, Salisbury*. pp. 357–65.

Ford, J. (1971). *The Role of the Trypanosomiasis in African Ecology: A Study of the Tsetse Fly Problem*. Oxford: Clarendon Press.

Fordyce, B. (1980). The prehistory of Nylsvley. Report to the National Programme for Environmental Sciences, CSIR, Pretoria. Typescript, 10pp.

Frost, P. G. H. (1984). The responses and survival of organisms in fire-prone environments. In *Ecological Effects of Fire in South African Ecosystems*, ed. P. de V. Booysen & N. M. Tainton, pp. 273–309. *Ecological Studies* 48. Berlin: Springer-Verlag.

Frost, P. G. H. (1985a). Organic matter and nutrient dynamics in a broadleafed African savanna. In *Ecology and Management of the World's Savannas*, ed. J. C. Tothill & J. J. Mott, pp. 200–6. Canberra: Australian Academy of Science.

Frost, P. G. H. (1985b). The responses of savanna organisms to fire. In *Ecology and Management of the World's Savannas*, ed. J. C. Tothill & J. J. Mott, pp. 232–7. Canberra: Australian Academy of Science.

Frost, P. G. H. (1987). The regional landscape: Nylsvley in perspective. *South African National Scientific Programmes Report* 133. CSIR, Pretoria. 30 pp.

Frost, P. G. H., Menaut, J. C., Walker, B. H., Medina, E., Solbrig O. T. & Swift, M. (1986). Responses of savannas to stress and disturbance. *Biology International Special Issue* 10.

Frost, P. G. H. & Robertson F. (1987). The ecological effects of fire in savannas. In *Determinants of Tropical Savannas*, ed. B. H. Walker, pp. 93–140. Eynsham, Oxford: IRL Press.

Frost, S. K. (1981). Food selection among young naive impala *Aepyceros melampus*. *South African Journal of Zoology* **16**, 123–4.

Frost, S. K. (1987). Factors affecting habitat separation in the Pallid Flycatcher *Melaenornis pallidus* and Marico Flycatcher *Melaenornis mariquensis*. MSc thesis, University of Cape Town.

Furniss, P. R., Ferrar, P., Morris, J. W. & Bezuidenhout, J. J. (1982). A model of savanna litter decomposition. *Ecological Modelling* **17**, 33–51.

Galpin, E. E. (1926). Botanical Survey of the Springbok Flats. *Memoirs of the Botanical Survey of South Africa* 12.

Gandar, M. V. (1980). The short-term effects of the exclusion of large mammals and of insects in a broad-leafed savanna. *South African Journal of Science* **76**, 29–31.

Gandar, M. V. (1982a). Trophic ecology and plant/herbivore energetics. In *Ecology of Tropical Savannas*, ed. B. J. Huntley & B. H. Walker, pp. 514–34. *Ecological Studies* 42. Berlin: Springer Verlag.

Gandar, M. V. (1982b). The dynamics and trophic ecology of grasshoppers (Acridoidea) in a South African savanna. *Oecologia* **54**, 370–8.

Gandar, M. V. (1982c). Description of a fire and its effects in the Nylsvley Nature Reserve: A synthesis report. *South African National Scientific Programmes Report* 63. CSIR, Pretoria. 35 pp.

Gandar, M. V. (1983). Ecological notes and annotated checklist of the grasshoppers (Orthoptera: Acridoidae) of the Savanna Ecosystem Project Study Area, Nylsvley. *South African National Scientific Programmes Report* 74. CSIR, Pretoria. 37 pp.

Gandar, M. V. (1986). Some social and environmental aspects of the use of fuelwood. Report to the National Programme for Environmental Sciences, CSIR, Pretoria.

Gardner, W. R. & Hillel, D. I. (1962). The relation of external evaporative conditions to the drying of soils. *Journal of Geophysical Research* **67**, 4319–25.

Gerdemann, J. W. (1968). Vesicular arbuscular mycorrhizae and plant growth. *Annual Review of Phytopathology* **6**, 397–418.

Gertenbach, W. P. D. (1983). Landscapes of the Kruger National Park. *Koedoe* **26**, 9–122.

Gillon, Y. (1973). Bilan energetique de la population d'*Orthoctha brachynemis* (Karsh) principale espèce Acridienne de la savane de Lamto (Côte d'Ivoire). *Annales University Abidjan (Ecology)* **6**, 105–25.

Gillon, D. (1983). The fire problem in tropical savannas. In *Tropical Savannas*, ed. F. Bourlière, pp. 617–41. *Ecosystems of the World* 13. Amsterdam: Elsevier.

Goodman, P. S. (1985). Multispecies wild herbivore systems vs. domestic single species systems: a comparison of net animal productivity. *Journal of the Grassland Society of southern Africa* **2**, 13–16.

Grant, K. (1983). The role of competition on the structure of a *Burkea africana* savanna. BSc Honours report, Dept of Botany, University of the Witwatersrand, Johannesburg.

Grant, S. (1984). The influence of ants on the insect fauna of broad-leafed savanna trees. MSc thesis, Rhodes University, Grahamstown.

Green, G. C. (1969). Variability and probability of rainfall in relation to coefficients of variation of rainfall series. *Agrochemophysica* **1**, 1–8.

Grei, E. (1990). An annotated checklist of the butterflies, hawk moths (Sphingidae) and emperor moths (Saturnidae) of the Nylsvley Nature Reserve. Occasional Report of the Ecosystems Programme. CSIR, Pretoria.

Griffioen, C. & O'Connor T. G. (1990). The influence of trees and termite mounds on the soils and herbaceous composition of a savanna grassland. *South African Journal of Ecology* **1**, 18–26.

Grobbelaar, N. & Rösch, M. W. (1981). Biological nitrogen fixation in a Northern Transvaal savanna. *Journal of South African Botany* **47**: 493–506.

Grossman, D. (1980). Studies on the herbaceous layer. MSc thesis, University of Pretoria.

Grossman, D., Grunow, J. O. & Theron, G. K. (1980). Biomass cycles, accumulation rates and nutritional characteristics of grass layer plants in canopied and uncanopied subhabitats of *Burkea* savanna. *Proceedings of the Grassland Society of southern Africa* **15**, 157–61.

Grossman, D., Grunow, J. O. & Theron, G. K. (1981). The effect of fire with and without subsequent defoliation on the herbaceous layer of *Burkea africana* savanna. *Proceedings of the Grassland Society of southern Africa* **16**, 117–20.

Grundlehner, S. (1989). The effects of different patterns of defoliation imposed by cattle and impala on savanna grasses. MSc thesis, University of the Witwatersrand, Johannesburg.

Grunow, J. O. (1974). Savanna Ecosystem Research Project: General information on Nylsvley farm. Report to the National Programme for Environmental Sciences, CSIR. 6 pp.

Grunow, J. O. & Bosch, O. J. H. (1978). Above ground annual dry matter dynamics of the grass layer in a tree savanna ecosystem. *Proceedings of the 1st International Rangeland Congress*, pp. 229–33.

Grunow, J. O., Groeneveld, H. T. & du Toit, S. H. C. (1980). Above-ground dry matter dynamics of the grass layer of a South African tree savanna. *Journal of Ecology* **68**, 877–89.

Gunton, C. (1981). Seasonal changes in nitrogen in the soils and selected plant species in *Burkea africana* (Hook) savanna. MSc thesis, University of the Witwatersrand, Johannesburg.

Hall, B. P. & Moreau, R. E. (1970). *An Atlas of Speciation in African Passerine Birds*. London: Trustees of the British Museum (Natural History).

Harmse, H. J. von M. (1977). Grondsoorte van die Nylsvley natuurreservaat. *South African National Scientific Programmes Report* 16, CSIR, Pretoria. 64 pp.

Harrison, T. D. (1984). A study of the climate of Nylsvley and aspects of the physical environments of selected habitats of *Burkea africana* savanna. PhD thesis, University of the Orange Free State, Bloemfontein.

Heidger, C. (1988). Ecology of spiders inhabiting abandoned mammal burrows in South African savanna. *Oecologia* **76**, 303–6.

Henderson, L. (1979). The age structure and dynamics of a *Burkea africana* community at Nylsvley. BSc Honours dissertation, University of the Witwatersrand, Johannesburg.

Henning, J. (1980). Hydrological characterisation of soils under selected sub-habitats in a *Burkea-Eragrostis* ecosystem. MSc thesis, University of South Africa.

Hide, J. C. (1954). Observations on factors influencing the evaporation of soil moisture. *Soil Science Society of America Proceedings* **18**, 234–9.

Hide, J. C. (1958). Soil moisture conservation in the Great Plains. *Advances in Agronomy* **10**, 23–6.

Holm, E., Kirsten, J. F. & Scholtz, C. (1976). A general survey of feeders on woody vegetation: diversity, abundance and crude energetics. Report to the National Programme for Environmental Sciences, CSIR, Pretoria.

Holmes, S. (1982). A preliminary survey of the soil microfungi of the Nylsvley Provincial Nature Reserve (2 vols). BSc Honours dissertation, University of the Witwatersrand, Johannesburg.

Horak, I. G. (1978a). Parasites of domestic and wild animals in South Africa. X. Helminths in impala. *Onderstepoort Journal of Veterinary Research* **45**, 221–8.

Horak, I. G. (1978b). Parasites of domestic and wild animals in South Africa. XI. Helminths in cattle on natural pastures in the Northern Transvaal. *Onderstepoort Journal of Veterinary Research* **45**, 229–34.

Horak, I. G. (1980a). The control of parasites in antelope in small game reserves. *Journal of the South African Veterinary Association* **51**, 17–19.

Horak, I. G. (1980b). Host specificity and the distribution of helminth parasites of sheep, cattle, impala and blesbok according to climate. *Journal of the South African Veterinary Association* **52**, 201–6.

Horak, I. G. (1981). The seasonal incidence of major nematode genera recovered from sheep, cattle, impala and blesbok in the Transvaal. *Journal of the South African Veterinary Association* **52**, 213–23.

Horak, I. G. (1982). Parasites of domestic and wild animals in South Africa. XV. The seasonal prevalence of ectoparasites on impala and cattle in the Northern Transvaal. *Onderstepoort Journal of Veterinary Research* **49**, 85–93.

Horak, I. G. (1983). Helminth, arthropod and protozoan parasites of mammals in African savannas. In *Tropical Savannas*, ed. F. Bourlière, pp. 563–81. *Ecosystems of the World* 13. Amsterdam: Elsevier.

Houerou, H. N. & Hoste, C. H. (1977). Rangeland production and annual rainfall relations in the Mediterranean basin and the African Sahelo-Sudanian zone. *Journal of Range Management* **30**, 181–9.

Howe, H. F. & Westley, L. C. (1988). *Ecological Relationships of Plants and Animals*. New York: Oxford University Press.

Huffman, T. N. (1986). Iron Age settlement patterns and the origins of class distinction in southern Africa. *Advances in World Archaeology*, pp. 219–388.

Huntley, B. J. (1982). Southern African savannas. In *Ecology of Tropical Savannas*, ed. B. J. Huntley & B. H. Walker, pp. 101–19. *Ecological Studies* 42. Berlin: Springer-Verlag.

Huntley, B. J. (1987). Ten years of cooperative ecological research in South Africa. *South African Journal of Science* **83**, 72–9.

Huntley, B. J. & Morris, J. W. (1978). Savanna ecosystem project: Phase I summary and Phase II progress. *South African National Scientific Programmes Report* 29. CSIR, Pretoria. 50 pp.

Huntley, B. J. & Walker, B. H. (ed.) (1982). *Ecology of Tropical Savannas. Ecological Studies* 42. Berlin: Springer Verlag.

Ingram, J. S. I. & Swift, M. J. (1989). Tropical Soil Biology and Fertility (TSBF) Programme: Report of the fourth interregional workshop. *Biology International* Special Issue 20. 64 pp.

Inskeep, R. R. (1979). *The Peopling of Southern Africa*. New York: Harper & Row.

Irvine, L. O. F. (1941). The major veld types of the Northern Transvaal and their grazing management. DSc thesis, University of Pretoria.

Jacobsen, N. G. H. (1977). An annotated checklist of the amphibians, reptiles and mammals of the Nylsvley Nature Reserve. *South African National Scientific Programs Report* 21. CSIR, Pretoria. 65 pp.

Janse van Rensburg, D. (1982). 'n Outecologiese studie van enkele plantsoorte op die Nylsvley-natuurreservaat. MSc thesis, University of Pretoria.

Jarman, P. J. (1974). The social organisation of antelope in relation to their ecology. *Behaviour* **48**, 215–68.

Jarvis, P. G. & McNaughton, K. G. (1986). Stomatal control of transpiration: scaling up from the leaf to the region. *Advances in Ecological Research* **15**, 1–45.

Johnson, I. R. & Thornley, J. H. M. (1984). A model of instantaneous and daily canopy photosynthesis. *Journal of Theoretical Biology* **107**, 531–45.

Johnson, R. W. & Tothill, J. C. (1985). Definition and broad geographical outline of savanna lands. In *Ecology and Management of the World's Savannas*, ed. J. C. Tothill & J. J. Mott, pp. 1–13. Canberra: Australian Academy of Science.

Johnstone, P. A. (1975). Evaluation of a Rhodesian game ranch. *Journal of the southern African Wildlife Management Association* **5**, 43–51.

Jones, C. L., Smithers, N. L., Scholes, M. C. & Scholes, R. J. (1990). The effect of fire frequency on the organic components of a basaltic soil in the Kruger National Park. *South African Journal of Plant And Soil* **7**, 236–8.

Josens, G. (1983). The soil fauna of tropical savannas. II. The termites. In *Tropical Savannas*, ed. F. Bourlière, pp. 505–24. *Ecosystems of the World* 13. Amsterdam: Elsevier.

Kapustka, L. A. & Rice, E. L. (1978). Symbiotic and asymbiotic N_2-fixation in a tall grass prairie. *Soil Biology and Biochemistry* **10**, 553–4.

Keay R. W. J. (1959). *Vegetation map of Africa south of the Tropic of Cancer. Explanatory notes*. Oxford: Oxford University Press.

Kellman, M. (1979). Soil enrichment by neotropical savanna trees. *Journal of Ecology* **67**, 565–77.

Kemper, N. (1984). Competition between two grass species – *Eragrostis pallens* and *Digitaria eriantha* – and the effects of defoliation and water stress. BSc Honours report, Dept of Botany, University of the Witwatersrand, Johannesburg.

Kennan, T. S. D. (1972). The effects of fire in two vegetation types at Matopos, Rhodesia. *Proceedings of the Tall Timbers Fire Ecology Conference* **11**, 53–98.

Kennard, D. G. & Walker, B. H. (1973). Relationships between tree canopy and *Panicum maximum* in the vicinity of Fort Victoria. *Rhodesian Journal of Agricultural Research* **11**, 145–53.

King, J. M., de Moor, F. C. & Chutter, F. M. (1992). Alternative ways of classifying rivers in South Africa. In *River Conservation and Management*, ed. P. J. Boon, P. Calow & G. E. Petts, pp. 213–28. New York: Wiley.

Kirsten, J. F. (1978). Ecological energetics of social Hymenoptera in a savanna ecosystem with notes on their biology. MSc thesis, University of Pretoria.

Knoop, W. T. (1982). Interactions of herbaceous and woody vegetation in two savanna communities at Nylsvley. MSc thesis, University of the Witwatersrand, Johannesburg.

Knoop, W. T. & Walker, B. H. (1984). Interactions of woody and herbaceous vegetation in two savanna communities at Nylsvley. *Journal of Ecology* **73**, 235–53.
Körn, H. (1986a). A case of daily turpor in the Golden Mole, *Amblysomus hottentotus* (insectivora) from the Transvaal highveld, South Africa. *Säugertierkundliche Mitteilungen Band.* **33**, 86–7.
Körn, H. (1986b). Populations- und verhaltensstudien an freilebenden europäischen, afrikanischen und nordamerikanischen Nagetieren während der ungünstigen Jahrezeit. Dissertation, Philipps-Universität, Marburg.
Körn, H. (1987a). Effects of live trapping and toe clipping on body weight of European and African rodent species. *Oecologia* **71**, 597–600.
Körn, H. (1987b). Densities and biomass of non-fossorial southern African savanna rodents during dry season. *Oecologia* **72**, 410–13.
Körn, H. & Braak, L. E. O. (1987). Survival of wild gerbils (Mammalia: Rodentia) parasitised by larvae of the blowfly *Cordylobia anthropophaga* (Insecta: Diptera). *Säugertierkundliche Mitteilungen Band.* **52**, 56–7.
Langkamp, P. J., Farnell, J. K. & Dalling, M. J. (1982). Nutrient cycling in a stand of *Acacia holosericea* A.Cunn. ex G.Don. I. Measurements of precipitation interception, seasonal acetylene reduction, plant growth and nitrogen requirements. *Australian Journal of Botany* **30**, 87–106.
Lavelle, P. (1983). The soil fauna of tropical savannas. I. The community structure. In *Tropical Savannas*, ed. F. Bourlière, pp. 477–84. *Ecosystems of the World* 13. Amsterdam: Elsevier.
Levey, B. (1977). *Detailed quantitative ecology of dominant leaf and seed eating insects on woody vegetation (excluding Lepidoptera)*. Final Report to the SA Savanna Ecosystem Programme, CSIR, Pretoria.
Lindeman, R. L. (1942). The trophic–dynamic aspect of ecology. *Ecology* **23**, 399–418.
Lobert, J. M., Scharffe, D. H., Hao, W. M. & Crutzen, P. J. (1990). Importance of biomass burning in the atmospheric budgets of nitrogen-containing gases. *Nature* **346**, 552–4.
Londt, J. G. H., Horak, I. G. & de Villiers, I. L. (1979). Parasites of domestic and wild animals in South Africa XIII. The seasonal incidence of adult ticks (Acarine: Ixodidae) on cattle in the northern Transvaal. *Onderstepoort Journal of Veterinary Research* **46**, 31–9.
Long, S. P., Garcia-Moya, E., Imbamba, S. K., Kamnalrut, A., Piedade, M. T. F., Scurlock, J. M. O., Shen, Y. K. & Hall, D. O. (1989). Primary productivity of natural grass ecosystems of the tropics: a reappraisal. *Plant and Soil* **115**, 155–66.
Loots, G. C. (1974). The functional morphology of the gnathosoma of the predaceous and parasitic mites (Acari). *Proceedings of the 4th International Congress of Acarology*. Hungarian Academy of Sciences, pp. 581–4.
Lubke, R. A. (1976). A re-assessment of the woody vegetation of the Nylsvley study area. Report to the National Programme for Environmental Sciences, CSIR, Pretoria. 5pp.
Lubke, R. A. (1977). The woody vegetation (1974, 1975, 1976) of the Nylsvley Study area. Report to the National Programme for Environmental Sciences, CSIR, Pretoria. 3pp.
Lubke, R. A. (1985). Research techniques implemented in a quantitative survey of the woody vegetation of the Nylsvley Savanna Ecosystem study area. Occasional Report 3, Ecosystem Programmes, CSIR, Pretoria. 25 pp.
Lubke, R. A., Clinning, D. F. & Smith, F. R. (1975). A quantitative ecological survey of the woody vegetation of the Nylsvley Study Area. Report to the National Programme for Environmental Sciences, CSIR, Pretoria. Typescript, 123 pp.
Lubke, R. A., Clinning, C. F. & Smith, F. R. (1976). The pattern of the woody species of the Nylsvley savanna ecosystem project area. *Proceedings of the Grassland Society of Southern Africa* **11**, 29–35.

Lubke, R. A., Morris, J. W., Theron, G. K. & van Rooyen, N. (1983). Diversity, structure and pattern in Nylsvley vegetation. *South African Journal of Botany* **2**, 26–41.
Lubke, R. A. & Thatcher, F. M. (1983). Short-term changes in the *Burkea* savanna. *South African Journal of Botany* **2**, 85–97.
Ludlow, A. E. (1987). A developmental study of the anatomy and fine structure of the leaves of *Ochna pulchra* Hook. PhD thesis, University of the Witwatersrand, Johannesburg.
Ludlow, M. M. & Wilson, G. L. (1971). Photosynthesis of tropical pasture plants. *Australian Journal of Biological Science* **24**, 449–70.
MacVicar, C. N., de Villiers, J. M. Loxton, R. F., Verster, E., Lambrechts, J. J. N., Merryweather, F. R., le Roux, J., van Rooyen, T. H. & Harmse, H. J. von M. (1977). *Soil Classification: A Binomial System for South Africa*. Department of Agriculture, Pretoria. 150 pp.
Malaisse, F. (1978). High termitaria. In *The Biogeography and Ecology of Southern Africa*, ed. M. J. A. Werger, pp. 1279–1300. *Monographs in Biology* 31. The Hague: Junk.
Malaisse, F., Freson, R., Geoffinet, G. & Malaisse-Mousset, M. (1975). Litterfall and litter breakdown in miombo. In *Tropical Ecological Systems: Trends in Terrestrial and Aquatic Research*, ed. F. B. Golley & E. Medina, pp. 137–55. *Ecological Studies* 11. Berlin: Springer-Verlag.
Martin, H. (1961). Hydrology and water balance of some regions covered by Kalahari sands in South West Africa. In *Inter-African Conference on Hydrology, Nairobi*. Commission for Technical Cooperation in Africa South of the Sahara.
Mason, R. J. (1962). *Prehistory of the Transvaal*. Johannesburg: Witwatersrand University Press. 498 pp.
Mason, R. J. (1988). Cave of Hearths, Makapansgat, Transvaal. Occasional paper 21, Archaeological Research Unit, University of the Witwatersrand, Johannesburg.
McMillan, W. D. & Burgy, R. H. Interception loss from grass. *Journal of Geophysical Research* **65**, 2389–94.
McNaughton, S. J. (1979). Grazing as an optimization process: grass–ungulate relationships in the Serengeti. *The American Naturalist* **113**, 691–703.
Meissner, H. H. (1982). Theory and application of a method to calculate forage intake of wild southern African ungulates for purposes of estimating carrying capacity. *South African Journal of Wildlife Research* **12**, 41–7.
Melillo, J. M., Aber, J. D. & Muratore, J. F. (1982). Nitrogen and lignin control of hardwood leaf litter decomposition dynamics. *Ecology* **63**, 621–6.
Menaut, J. C., Abbadie, L., Lavenu, F., Loudjani, Ph. & Podaire, A. (1991). Biomass burning in West African savannas. In *Global Biomass Burning*, ed. J. S. Levine. MIT Press.
Meredith, F. (1987). The effect of soil and vegetation disturbance on the movement of selected nutrients in savanna soils. MSc thesis, University of the Witwatersrand, Johannesburg. 101 pp.
Mes, M. G. (1958). The influence of veld burning or mowing on the water, nitrogen and ash content of grasses. *South African Journal of Science* **54**, 83–6.
Michelmore, A. P. G. (1939). Observations on tropical African grasslands. *Journal of Ecology* **27**, 282–312.
Milne, G. (1936). Some suggested units of classification and mapping, particularly for East African soils. *Soils Research* **4**, 183–98.
Milton, S. J. (1988). The effects of pruning on shoot production and basal increment of *Acacia tortilis*. *South African Journal of Botany* **54**, 109–17.
Molz, F. J. (1981). Simulation of plant-water uptake. In *Modelling Wastewater Renovation Land Treatment*. ed. I. K. Iskander, pp. 69–91. New York: Wiley.

Monro, R. (1978). A summary of the effects of fire on the populations of impala at Nylsvley. Report to the National Programme for Environmental Sciences. Typescript, 5 pp.

Monro, R. (1979). A study on the growth, feeding and body condition of impala, *Aepyceros melampus* (Lichtenstein 1812). MSc thesis, University of Pretoria.

Monro, R. H. & Skinner, J. D. (1979). A note on condition indices for adult male impala, *Aepceros melampus*. *South African Journal of Animal Science* **9**, 47–51.

Monro, R. H. (1980). Observations on the feeding ecology of impala. *South African Journal of Zoology* **15**, 107–10.

Monson, R. K. (1989). On the evolutionary pathways resulting in C_4 photosynthesis and Crassulacean Acid Metabolism. *Advances in Ecological Research* **19**, 57–110.

Moore, A. (1980). Waterbalans studies in geselekteerde subhabitatte van 'n *Burkea*-savanna. MSc (Agric) thesis, University of the Orange Free State, Bloemfontein.

Moore, A. (1982). Herverspreiding van grondwater in drie subhabitatte van 'n *Burkea* savanna. *Proceedings of the Grassland Society of southern Africa* **17**, 112–15.

Moore, A. W. (1960). The influence of annual burning on a soil in the derived savanna of Nigeria. In *Transactions of the International Congress of Soil Science*, Madison, Wisconsin, pp. 257–64

Morris, J. W. (1983). A review of decomposition and reduction and of soil organic matter in tropical African biomes. *Journal of South African Botany* **49**, 65–78.

Morris, J. W., Bezuidenhout, J. J., Ferrar, P., Horne, J. C. & Judelman, M. (1978). A first savanna decomposition model. *Bothalia* **12**, 547–53.

Morris, J. W., Bezuidenhout, J. J. & Furniss, P. R. (1982). Litter decomposition. In *The Ecology of Tropical Savannas*. ed. B. J. Huntley & B. H. Walker, pp. 535–53. *Ecological Studies* 42. Berlin: Springer Verlag.

Mosier, A., Schimel, D., Valentine, D., Bronson, K. & Parton, W. (1991). Methane and nitrous oxide fluxes in native, fertilised and cultivated grassland. *Nature* **350**, 330–2.

Myers, J. H. (1988). Can a general hypothesis explain population cycles of forest lepidoptera. *Advances in Ecological Research* **18**, 179–242.

Naveh, Z. & Whittaker, R. H. (1979). Structural and floristic diversity of shrublands and woodlands in northern Israel and other mediterranean areas. *Vegetatio* **41**, 171–90.

Nelson, D. W. & Sommers, L. E. (1975). A rapid and accurate method for estimating organic carbon in the soil. *Proceedings of the Indiana Academy of Science* **84**, 456–62.

Nix, H. A. (1983). Climate of tropical savannas. In *Tropical Savannas*, ed. F. Bourlière. *Ecosystems of the World* 13. Amsterdam: Elsevier.

Norton, G. A. & Walker, B. H. (1985). A decision analysis approach to savanna management. *Journal of Environmental Management* **21**, 15–31.

Nunn, S. M. (1978). Development of a monitoring programme for key insect taxa in the *Burkea* savanna at Nylsvley. Report to the National Programme for Environmental Sciences, CSIR, Pretoria. 61 pp.

O'Connor, T. G. (1977). Habitat selection of impala *Aepyceros melampus* in the Nylsvley Nature Reserve. BSc Honours thesis, University of the Witwatersrand, Johannesburg.

O'Connor, T. G. (1985). A synthesis of field experiments concerning the grass layer in the savanna regions of southern Africa. *South African National Scientific Programmes Report* 114. CSIR, Pretoria.

Olivier, P. A. S. (1976). 'n Taksonomiese studie van die Prostigmata (Acari) in 'n savannebiotoop te Nylsvley. MSc thesis, Potchefstroom University for Christian Higher Education, Potchefstroom.

Olivier, P. A. S. & Theron, P. D. (1988). A new species of *Eutogenes* Baker, 1949 (Acari: Cheyletidae) from South Africa. *Phytophylactica* **20**, 253–6.

Olivier, P. A. S. & Theron, P. D. (1989a). A new species of *Speleorchestes* (Nanorchestidae: Prostigmata) from a savanna biotype in South Africa. *South African Journal of Zoology* **24**, 356–60.

Olivier, P. A. S. & Theron, P. D. (1989b). A new genus and species of Cheyletidae (Acari: Prostigmata) from South Africa. *Journal of the Entomological Society of southern Africa* **52**, 237–43.

Olsen, S. R. & Sommers, L. E. (1982). Phosphorus. In *Methods of Soil Analysis,* Part 2. *Chemical and Microbiological Properties.* (2nd edition, ed. A. L. Page, R. H. Miller & D. R. Keeney, pp. 403–30. Madison: American Society of Agronomy.

Olsvig-Whittaker, L. & Morris, J. W. (1982). Comparison of certain Nylsvley soils using a bioassay technique. *South African Journal of Botany* **1**, 91–96.

Opperman, D. P. J., Human, J. J. & Viljoen, M. F. Evapotranspiration studies on *Themeda triandra* Forsk. under field conditions. *Proceedings of the Grassland Society of southern Africa* **12**, 71–6.

Owen-Smith, R. N. (1982). Factors influencing the consumption of plant products by large herbivores. In: *Ecology of Tropical Savannas*, ed. B. J. Huntley & B. H. Walker, pp. 359–404. *Ecological Studies* 42. Berlin: Springer-Verlag.

Owen-Smith, R. N. (ed.) (1983). *Management of Large Mammals in African Conservation Areas*. Pretoria: Haum.

Owen-Smith, R. N. (1985a). The ecological potential of the kudu for commercial production in savanna regions. *Journal of the Grassland Society of southern Africa* **2**, 7–10.

Owen-Smith, R. N. (1985b). Niche separation among African ungulates. In *Species and speciation*, ed. E. S. Vrba, pp. 167–71. Transvaal Museum Monograph No. 4, Transvaal Museum, Pretoria.

Owen-Smith, R. N. (1988). *Megaherbivores: The Influence of Very Large Body Size on Ecology*. Cambridge: Cambridge University Press.

Owen-Smith, R. N. & Cooper, S. M. (1985). Comparative consumption of vegetation components by kudus, impalas and goats in relation to their commercial potential as browsers in savanna regions. *South African Journal of Science* **81**, 72–6.

Owen-Smith, N. & Cooper, S. M. (1987a). Palatability of woody plants to browsing ruminants in a South African savanna. *Ecology* **68**, 319–31.

Owen-Smith, N. & Cooper, S. M. (1987b). Assessing food preferences of ungulates by acceptability indices. *Journal of Wildlife Management* **51**, 372–8.

Owen-Smith, N. & Cooper, S. M. (1989). Nutritional ecology of a browsing ruminant, the kudu (*Tragelaphus strepsiceros*) through the seasonal cycle. *Journal of Zoology, (London)* **219**, 29–43.

Owen-Smith, R. N., Cooper, S. M. & Novellie, P. A. (1983). Aspects of the feeding ecology of a browsing ruminant: the kudu. *South African Journal of Animal Science* **13**, 35–8.

Owen-Smith, N & Novellie, P. (1982). What should a clever ungulate eat? *The American Naturalist* **119**, 151–78.

Parton, W. J. (1978). Abiotic section of ELM. In *Grassland Simulation Model*, ed. G. S. Innes, pp. 31–53. *Ecological Studies* 26. New York: Springer-Verlag.

Partridge, T. C. (1986). Paleoecology of the Pliocene and lower Pleistocene hominids of southern Africa: how good is the chronological and paleoenvironmental evidence? *South African Journal of Science* **82**, 80–3.

Peakin, G. J. & Josens, G. (1978). Respiration and energy flow. In *Production Ecology of Ants and Termites*, ed. M. V. Brian, pp. 111–63. Cambridge: Cambridge University Press.

Pendle, B. G. (1982). The estimation of water use by the dominant species in the *Burkea africana* savanna. MSc thesis, University of the Witwatersrand, Johannesburg.

Penman, H. L. (1948). Natural evaporation from open water, bare soil and grass. *Proceedings of the Royal Society of London, Ser.* A **193**, 120–45.

Penridge, L. K. & Walker, J. (1986). Effect of neighbouring trees on eucalypt growth in a semi-arid woodland in Australia. *Journal of Ecology* **74**, 925–36.

Phillips, J. F. V. (1965). Fire – as master and servant, its influence in the bioclimatic regions of trans-Saharan Africa. *Proceedings of the Tall Timbers Fire Ecology Conference* **4**, 7–109.

Phillips, J. F. V. (1968). The influence of fire in trans-Saharan Africa. *Acta Phytogeographica Suecica* **54**, 13–29.

Phillipson, D. W. (1977). *The Later Prehistory of Eastern and Southern Africa*. London: Heinemann.

Phillipson, J. (1973). The biological efficiency of protein production by ecosystems. In *The Biological Efficiency of Protein Production*, ed. J. G. W. Jones, pp. 217–35. Cambridge: Cambridge University Press.

Pratt, D. J., Greenway, P. J. & Gwynne, M. D. (1966). A classification of East African rangeland, with an appendix on terminology. *Journal of Applied Ecology* **3**, 369–82.

Preston-Whyte, R. A. & Tyson, P. D. (1988). *The Atmosphere and Weather of Southern Africa*. Cape Town: Oxford University Press.

Purchase, B. S. (1974). The influence of phosphate deficiency on nitrification. *Plant and Soil* **41**, 541–7.

Pyke, G. H., Pulliam, H. R. & Charnov, E. L. (1977). Optimal foraging: a selective review of theory and tests. *Quarterly Review of Biology* **52**, 137–53.

Raich, J. W. & Nadelhoffer, K. J. (1989). Belowground carbon allocation in forest ecosystems: global trends. *Ecology* **70**, 1346–54.

Rains, A. B. (1963). Grassland research in northern Nigeria 1952–1962. *Miscellaneous Papers of the Institute of Agricultural Research, Samaru* **11**, 1–67.

Raison, R. J., Connell, M. H. & Khanna, P. K. (1987). Methodology for studying fluxes of mineral-N *in situ*. *Soil Biology and Biochemistry* **19**, 521–30.

Randall, L. A. (1983). Factors influencing photosynthesis and above-ground production in savanna grasses. MSc thesis, University of the Witwatersrand, Johannesburg.

Randall, L. A. & Cresswell, C. F. (1983). Growth analysis and photosynthetic rates in three selected grass species in the *Burkea–Eragrostis* savanna. *Proceedings of the Grassland society of southern Africa* **18**, 120–3.

Rasmussen, R. A. & Khalil, M. A. K. (1983). Global production of methane by termites. *Nature* **301**, 700–2.

Raven, P. H. & Axelrod, D. I. (1974). Angiosperm biogeography and past continental movements. *Annals of the Missouri Botanical Garden* **61**, 539–673.

Reuss, J. O. & Copley, P. W. (1971). Soil nitrogen investigation on the Pawnee site, 1970. *US/IBP Grassland Biome Technical Report* 106. Colorado State University, Fort Collins. 44 pp.

Rey, M. E. C. (1982). Epidemiological, morphological and physiological studies of selected plant diseases at Nylsvley. PhD thesis, University of the Witwatersrand, Johannesburg.

Rhoades & Cates (1976). Towards a general theory of plant antiherbivore chemistry. *Recent Advances in Phytochemistry*, **19**, 168–213.

Ripley, E. A. & Saungier, B. (1978). Biophysics of a natural grassland: evaporation. *Journal of Applied Ecology* **15** 459–79.

Risser, P. G. (1989). South African Savanna Biome Programme: review. *Report to the Committee for Terrestrial Ecosystems and Steering Committees*. Foundation for Research Development, Pretoria. 14 pp.

Robertson, G. P. (1987). Geostatistics in ecology: interpolating with known variance. *Ecology* **68**, 744–8.
Rose-Innes, R. (1972). Fire in West African vegetation. *Proceedings of the Tall Timbers Fire Ecology Conference* **11**, 147–73.
Rosenthal, G. A. & Jansen, D. H. (ed.) (1979). *Herbivores: Their Interaction with Secondary Plant Metabolites*. New York: Academic Press.
Rutherford, M. C. (1979). Aboveground biomass subdivisions in woody species of the savanna ecosystem project study area, Nylsvley. *South African National Scientific Programmes Report* 36. CSIR, Pretoria. 30 pp.
Rutherford, M. C. (1980). Annual plant production–precipitation relations in arid and semi-arid regions. *South African Journal of Science* **76**, 53–6.
Rutherford, M. C. (1981). Survival, regeneration and leaf biomass changes in woody plants following spring burns in *Burkea africana–Ochna pulchra* savanna. *Bothalia* **13**, 531–52.
Rutherford, M. C. (1983). Growth rates, biomass and distribution of selected woody plant roots in *Burkea africana–Ochna pulchra* savanna. *Vegetatio* **52**, 45–63.
Rutherford, M. C. (1984). Relative allocation and seasonal phasing of growth of woody plant components in a South African savanna. *Progress in Biometeorology* **3**, 200–21.
Rutherford M. C. & Curran, B. (1981). A root observation chamber for replicated use in a natural plant community. *Plant and Soil* **63**, 123–9.
Rutherford, M. C. & Kelly, R. D. (1978). Woody plant basal area and stem increment in *Burkea africana–Ochna pulchra* woodland. *South African Journal of Science* **74**, 307–8.
Rutherford, M. C. & Panagos, M. D. (1982). Seasonal woody plant shoot growth in *Burkea africana–Ochna pulchra* savanna. *South African Journal of Botany* **1**, 104–16.
Sanchez, P. A. (1976). *Properties and Management of Soils in the Tropics*. New York: Wiley. 618 pp.
Sarmiento, G. (1984). *The Ecology of Neotropical Savannas*. Harvard University Press.
Scholes, M. C. & Scholes, R. J. (1989). Phosphorus mineralisation and immobilisation in savannas. *Proceedings of the Phosphorus Symposium* 1988, pp. 101–3. Soils and Irrigation Research Institute, Private Bag X79, Pretoria.
Scholes, R. J. (1987). Response of three semi-arid savannas on contrasting soils to the removal of the woody component. PhD thesis, University of the Witwatersrand, Johannesburg.
Scholes, R. J. (1990a). The regrowth of *Colophospermum mopane* following clearing. *Journal of the Grassland Society of southern Africa* **7**.
Scholes, R. J. (1990b). The influence of soil fertility on the ecology of southern African savannas. *Journal of Biogeography* **17**, 417–19.
Scholtz, C. H. (1976). Biology and ecological energetics of Lepidoptera larvae associated with woody vegetation in a savanna ecosystem. MSc thesis, University of Pretoria.
Scholtz, C. H. (1982). Trophic ecology of Lepidoptera larvae associated with woody vegetation in a savanna ecosystem. *South African National Scientific Programmes Report* 55. CSIR, Pretoria.
Scott, J. A. (1979). An ecosystem-level trophic-group arthropod and nematode bioenergetics model. In *Perspectives in Grassland Ecology*, ed. N. French, pp. 107–16. *Ecological Studies* 32. Berlin: Springer-Verlag.
Scott, L. (1982). A late Quaternary pollen record from the Transvaal Bushveld, South Africa. *Quaternary Research* **17**, 339–70.
Seastedt, T. R., Ramundo, R. A. & Hayes, D. C. (1988). Maximisation of densities of soil animals by foliage herbivory: empirical evidence, graphical and conceptual models. *Oikos* **51**, 243–8.

Seiler, W., Conrad, R. & Scharfe, D. (1984). Field studies of methane emission from termite nests into the atmosphere and measurements of methane uptake by tropical soils. *Journal of Atmospheric Chemistry* **1**, 171–86.

Sibbensen, E. (1978). An investigation of the anion-exchange resin method for soil phosphate extraction. *Plant and Soil* **50**, 305–21.

Sinclair, A. R. E. (1975). The resource limitation of trophic levels in tropical grassland ecosystems. *Journal of Animal Ecology* **44**, 497–520.

Sinclair, A. R. E. & Norton-Griffiths, M. (ed.) (1979). *Serengeti: Dynamics of an Ecosystem*. Chicago: University of Chicago Press.

Singh, F. S. & Gupta, S. R. (1977). Plant decomposition and soil respiration in terrestrial ecosystems. *Botanical Review* **43**, 449–528.

Skinner, J. D., Monro, R. H. & Zimmermann, I. (1984). Comparative food intake and growth of cattle and impala on mixed tree savanna. *South African Journal of Wildlife Research* **14**, 1–9.

Sloff, D. (1982). A study of the effects of vesicular-arbuscular mycorrhizal infection on two grass species, *P. maximum* and *E. curvula*, grown on two soil types from the Nylsvley Nature Reserve. BSc Honours dissertation, University of the Witwatersrand, Johannesburg.

Smith, F. R. (1974). A study of the structural analysis of the woody species in the Nylsvley ecosystem study area, Naboomspruit. BSc Honours Thesis, University of the Witwatersrand, Johannesburg.

Smith, T. M. & Walker, B. H. (1983). The role of competition in the spacing of savanna trees. *Proceedings of the Grassland Society of southern Africa* **18**, 159–64.

Smithers, R. H. N. (1983). *The Mammals of the Southern African Subregion*. University of Pretoria.

Smucker, A. J. M., McBurney, S. L. & Srivastava, A. K. (1982). Quantitative separation of roots from compacted soil profiles by the hydropneumatic elutriation system. *Agronomy Journal* **74**, 500–3.

Soil Survey Staff (1975). *Soil Taxonomy: a basic system of soil classification for making and interpreting soil surveys. Agriculture Handbook* 436. Soil Conservation Service, United States Dept of Agriculture, Washington. 754 pp.

South African Geological Survey (1978). 1:250 000 Geological Series map 2428 Nylstroom. Pretoria: Government Printer.

Stebbins, G. L. (1981). Coevolution of grasses and herbivores. *Annals of the Missouri Botanical Garden* **68**, 75–86.

Steudler, P. A., Bowden, R. D., Melillo, J. M. & Aber, J. D. (1989). Influence of nitrogen fertilization on methane uptake in temperate forest soils. *Nature* **341**, 314–15.

Strang, R. M. (1974). Some man-made changes in successional trends on the Rhodesian highveld. *Journal of Applied Ecology* **11**, 249–63.

Stuart-Hill, G. C., Tainton, N. N. & Barnard, H. J. (1987). The influence of an *Acacia karroo* tree on grass production in its vicinity. *Journal of the Grassland Society of southern Africa* **4**, 83–8.

Swift, M. J., Heal, O. W. & Anderson, J. M. (1979). *Decomposition in Terrestrial Ecosystems. Studies in Ecology* 5. Berkeley: University of California Press.

Tabatabai, M. A. (1982). Sulphur. In *Methods of Soil Analysis*, Part 2. *Chemical and Microbiological Properties*. (2nd edition), ed. A. L. Page, R. H. Miller & D. R. Keeney, pp. 501–38. Madison: American Society of Agronomy.

Tainton, N. M., Groves, R. H. & Nash, R. C. (1977). Time of mowing and burning veld: short term effects on production and tiller development. *Proceedings of the Grassland Society of southern Africa* **12**, 59–64.

Talma, A. S., Vogel, J. C. & Partridge, T. C. (1974). Isotopic contents of some Transvaal speleotherms and their paleoclimatic significance. *South African Journal of Science* **70**, 135–40.

Tarboton, W. R. (1977). A checklist of the birds of the Nylsvley Nature Reserve. *South African National Scientific Programmes Report* 15. CSIR, Pretoria. 14 pp.

Tarboton, W. R. (1979). The ecology of a savanna bird community. MSc dissertation, University of the Witwatersrand, Johannesburg.

Tarboton, W. R. (1980). Avian populations in a Transvaal savanna. *Proceedings of the 4th Pan African ornithological conference*, pp. 113–24.

Theron, G. K. (1973). 'n Ekologiese studie van die plantegroei van die Loskopdamnatuurreservaat. DSc thesis, University of Pretoria.

Theron, G. K., Morris, J. W. & van Rooyen, N. (1984). Ordination of the herbaceous stratum of savanna in the Nylsvley Nature Reserve, South Africa. *South African Journal of Botany* **3**, 22–32.

Theron, P. D. (1974). The functional morphology of the gnathosoma of some liquid and solid feeders in the Trombidiformes, Cryptostigmata and Astigmata (Acarina). *Proceedings of the 4th International Congress of Acarology*, pp. 575–79. Hungarian Academy of Sciences.

Thom, A. S. (1975). Momentum, mass and heat exchange of plant communities. In *Vegetation and the Atmosphere*, Vol. 1, ed. J. L. Monteith, pp. 57–109. London: Academic Press.

Tidmarsh, C. E. M. & Havenga, C. M. (1955). The wheelpoint method of survey and measurement of open and semi-open savanna vegetation. *Memoirs of the Botanical Survey of South Africa* 29.

Tinker, P. B. (1975). Soil chemistry of phosphate and mycorrhizal effects on plant growth. In *Endomycorrhizas*, ed. F. E. Sanders, B. Mosse & P. B. Tinker. London: Academic Press.

Tinley, K. L. (1981). Salient landscape features of the Nylsvley area and its regional environs. Report to the National Programme for Environmental Sciences, CSIR, Pretoria. 4 pp.

Tothill, J. C. & Mott, J. J. (1985). *Ecology and Management of the World's Savannas*. Canberra: Australian Academy of Science.

Trapnell, C. G. (1959). Ecological results of woodland burning experiments in Northern Rhodesia. *Journal of Ecology* **47**, 129–68.

Trapnell, C. G., Friend, M. T., Chamberlain, G. T. & Birch, H. F. (1976). The effect of fire and termites on a Zambian woodland soil. *Journal of Ecology* **64**, 577–88.

Trollope, W. S. W. (1980). Controlling bush encroachment with fire in the savanna area of South Africa. *Proceedings of the Grassland Society of southern Africa* **15**, 173–7.

Trollope, W. S. W. (1984). Characteristics of fire behaviour. In *Ecological Effects of Fire in Southern African Ecosystems*, ed. P. de V. Booysen & N. M. Tainton. *Ecological Studies* 48. Berlin: Springer Verlag.

Trollope, W. S. W. & Potgieter, A. L. F. (1985). Fire behaviour in the Kruger National Park. *Journal of the Grassland Society of southern Africa* **2** (2), 17–23.

Truswell, J. F. (1970). *An Introduction to the Historical Geology of South Africa*. Cape Town: Purnell.

Turner, G. L. & Gibson, A. H. (1980). Measurement of nitrogen fixation by indirect means. In *Methods for Evaluating Biological Nitrogen Fixation*. ed. F. J. Bergsen. New York: Wiley.

Tyson, P. D. (1986). *Climatic Change and Variability in Southern Africa*. Cape Town: Oxford University Press.

Tyson, P. D., Kruger, F. J. & Louw, C. W. (1988). Atmospheric pollution and its implications in the Eastern Transvaal Highveld. *South African National Scientific Programmes Report* 150, CSIR, Pretoria. 111 pp.

Ueckermann, E. A. (1978). 'n Faunistiese studie van die Acari in 'n savanna-biotoop. MSc thesis, Potchefstroom University for Christian Higher Education, Potchefstroom.

van der Meulen, F. & Werger, M. J. A. (1984a). Crown characteristics, leaf size and light throughfall of some savanna trees in southern Africa. *South African Journal of Botany* **3**, 208–18.

van der Meulen, F. & Werger, M. J. A. (1984b). Savanna tree crowns and light interception in some climatically different regions of southern Africa. *Progress in Biometeorology* **3**, 222–8.

van Genuchten, M. Th. (1980). A closed form equation for predicting the hydraulic conductivity of saturated soils. *Soil Science Society of America Journal* **44**, 892–8.

van Hoven, W. & Boomker, E. A. (1983). The influence of inoculum source on *in vitro* digestibility. *South African Journal of Animal Science* **13**, 207–9.

van Rooyen, N. & Theron, G. K. (1982). 'n Kwantitatiewe analise van die kruidstratum van die *Eragrostis pallens–Burkea africana* boomsavanna op die Nylsvley-natuurreservaat. *South African Journal of Science* **78**, 116–21.

van Vegten, J. A. (1984). Thornbush invasion in a savanna ecosystem in eastern Botswana. *Vegetatio* **56**, 3–7.

van Wyk, J. J. P. (1977). 'n Studie van die ondergrondse biomassa van die *Eragrostis pallens–Burkea africana* savanne op Nylsvley. Report to the South African National Programme for Environmental Sciences, CSIR, Pretoria. 13 pp.

Venter, H. J. T. (1976). Grond-saadreserwes van die *Eragrostis pallens–Burkea africana*-veld. Report to the National Programme for Environmental Sciences, CSIR, Pretoria.

Verhagen, B. Th. (1991). Isotope hydrology of the Kalahari: recharge or no recharge? *Paleoecology of Africa* **21**, 143–60.

Vlassak, K., Paul, E. A. & Harris, R. (1973). Assessment of biological nitrogen fixation in grassland and associated sites. *Plant and Soil* **38**, 637–49.

Vogt, K. A., Grier, C. C., Gower, S. T., Sprugel, D. G. & Vogt, D. J. (1986). Overestimation of root production: a real or imaginary problem? *Ecology* **67**, 577–9.

von dem Bussche, G. (1988). Silvicultural, economic and socio-economic aspects of woodlot establishment and management in southern Africa. Technical Report, SARCCUS, Pretoria. 19 pp.

von Maltitz, G. P. (1984). Nutrient translocation and leaching from leaves approaching senescence: a comparison between two savanna communities. BSc Honours dissertation, University of the Witwatersrand, Johannesburg.

von Maltitz, G. P. (1990). The effect of spatial scale of disturbance on patch dynamics. MSc thesis, University of the Witwatersrand, Johannesburg.

Vrba, E. S. (1974). Chronological and ecological implications of the fossil Bovidae at the Sterkfontein Australopithecene site. *Nature* **250**, 19–23.

Vrba, E. S. (1975). Some evidence of chronology and paleoecology of Sterkfontein, Swartkrans and Kromdraai from the fossil Bovidae. *Nature* **254**, 301–4.

Vrba, E. S. (1980). The significance of Bovid remains as indicators of environment and predation patterns. In *Fossils in the Making*, ed. A. K. Behrensmeyer & A. P. Hill, pp. 247–71. Chicago: University of Chicago Press.

Vrba, E. S. (1985). African Bovidae: evolutionary aspects since the Miocene. *South African Journal of Science* **81**, 263–6.

Wagenet, R. J. & Hutson, J. L. (1987). LEACHIM: Leaching estimation and chemistry model. A process-based model of water and solute movement, transformations, plant

uptake and chemical reactions in the unsaturated zone. *Continuum* 2. Water Resources Institute, Cornell University, Ithaca.
Walker, B. H. (1979). Game ranching in Africa. In *Management of Semi-Arid Ecosystems*, ed. B. H. Walker. Amsterdam: Elsevier.
Walker, B. H. (1985). Structure and function of savannas: an overview. In *Ecology and Management of the World's Savannas*, ed. J. C. Tothill & J. J. Mott, pp. 83–91. Canberra: Australian Academy of Science.
Walker, B. H. (ed.) (1987). *Determinants of Tropical Savannas. IUBS Monograph* 3. Oxford: IRL Press.
Walker, B. H., Emslie, R. H., Owen-Smith, R. N. & Scholes, R. J. (1987). To cull or not to cull: lessons from a southern African drought. *Journal of Applied Ecology* **24**, 381–401.
Walker, B. H. & Knoop, W. T. (1987). The response of the herbaceous layer in a dystrophic *Burkea africana* savanna to increased levels of nitrogen, phosphate and potassium. *Journal of the Grassland Society of southern Africa* **4**, 31–4.
Walker, B. H., Ludwig, D., Holling, C. S. & Peterman, R. S. (1981). Stability of semi-arid savanna grazing systems. *Journal of Ecology* **69**, 473–98.
Walker, B. H., Norton, G. A., Conway, G. R., Comins, H. N. & Birley, M. (1978). A procedure for multidisciplinary research: with reference to the South African Savanna Ecosystem Project. *Journal of Applied Ecology* **15**, 481–502.
Walker, B. H. & Noy-Meir, I. (1982). Aspects of the stability and resilience of savanna ecosystems. In *Ecology of Tropical Savannas*, ed. B. J. Huntley & B. H. Walker, pp. 556–609. *Ecological Studies* 42. Berlin: Springer Verlag.
Walker, B. H., Stone, L., Henderson, L. & Vernede, M. (1986). Size structure analysis of the dominant trees in a South African savanna. *South African Journal of Botany* **52**, 397–402.
Walter, D. E. (1988). Predation and mycophagy by endostigmatid mites (Acariformes: Prostigmata). *Experimental and Applied Acarology* **4**, 159–66.
Walter, H. (1939). Grasland, Savanne und Busch der ariden Teile Africas in ihrer ökologischen Bedingtheid. *Jb. Wiss. Bot.* **87**, 750–860.
Walter, H. (1971). *Ecology of Tropical and Subtropical Vegetation*. Edinburgh: Oliver & Boyd.
Warren Wilson, J. (1963). Estimation of foliage density and foliage angle by inclined point quadrats. *Australian Journal of Botany* **11**, 95–105.
Weinmann, H. (1955). The nitrogen content of rainfall in Southern Rhodesia. *South African Journal of Science* **51**, 82–4.
Weltzin, J. F. & Coughenour, M. B. (1990). Savanna tree influence on understory vegetation and soil nutrients in northwestern Kenya. *Journal of Vegetation Science* **1**, 325–34.
Werger, M. J. A. & Coetzee, B. J. (1978). Biogeographical division of southern Africa. In *Biogeography and Ecology of Southern Africa*, Vol. 1, ed. M. J. A. Werger, pp. 301–462. The Hague: Junk.
Werner, P. A. (1991). *Savanna ecology and management*. Special edition of the Journal of Biogeography, Blackwell Scientific.
West, O. (1965). *Fire in vegetation and its use in pasture management with special reference to tropical and subtropical Africa*. Commonwealth Bureau of Pastures and Crops, Farnham Royal.
White, F. (1976). The underground forests of Africa: a preliminary review. *Gardens Bulletin (Singapore)* **29**, 55–71.

White, F. (1980). *The Vegetation of Africa. Natural Resources Research* XX. Paris: Unesco. 356 pp.
Whittaker, R. H., Morris, & Goodman, D. (1984). Pattern analysis in savanna woodlands at Nylsvley, South Africa. *Memoirs of the Botanical Survey of South Africa* **49**. 51 pp.
Whyte, R. O. (1962). The myth of tropical grasslands. *Tropical Agriculture (Trinidad)* **39**, 1–12.
Wiegert, R. G. & Coleman, D. C. (1970). Ecological significance of low oxygen consumption and high fat accumulation by *Nasutitermes costalis* (Isoptera: Termitidae). *Bioscience* **20**, 663–5.
Wilson, J. R. (1975). Influence of temperature and nitrogen on growth, photosynthesis and accumulation of non-structural carbohydrate in a tropical grass, *Panicum maximum* var. *trichoglume*. *Netherlands Journal of Agricultural Science* **23**, 48–61.
Wolfson, M. M. (1988). The effect of inorganic nitrogen on growth, morphology and some aspects of the physiology of *Digitaria eriantha* (Steud). PhD thesis, University of the Witwatersrand, Johannesburg.
Wolfson, M. M. & Cresswell, C. F. (1984). The effect of nitrogen on the development and photosynthetic activity of *Digitaria eriantha* Steud. subsp *eriantha*. *Journal of the Grassland Society of southern Africa* **1**, 33–6.
Wood, T. G. (1976). The role of termites (Isoptera) in decomposition processes. In *The Role of Terrestrial and Aquatic Organisms in Decomposition Processes*, ed. J. M. Anderson & A. Macfadyen, pp. 145–68. Oxford: Blackwell.
Wood, T. G. (1978). Food and feeding habits of termites. In: *Production Ecology of Ants and Termites*, ed. M. V. Brian, pp. 55–80. Cambridge: Cambridge University Press.
Wood, T. G. & Sands, (1978). The role of termites in ecosystems. In *Production Ecology of Ants and Termites*, ed. M. V. Brian, pp. 245–92. Cambridge: Cambridge University Press.
Woods, P. V. & Raison, R. J. (1982). An appraisal of techniques for the study of litter decomposition in eucalypt forests. *Australian Journal of Ecology* **7**, 215–25.
Wullstein, L. H., Bruening, M. L. & Bollen, W. B. (1979). Nitrogen fixation associated with sand-grain root sheaths (rhizosheaths) of certain xeric grasses. *Physiologia Plantarum* **46**, 1–4.
Yeaton, R. I. (1988). Porcupines, fire and the dynamics of the tree layer of the *Burkea africana* savanna. *Journal of Ecology* **76**, 1017–29.
Yeaton R. I., Frost, S. & Frost, P. G. H. (1986). A direct gradient analysis of grasses in a savanna. *South African Journal of Science* **82**, 482–6.
Yeaton, R. I., Frost, S. & Frost, P. G. H. (1988). The structure of a grass community in *Burkea africana* savanna during recovery from fire. *South African Journal of Botany* **54**, 367–71.
Zietsman, P. C., Grobbelaar, N. & van Rooyen, N. (1988). Soil nitrogenase activity of the Nylsvley Nature Reserve. *South African Journal of Botany* **54**, 21–7.
Zimmermann, I. (1978). The feeding ecology of Africander steers (*Bos indicus*) on mixed bushveld at Nylsvley Nature Reserve, Transvaal. MSc thesis, University of Pretoria.
Zimmermann, I. (1979). The efficiency of Africander steers (*Bos indicus*) on a mixed tree savanna. *Journal of the South African Biological Society* **20**, 66–78.
Zimmermann, I. (1980a). Predicting diet quality from measurement of nitrogen and moisture in cattle dung. *South African Journal of Wildlife Management* **10**, 56–60.
Zimmermann, I. (1980b). Factors influencing the feed intake and liveweight change of beef cattle on a mixed tree savanna in the Transvaal. *Journal of Range Management* **33**, 132–6.
Zimmerman, P. R., Greenberg, J. P. & Darlington, J. P. E. C. (1984). Reply to Collins & Wood (1984). *Science* **224**, 86.

Zimmerman, P. R., Greenberg, J. P., Wandiga, S. O. & Crutzen, P. J. (1982). Termites: a potentially large source of atmospheric methane, carbon dioxide and molecular hydrogen. *Science* **218**, 563–5.
Zucker, W. V. (1983). Tannins: Does structure determine function? *American Naturalist* **121**, 335–65.

INDEX

Printed in the United States
By Bookmasters

Printed in the United States
By Bookmasters